LOTUS
Racing Cars

Club Racers to World Champions 1948–1968

JOHN TIPLER

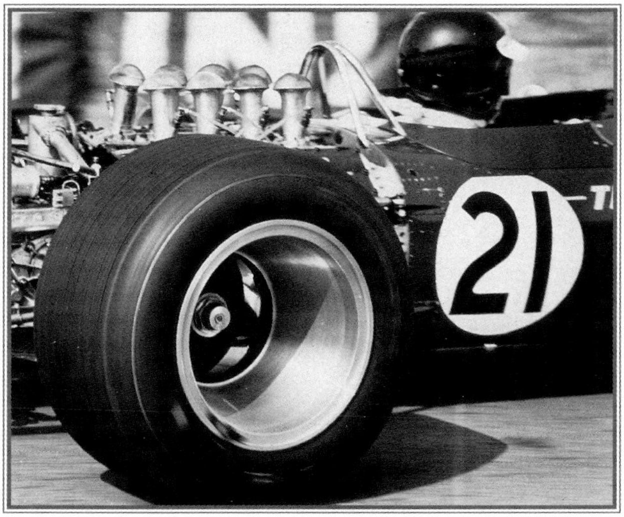

SUTTON PUBLISHING LIMITED

Sutton Publishing Limited
Phoenix Mill · Thrupp · Stroud
Gloucestershire · GL5 2BU

First published 2000

Reprinted 2001

Title page photograph: No driver was as easy on his tyres as Jim Clark.

British Library Cataloguing in Publication Data
A catalogue record for this book is available from the British Library.

ISBN 0-7509-2389-X

Typeset in 10.5/13 Photina.
Typesetting and origination by
Sutton Publishing Limited.
Printed and bound in England by
J.H. Haynes & Co. Ltd, Sparkford.

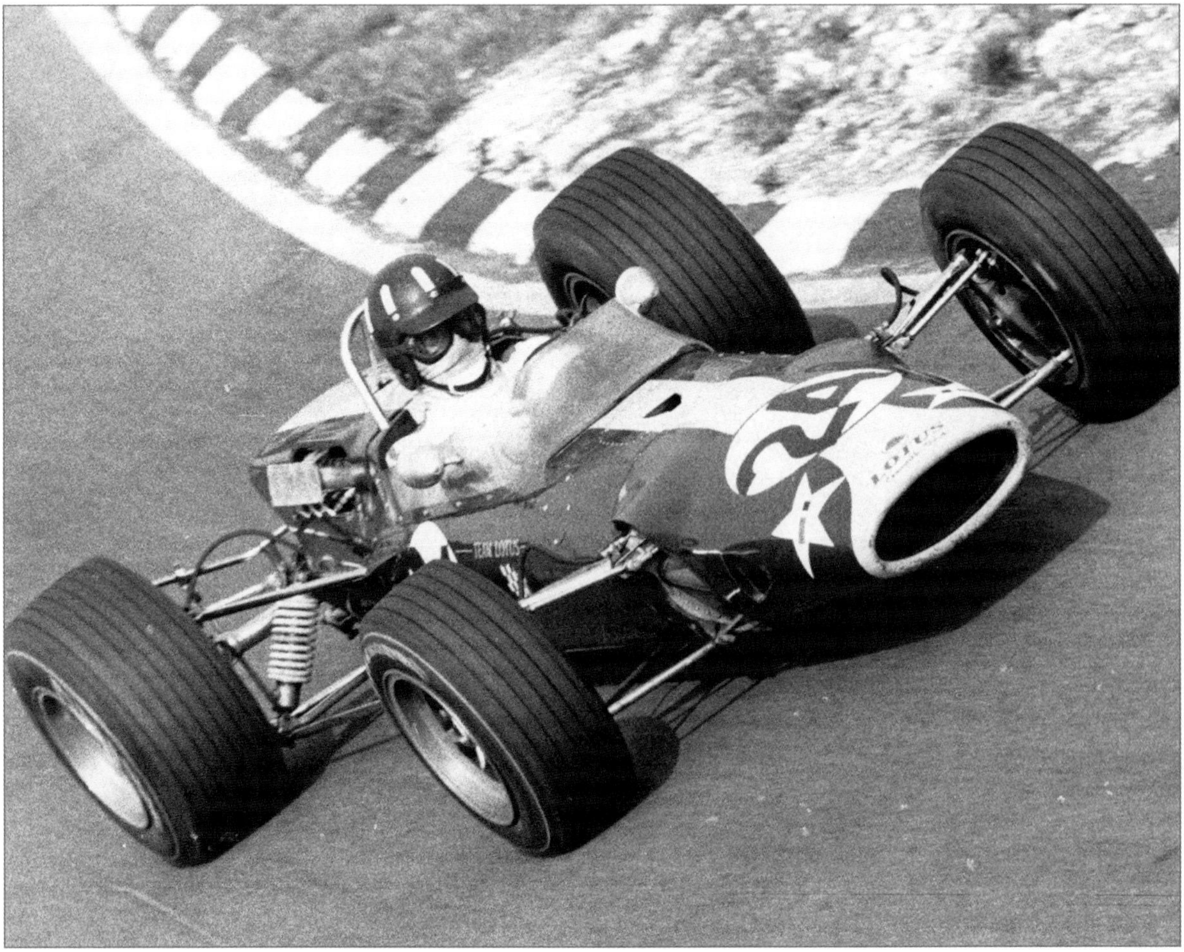

August Bank Holiday Monday at Brands Hatch in 1967 was an F2 festival, with two ten-lap heats and a forty-lap final. All the top names of the day – except Clark – were present. While Rindt and Stewart dominated the proceedings in heat one and the final, Graham Hill (above) won the second heat in the Team Lotus 48 and was seventh in the final. Jackie Oliver was second in heat two and sixth in the final. Oliver also won the unruly saloon car race in the DR Fabrications Mustang in the interim, with Ickx second in a Lotus-Cortina FVA.

Contents

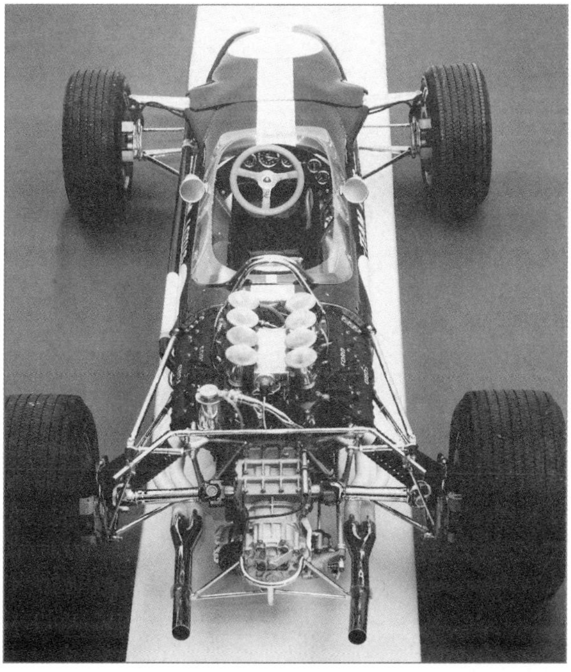

The Cosworth DFV-powered 49 ushered in the modern F1 era.

INTRODUCTION

The history and evolution of Lotus racing cars is inextricably linked to the innovative and blustery life of Colin Chapman. Building and racing cars was more than a career, it was his very existence. Not only was he the Team owner, he was also its Chief Engineer. Where he led, others followed, and he was always surrounded by a coterie of brilliant engineers, designers, mechanics and administrators.

The company was founded in 1948, and its episodic progress built in a crescendo during the 1950s to its first peak in the mid-'60s with Jim Clark's first World Championship victory and the legendary Indianapolis success. A series of peaks and troughs followed during the next two decades, with a slow but gradual tailing off after Chapman died in 1982, to the Team's final demise in 1994. Team Lotus successes were many and included seven World Constructors' titles and six Drivers' World Championships. There were seventy-nine Grand Prix victories between 1957 and 1994, plus numerous non-championship F1 wins and Lotus successes in Formula 2, Formula 3, Formula Junior and Formula Ford, Indianapolis, the World Championship for Makes, including class wins at Le Mans, as well as GT1 and GT2

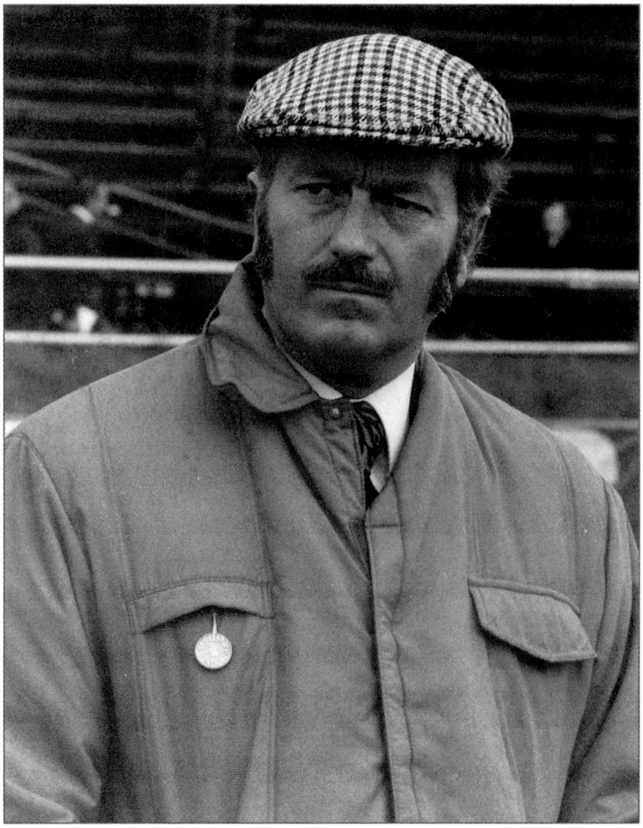

For the first thirty years or so, the development of Team Lotus and the racing cars built by Lotus Components revolved around Colin Chapman, pictured here around 1974.

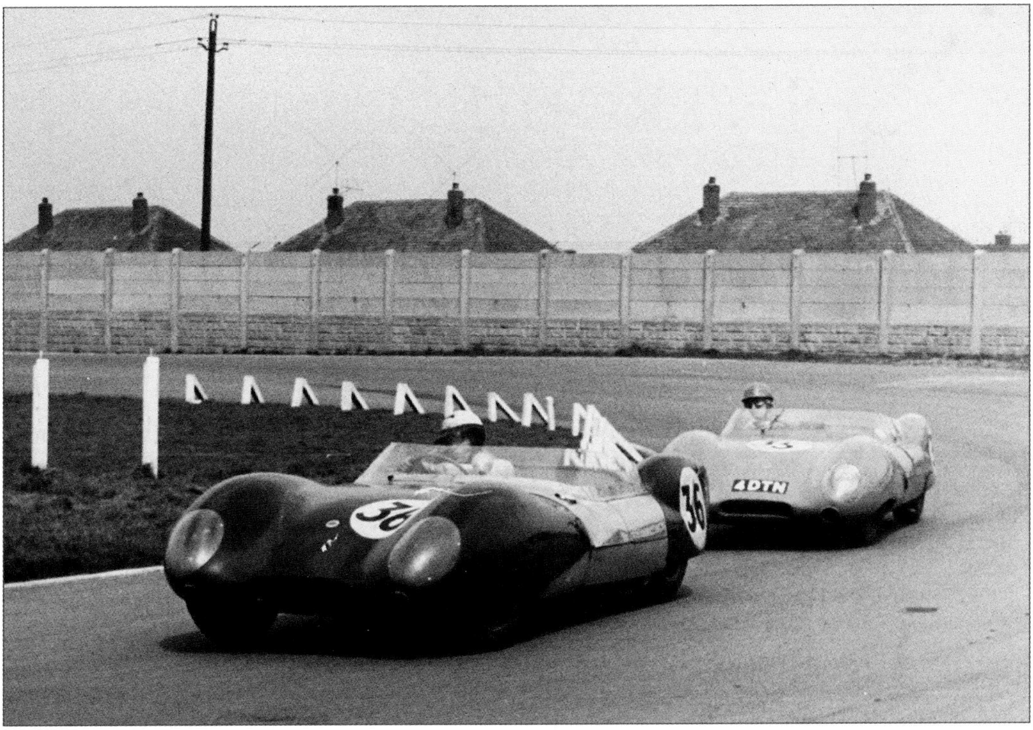

Lotus was always an innovator, and led the way with lightweight cars that had highly streamlined bodies. One of the first to appear was the Eleven, introduced in 1956. Here, the Elevens of Alan Stacey and Keith Hall contest the lead of the sports car race at the *Daily Mail* International 200 at Aintree on 19 April 1958. Stacey, fastest in a wet practice at 1 minute 19.5 seconds, put in a best lap of 1 minute 9.5 seconds in the race, leading for nine laps until his engine overheated, dropping him to third and leaving the win to Hall, with Tom Dickson's Eleven second.

in more recent times. Chapman brought about major changes in the sport, notably in sports car racing in the 1950s, Formula 1 and Indianapolis, and on occasion he ingeniously circumvented the rules to get his cars to the grid.

The year 1968 marked a turning point in his attitude to the sport, and his business. That's why the two volumes of *Lotus Racing Cars* are divided at this point. The pivotal event was the death of Chapman's master driver Jim Clark, followed swiftly by that of Mike Spence. Other Team Lotus drivers had died in the past – Mac Fraser, Alan Stacey – but the Chapman-Clark relationship had been far deeper, and he was severely shaken. The future of the racing team hung in the balance for some weeks, and it is possible that had it not been for the strength and cohesiveness of Graham Hill's personality – not to mention a win at Jarama – there would not have been the will to continue. At least, not without a break of some kind.

No subsequent Team Lotus driver enjoyed the same rapport with Colin as Jim had. Graham and Colin went back an awful long way, so that relationship was a different matter. Chapman, always in a hurry and keen to move on to the next thing, might have espoused powerboats or aircraft at this point. He had already tasted the first fruits of major commercial sponsorship. And this is the other reason for splitting the two volumes at 1968. Clearly, Team Lotus racing cars up to that point had remained true to the traditional British Racing Green – well, green at least, given a yellow

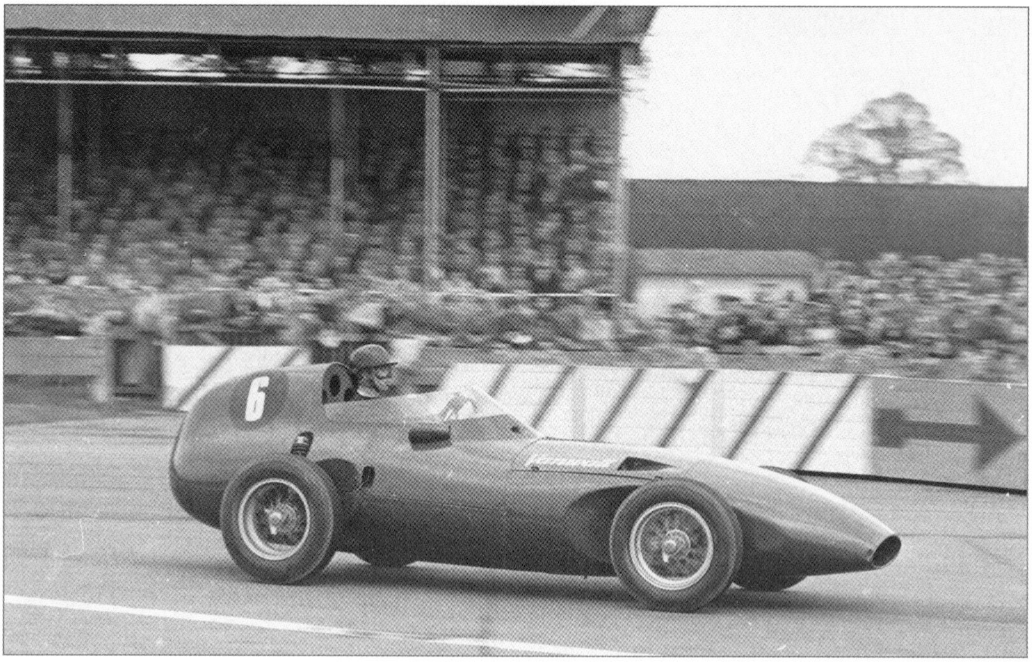

Colin Chapman's expertise was sought by other racing teams, and in 1955 he was asked to redesign the Formula 1 Vanwall's chassis and rear suspension, while Frank Costin styled the rotund bodywork. Chapman was nominated as Vanwall's third driver for the 1956 French Grand Prix as team-mate to Schell and Hawthorn, but his car was sidelined by a practice incident. Here, Tony Brooks is pictured in the Vanwall at Goodwood during the Easter weekend 1957.

racing stripe or two. Post-1968 of course, Lotuses were liveried in commercial brands of one sort or another. This was typical of Chapman – to have the first racing team outside the USA to endorse the products of non-motor industry companies, and to do it thoroughly, too.

Chapman was very much his own man, did not suffer fools at all, and was a swift and mostly accurate judge of character. He could thus command wholehearted and unquestioning loyalty from his chosen employees. On a good day, he was excellent company. On a bad day, it was best to give him as wide a berth as possible. In the mid-1970s I worked on the fringe of Team Lotus, running the JPS press office alongside Grand Prix statistician Mike Doodson, and the Travco motorhome that served as Team support vehicle was not a good environment to be in if things were going badly. Mr Chapman would be present during drivers' briefings, and it was our job to be on hand to catch the appropriate form of words regarding the Team's fortunes to broadcast to the world's motoring media. They were heady times, of course, and although the stakes were high and only the sharpest witted survived, the atmosphere was nowhere near as up-tight, sanitised and in thrall to Mammon as it is today. Go to the Goodwood Revival meeting to catch a glimpse of how it used to be, even in the 1970s, when the big-time sponsors were on the scene.

As a race fan during childhood in the '50s and '60s, it was impossible not to be a Lotus enthusiast. One knew little of the fragility of Lotuses and the ethos behind Chapman's methods then, despite being a fairly regular reader of the motor racing press. But from 1960 onwards, Lotuses were always up there in the big time doing the

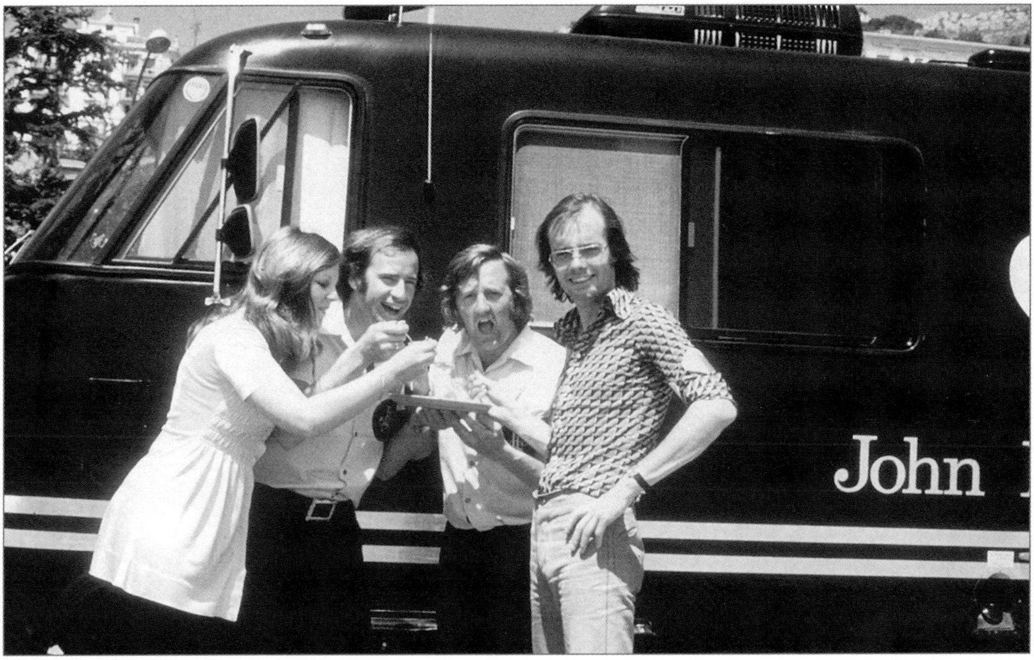

Big time sponsorship brought better conditions for the Team Lotus mechanics, including cooked meals in the paddock. Here at Monaco in 1973, the cooks Anne and Mike Murphy demonstrate their cuisine to Player's executive Ian Catt and JPS Press Officer John Tipler, right, outside the Dodge Travco motorhome that served as the Team's hospitality vehicle.

winning, even though they might have fallen apart as they crossed the line, or before that in some unfortunate cases. Come to think of it, one of my first Scalextric cars was a Lotus 16 – hardly a winner exactly, but so aesthetically pretty it couldn't be ignored. From Stirling and Innes to Jimmy and Trevor, Lotus could do no wrong. The attrition rate was high, and not necessarily because the cars broke. One of Team Lotus's first casualties was the highly promising Alan Stacey, killed at Spa in 1960 when he was struck in the face by an errant pheasant while travelling at top speed. I suppose that, like Chapman, one remained saddened by the demise of a hero, but moved swiftly onto the next one.

For the next seven years, then, Jimmy Clark wore the Grand Prix winner's laurel wreath more often than not, to the point that his victories were almost predictable. Statistics show otherwise, but that was how it seemed, and our interest veered away from F1 to tin-tops, CanAm and World Sports prototypes. However, there were always teenage kicks to be had watching the antics of John Miles in a Lotus Elan or Sir John Whitmore in a Lotus-Cortina at Snetterton or Brands Hatch, our closest venues. I graduated to reporting on minor club events for *Autosport* and *Motoring News* in the late 1960s at places like Lydden Hill, where it was still possible to find highly modified Lotus Sevens to the fore in Clubmans' events. On Sundays I worked for Brands Hatch Press Officer Graham Macbeth, handing out passes and photographers' armbands mostly, but it was my first taste of being 'on the inside'. One of the support races for the Grand Prix was a celebrity race in Escort Mexicos, and on one such occasion I recall Chapman swapping paint with the Grand Prix

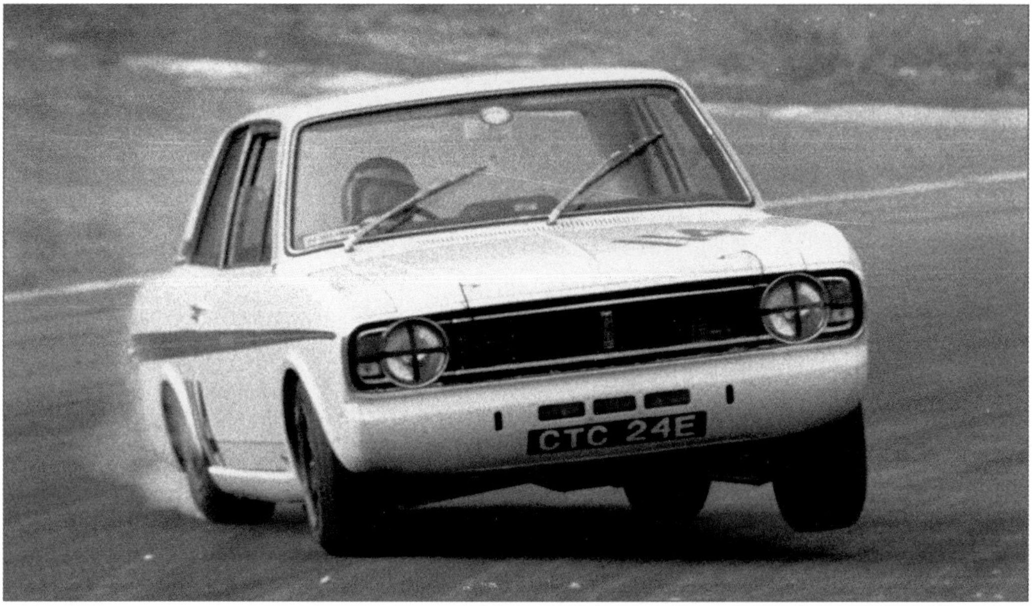

Jacky Ickx drifts the works Lotus-Cortina Mk 2 through South Bank bend (as it was known) at Brands Hatch on 28 August 1967 on his way to a class win and second overall. As a John Player Team Lotus F1 driver in 1974, Jacky gave rides to journalists around the host GP circuit to promote the Grand Prix, and to sit alongside him in an Elan was to be in the presence of a supreme expert. Jacky once treated me to a few laps in the Chevrolet Corvette of US journalist Pete Lyons, which was equally exhilarating.

greats in outrageous fashion, to the delight of the assembled media personnel. After all, Chapman was no mean driver in his own right.

From a weekend rôle in the Brands Hatch press box it was a short step to the East End offices of Stanbury Foley, the motor racing consultancy that handled JPS Lotus publicity, among others. Noel Stanbury subsequently became Team Lotus's Commercial Manager in the 1980s, while Barry Foley was more noted for his long-running 'Catchpole' cartoon in *Autosport* magazine. Back in the early '70s our office hack was a black and gold Lotus Elan Plus 2, and my own transport was an Elan S4 SE. We virtually lived at race tracks, and were certainly present at some event or other every weekend. One of our functions was to obtain publicity for the John Player-funded British Grand Prix, and the Team Lotus drivers of the day, Emerson, Ronnie and Jacky Ickx were engaged to give the world's motoring journalists rides around Brands Hatch or Silverstone aboard Lotus-supplied road cars. After the hacks had struggled, knees knocking, to the bar to regain their equilibrium, those of us on the periphery could catch a ride as well. It was a glimpse into another world, populated by gods whose skills were so innate as to be unattainable and indecipherable. I say this mainly to illustrate what calibre of talent you find at the wheel of a Lotus racing car. If the vehicles themselves were, without exception, pretty special, then so were the stars that raced them.

CHRONOLOGY

1947 *Mk 1:* Colin Chapman's first car. Trials special, based on 1930 Austin 7 saloon chassis fitted with plywood and alloy body.

1949 *Mk 2:* Trials special, based on Austin 7 chassis with extra tubular bracing. Ford 10 engine. Later used for racing.

1951 *Mk 3:* Racing car built to 750 Formula with modified Austin 7 chassis and engine featuring Chapman's divided inlet manifolding and ports. Aluminium body. First car to be sought by other enthusiast drivers. Lotus Engineering formed on 1 January 1952, new premises at Tottenham Lane, Hornsey.

1952 *Mk 4:* Road-going and Sprint car built on Austin 7 chassis. Ford 10 engine, Ford 8 gearbox and Austin 7 rear axle. Aluminium body. ***Mk 5:*** Not made. Intended as Austin 7-powered sports-racing car based on Mk 3.

1953–5 *Mk 6:* First production sports car, sold mainly in component form. Multi-tubular chassis. Aluminium body. Engine: 1172cc side-valve Ford 10. Became the car to have in British club racing in the mid-'50s. Lotus takes stand at 1955 Earls Court Show as members of the SMMT.

1954 *Mk 8:* First Lotus streamliner. Sports racing car with Frank Costin-designed aerodynamic aluminium bodywork. Tubular spaceframe chassis. Engines from 1100cc Coventry-Climax to 1500cc Connaught.

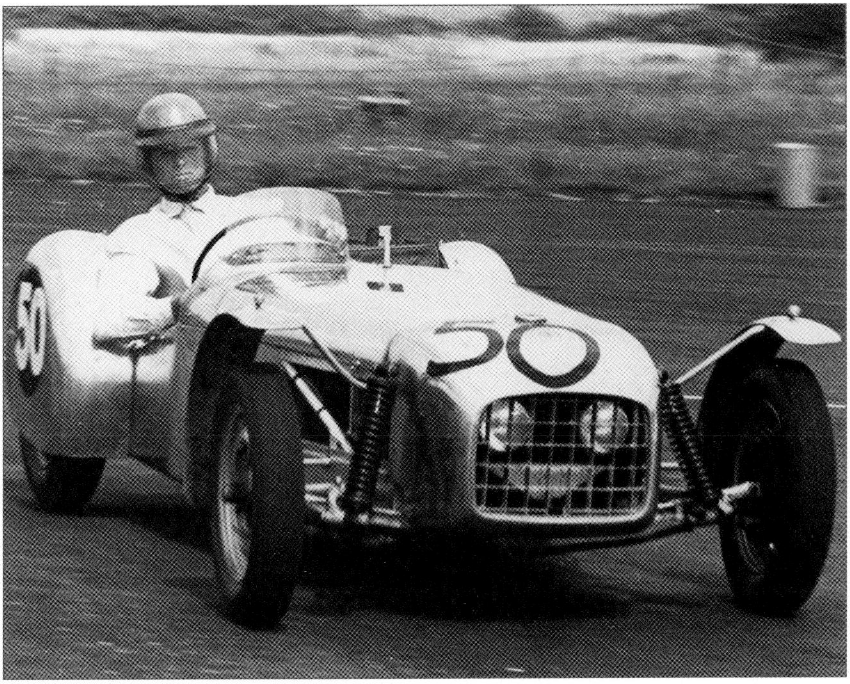

The Mk 6 was the first Lotus production car, made between 1953 and 1955 as a spaceframe component car, powered by a variety of engines including Ford 10, Ford Consul, MG and BMW, ranging from 750cc to 2.0-litres. They made 110 units, and this is Hazel Chapman at Silverstone – she once beat her husband in a hill climb in Dorset.

1955 *Mk 9:* Successor to Mk 8. Sports racing car with aerodynamic aluminium bodywork. Tubular spaceframe chassis. Engines from 1100cc Coventry-Climax to 1500cc MG. First Lotus appearance at Le Mans 24-Hours. Graham Hill gets first drive for Team Lotus. ***Mk 10:*** Larger-engined version of Mk 8. Sports racing car with aerodynamic aluminium bodywork. Tubular spaceframe chassis. Aluminium body. Engines: 2.0-litre Bristol or 2.0-litre Connaught.

1956–8 *Eleven:* Sports racing car, replacement for Mk 9. Spaceframe chassis. Series 2 had wishbone front suspension. Aerodynamic aluminium body. Engines: from 750cc Coventry-Climax, 1100cc Climax, 1172cc Ford, to 2.0-litre Maserati. class win and Index of Performance victory at Le Mans in 1957.

1957 *12:* First single-seater Lotus, Formula 2 car. Spaceframe chassis. Aluminium body. Engine: 1500cc Coventry-Climax. Grand Prix debut at Monaco 1958.

1957–61 *14 Elite:* First Lotus production car, sports GT model. Fibreglass monocoque construction reinforced with steel tubing and plates. Engine: 1216cc Coventry-Climax. Lotus's move to Cheshunt factory in June 1959 brings greater production capacity.

1957–73 *Seven:* Replacement for Mk 6. Multi-tubular spaceframe chassis. Aluminium body, later with fibreglass nosecone and clamshell wings, ultimately all fibreglass. Engines: various, ranging from 948cc BMC to 1600cc Lotus twin-cam.

1958–60 *15:* Sports racing car, bigger engined successor to Mk 10. Spaceframe chassis. Aluminium body. Engines: 1500cc to 2500cc Coventry-Climax.

1958–60 *16:* First Lotus Formula 1 car. Single-seater Formula 1/Formula 2 car. Successor to 12, with Costin-inspired 'Baby Vanwall' body. Spaceframe chassis. Aluminium and fibreglass body. Engines: 1500cc to 2500cc Coventry-Climax.

1959 *17:* Sports racing car, smaller than Eleven. Spaceframe chassis. Aluminium and glass-fibre body. Engine: 750cc or 1100cc Coventry-Climax.

1959–60 *18:* Rear-engined single-seater for Formulae 1, 2, and Junior. Tubular spaceframe chassis. Aluminium and fibreglass body. Engines: 1000cc Ford to 2500cc Coventry-Climax. Alan Stacey debuts FJ car at Boxing Day Brands Hatch meeting. Stirling Moss wins Monaco GP, Lotus's first World Championship success. Also Jim Clark's first GP.

1960–1 *19 Monte Carlo:* Rear-engined sports racing car, developed from 18. Spaceframe chassis. Aluminium and fibreglass body. Engines: 1998cc or 2500cc Coventry-Climax. Thirteen units built. **19B:** One-off sports-racer with revised suspension for Dan Gurney in 1962/3. Engine: Ford V8.

1961 *20:* Single-seater Formula Junior production car. Engine: 997cc or 1097cc Ford. 118 units built. ***20B:*** Single-seater Formule Libre or Formula B car. Disc-braked version of 20. Engine: 1498 cc Ford. ***21:*** Formula 1 car, spaceframe chassis, developed from 18. Engine: 1498cc Coventry-Climax. Eleven units built.

1962 *22:* Single-seater Formula Junior car developed from 20. Altered suspension, canted engine. Seventy-seven units built. ***23:*** Sports-racing car developed from 20 and 22. Engines: 997cc, 1097cc and 1599cc Ford. 131 units built. ***23B:*** Sports-racing car with strengthened chassis. Engine: Lotus-Ford twin-cam – first use of this engine at Nürburgring 1000km 1962. All later 23s produced to this specification. ***24:*** Single-seater Formula 1, spaceframe chassis. Engine: 1498cc Coventry-Climax V8. Twelve units built, plus three externally by Reg Parnell Racing. ***25:*** The world's first monocoque Grand Prix car debuts at Zandvoort. Single-seater Formula 1, aluminium monocoque. Engine: 1.5 litre Climax V8. seven units built. Jim Clark and Team Lotus are 1963 F1 World Champions. ***26R Elan:*** Racing derivative of production

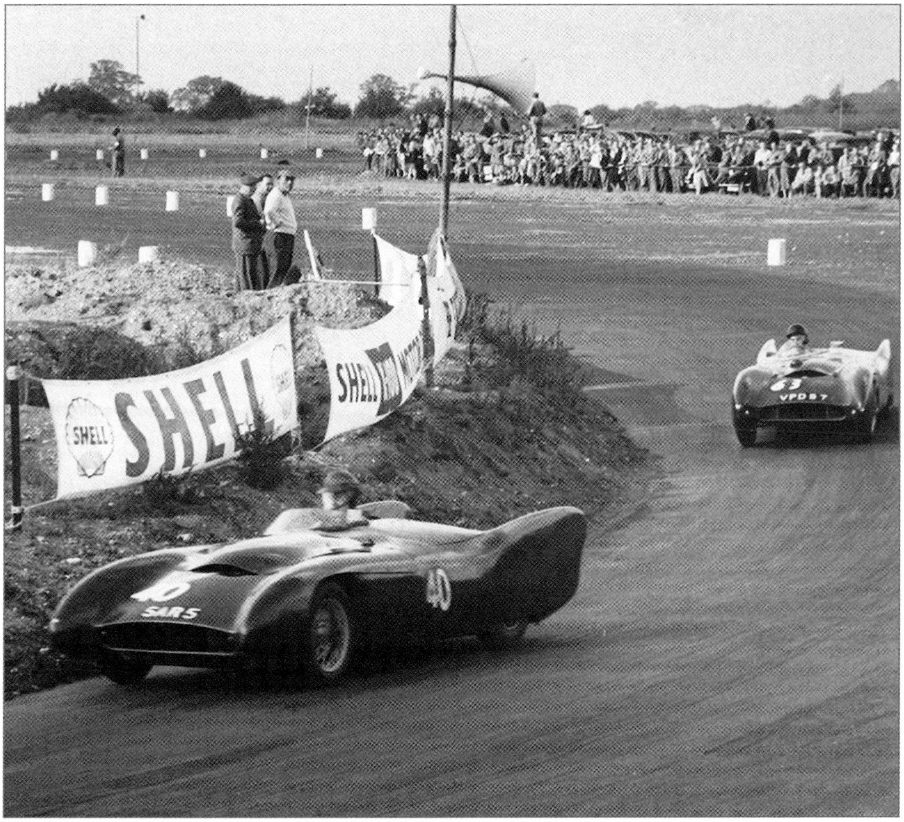

Colin Chapman's 1.5-litre MG-powered Mk 8 leads John Coombs's similar car through the Esses at Snetterton in 1954. This complex was much more worthy of the name in the days of the original circuit, and the trackside oil drums were fearsome objects. Crowd protection? Forget it!

sports car. Fabricated steel box-section backbone chassis. Engine: 1558cc Lotus-Ford twin-cam. Ninety-seven units built.

1963 ***27:*** Single-seater Formula Junior development of 25 monocoque chassis. Engine: 1097cc Ford. ***28 Lotus-Cortina:*** Two-door saloon homologated as Touring Car. Engine: BRM-built 1558cc Lotus-Ford twin-cam. Forty-nine units built by Lotus Components. Jim Clark is British Saloon Car Champion in 1964, Sir John Whitmore is European Touring Car Champion in 1965. ***29:*** First attempt at Indianapolis. Single-seater car, based on 25 monocoque with longer wheelbase, offset left suspension. Engine: 4260cc Ford V8 pushrod. Three units built. Jim Clark comes second in Indy 500.

1964 ***30:*** Sports-racing car for Group 7 category. Engine 4727cc Ford V8. Series 1: twenty-three units built. Series 2: ten units built. ***31:*** Single-seater Formula 3 car, developed from 22. Engine: 997cc Cosworth-Ford SCA. Twelve units built. ***32:*** Single-seater Formula 2/Formula B monocoque developed from 27. Engine: 997cc Cosworth-Ford SCA (Formula 2). Twelve units built. ***32B:*** Single-seater Tasman series derivative of 32. Engine: 2495cc four-cylinder Coventry-Climax FPF. One unit built. ***33:*** Single-seater Formula 1 monocoque developed from 25 with revised suspension and steering, wider wheels and tyres. Engine: 1.5-litre Coventry-Climax V8. Six units built. Jim Clark reigns as World Champion with six GP victories. ***34:*** Single-seater Indianapolis car developed from 29. Engine: 4200cc four-cam Ford V8. Three units built.

1965 ***35:*** Single-seater Formula 2/Formula 3/Formula B monocoque developed from 27 and 32. Engine: 997cc Cosworth-Ford SCA. Twenty-two units built. ***36:*** Elan Coupé. ***37:*** Seven-based 'Three-Seven' Clubman's Formula sports racing car. Engine: 1499cc Ford. One unit built. ***38:*** Jim Clark's victorious car at Indianapolis. Single-seater monocoque chassis with longer nosecone. Engine: 4200cc Ford V8. Five units built. ***39:*** Single-seater Formula 1/Tasman monocoque with tubular engine bay. Engine: intended for flat-sixteen Coventry-Climax; 2.5-litre Climax FPF four-cylinder engine fitted for 1966 Tasman series. One unit built. ***40:*** Sports racing car for Group 7 category, based on 30 with heavy-duty suspension. Engine: 5.8-litre Ford V8 engine. Three units built.

1966 ***41:*** Single-seater Formula 3 spaceframe chassis. Engine: 997cc Ford-Cosworth SCA. Sixty-one units built. ***43:*** Single-seater Formula 1 truncated monocoque. Engine: 3.0-litre BRM H16, acting as stressed chassis member. Two units built. ***44:*** Single-seater Formula 2 monocoque, developed from 35 fitted with Lotus 41 rear suspension. Engine: 997cc Cosworth-Ford SCA. Three units built. ***45:*** Series 3 Elan. ***46 Europa:*** Production mid-engined coupé. First car to be produced at the new Hethel factory in November 1966. ***47:*** Sports GT mid-engined coupé, based on Europa, for Group 4 category. Engine: 1558 Lotus-Ford twin-cam. Fifty-five units built.

1967 ***41B:*** Single-seater Formula 2 spaceframe chassis run by Lotus Components for Jackie Oliver. Type number also allocated to type 41 Formula B cars sold to the USA. ***41C:*** Single-seater Formula 3 spaceframe based on 41 with Europa-type front suspension. ***42:*** Single-seater Indianapolis monocoque. Engine: intended to be 4.2-litre version of BRM H16, subsequently fitted 4.2-litre Ford V8. Two units built. ***48:*** Single-seater Formula 2 monocoque with tubular frame engine bay. Engine: 1599cc Cosworth-Ford FVA. Four units built. Jim Clark perishes in his 48 at Hockenheim 7 April 1968. ***49:*** Single-seater Formula 1 truncated monocoque. Engine: 3.0-litre Cosworth-Ford DFV V8, acting as stressed chassis member. Twelve units built (including subsequent 1968/9 49Bs and 1970 49Cs). Jim Clark wins 1967/8 Tasman Championship, also takes his twenty-fifth GP win at Kyalami. Team Lotus adopts Gold Leaf livery of new sponsors John Player & Son. ***51:*** Single-seater Formula Ford spaceframe based on Lotus 22 and 31. Engine: 1599cc Ford pushrod. 218 units built.

1968 ***56:*** Single-seater Indianapolis turbine car. Wedge-shaped monocoque tub. Engine: Pratt & Whitney 5T6 turbine. Four units built. Mike Spence dies in Indy qualifying accident. Turbine project resurfaces in F1 in 1970.

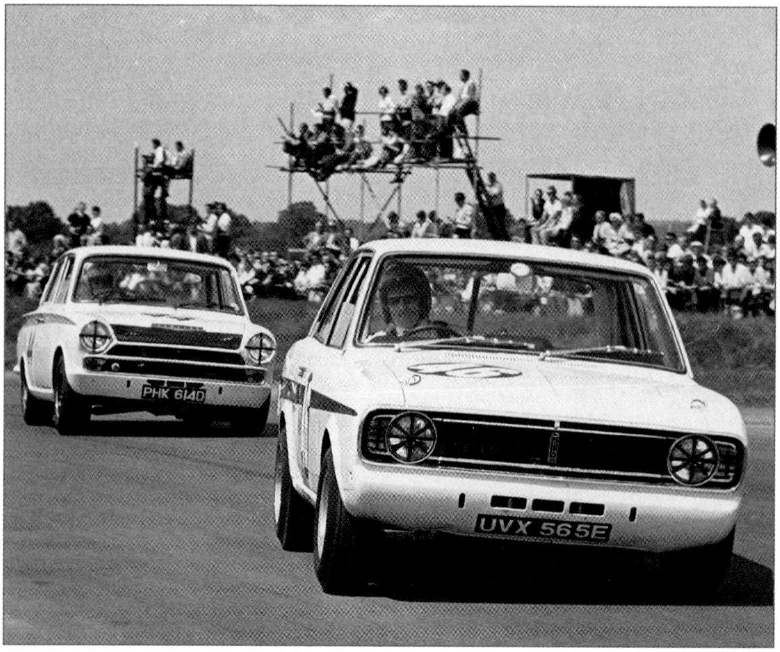

Mks 1 and 2 Lotus-Cortinas in action at Silverstone in 1967. This is the Grand Prix meeting on 15 July and the Mk 2 is in the hands of Lucien Bianchi, with John Miles driving the older model. Paul Hawkins drove the Team Lotus Mk 2 model to fourth overall and first in class, while Miles was sixth overall and third in class. Guest driver Lucien Bianchi's Mk 2 Lotus-Cortina was entered by the Ford Motor Company but retired. Elford's red Porsche 2.0-litre 911S was the team's principal opposition during 1967, and the Team Lotus cars went over to 210bhp Group 5 Cosworth FVA engines instead of the BRM twin-cams. They were now out-and-out racing cars rather than the 'family saloons' they'd started as. With suspension mounts altered and non-standard brackets fitted, they were good for 150mph.

CHAPTER ONE

GETTING THE SHOW ON THE ROAD

Colin Chapman's career as a racing car builder was founded on a structural engineering degree course at University College, London, in 1945. He augmented his student income by dabbling in second-hand cars. When post-war petrol rationing made this unviable, he turned his attention to transforming a 1930 Austin 7 into a sporting special. It was a tried and tested formula for enthusiastic motoring, but Chapman developed it successively into the Lotus Mk 6, which became the mainstay of the club racing fraternity in the early 1950s. In 1957 it metamorphosed into the Mk 7, better known as the Seven, which was arguably Lotus's most successful car, insofar as it lives on today in the Caterham Seven.

Chapman's Austin 7 special of 1948, registration number OX 9292 and named the Lotus Mk 1, was based on the chassis of a fabric-bodied Austin 7, clad in alloy panels over bonded plywood, which endowed it with greater rigidity. He extended the rear of the Mk 1 body to accommodate passengers or trialling ballast. Chapman discovered that inverting the rear leaf springs cured the Austin 7's propensity to oversteer. He also created an independent front suspension system by splitting the front axle beam and pivoting it in the centre, thereby ensuring that the wheels were vertical during cornering. The Austin 7 Special was assembled in a lock-up garage belonging to the father of Chapman's girlfriend Hazel Williams, and the pair campaigned it successfully in the mud and ruts of trials competitions all over England.

As a student, Chapman was a member of the London University air squadron, and designed the car along the principles of aircraft construction. He spent much of his time mulling over technical treatises at the Institute of Mechanical Engineers' library in order to ascertain what stresses cars were subject to on the limit, and what could be done to counteract them. This was not something that contemporary car designers devoted much thought to, and it accounted for much of the success that the Mks 6, Seven and Eleven enjoyed.

He spent his National Service in the RAF, and learned to fly a Harvard two-seat trainer. This skill became a passion that stood him in great stead when he commuted from home base to race circuit, often flying out for practice and then home again before going out once more on race day.

Meanwhile, back in 1950, Chapman had built the Mk 2 Lotus in his spare time. Based on another Austin 7 chassis and clad in an aluminium cigar-tube body, it was thus more sporty looking than the Mk 1, and its main chassis members were boxed in, with tubular braces replacing the existing cross members. It was powered by an engine from a Ford 8, subsequently replaced with a Ford 10 unit, and the Austin 7

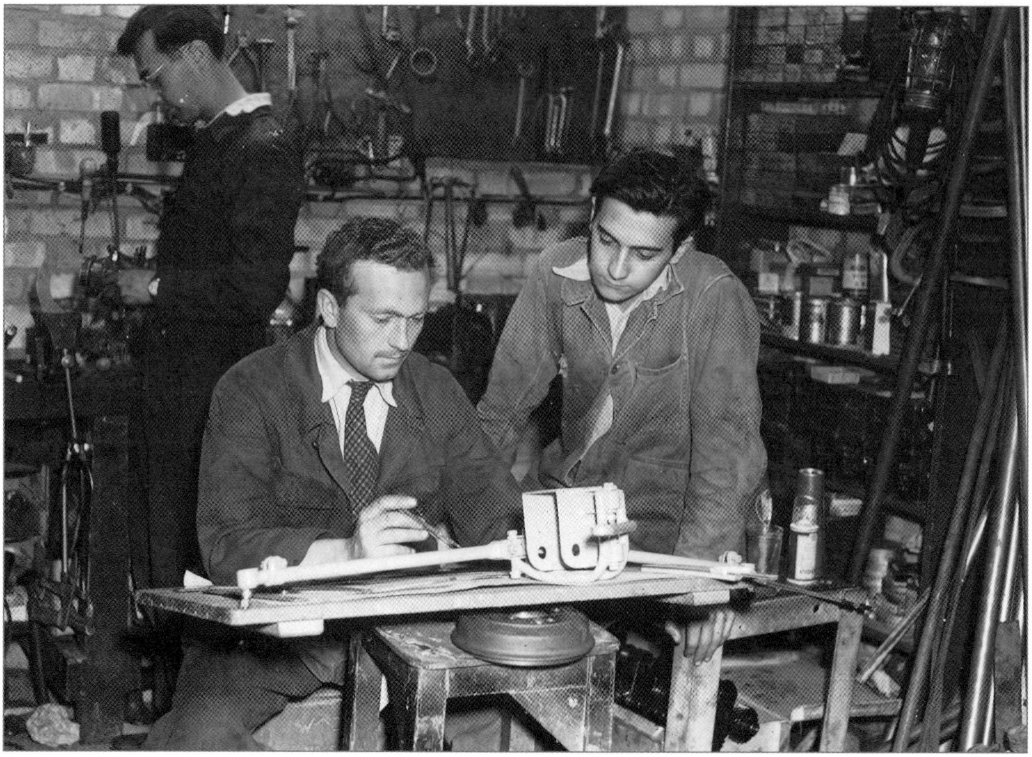

In 1951, Colin Chapman became friendly with 750 Motor Club members Nigel and Michael Allen who had a workshop in Wood Green. The brothers acquired a pair of Lotus Mk 3s – Colin and Michael deliberate over the front suspension on a makeshift drawingboard – and they were closely involved in the design of the Mk 6. While Chapman drove the works' car in 750 championship events, Michael Allen and Hazel – the future Mrs Chapman – drove it with no less gusto at minor meetings.

rear axle was retained. As an example of his dogged resourcefulness, Chapman obtained a 4.55:1 ratio final drive by pairing up an unmatched 42-tooth crown wheel with a 9-tooth pinion. He achieved this by filling the diff with metal polish and driving it for 50 miles. The bearings were ruined, but the crown wheel and pinion meshed perfectly.

Special building was still very much a hobby, and at this point Colin worked for British Aluminium as a construction engineer. He was nevertheless very much abreast of contemporary motor sport, and noted that Holly Birkett of the 750 Motor Club had just established the unsupercharged 750 Formula for Austin 7s in club racing. Chapman based his next creation on that formula. With the Mk 2, Chapman and Hazel were active most weekends in trials, rallies or some 750 Formula club event, and they had many successes. Chapman's first proper race in the Eight Clubs meeting at Silverstone on 3 June 1950, in which he beat Dudley Gahagan's Mk 37 Bugatti, spurred him to build a car that could win on the road rather than the less predictable terrain of a trials course.

To this end, he created the equally austere and spindly Mk 3 Lotus in 1951. The basis for the prototype was a 1930 Austin 7, the chassis of which was given the same beefing-up treatment with side-members boxed in and cross-members replaced with 14-gauge tubular beams. Every component was weighed prior to installation, and if deemed too heavy, was worked on or discarded for something lighter. Around this time Chapman became friends with the Allen brothers, Nigel and Michael. They were 750 Motor Club members who lived in Wood Green, and had reasonable workshop facilities. They were fired by Chapman's enthusiasm and built a pair of similar cars to the Mk 3 Lotus. They were also closely involved in the gestation of the seminally important Mk 6, which came out in 1953. In the chronology of Lotus models, the Mk 3 was superseded by the Mk 4, which was ostensibly intended as a trials machine, and the Mk 5, which was never completed, was intended to be a circuit racer.

There was sufficient interest at club racing level to find buyers for these cars, which were little more than one-off specials. Chapman's strategy for selling them depended on race successes to produce publicity and inspire potential buyers, and his prowess generated sufficient interest for him to consider productionising the concept, which is what led to the Mk 6. Chapman was therefore at the forefront of the kit-car industry, and as customers began to approach him for their own Mk 6, it wasn't long before he was selling the vehicles in component form.

He established two companies, one from which to dispense the kit car components, the other to manufacture complete cars. Lotus Components Ltd constructed racing cars for sale, and marketed Lotus Sevens in kit form. The company earned a modest revenue selling tuning equipment to members of the 750 Motor Club, but it wasn't enough to maintain the kind of business that Chapman had in mind. His workshop and drawing office was a former stable behind father Stan Kennedy-Chapman's Railway Arms Hotel, at 7 Tottenham Lane, Hornsey, in residential London N8.

After university, Chapman had served an apprenticeship at De Havilland aircraft factory at Hatfield, and in 1953 his day job was still as a structural engineer at British Aluminium. His personal transport was a Lotus Mk 6, which meant he was apt to arrive for work soaked to the skin. These were heady times in aviation, with the frontiers being thrust back all the time, and Chapman and his circle were similarly inspired. Three De Havilland design engineers – Mike Costin, Mac Macintosh and Peter Ross – frequently joined Chapman and his associate Michael Allen in their nocturnal activities, and were instrumental in honing Lotus chassis development. Macintosh was responsible for designing the early tubular space frames, and Ross, an alloys expert, did the castings. This background in aviation and familiarity with state-of-the-art technology manifested itself in Chapman's quest for lightness and evaluation of stresses in the chassis. Eventually this work would bear fruit in the monocoque chassis, and it was also born out by Chapman's adoption of refined aerodynamics. When the company moved to its present location at Hethel in Norfolk in 1966, Chapman the pilot had the factory's reception area built facing the airfield, so convinced was he that the aeroplane would be the inevitable mode of arrival. Even at this early stage in the development of the Lotus organisation, the charismatic Chapman commanded unswerving loyalty. This devotion would later be manifested in the willingness of Team Lotus mechanics to work 'all-nighters' – with no dinner – to fix crashed or blown-up cars, or simply to incorporate a component introduced by Chapman on a whim. The ability to inspire went back to his childhood, when as the leader of Hornsey Air Scouts he had been able to motivate his troop to construct and win soap-box derbies.

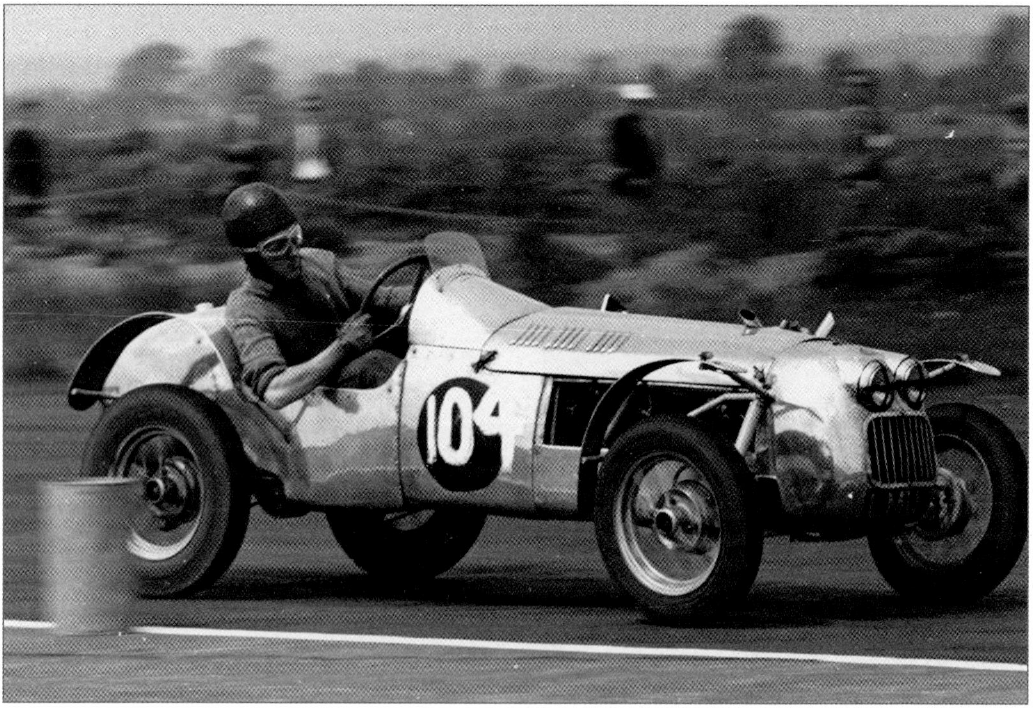

The Mk 3 was very successful in the 750 Motor Club events and Chapman received several orders for replicas, one of which was this 1951 car being driven in the 1953 AMOC meeting at Silverstone by J.B. Davidson. Lotus Engineering was founded in the wake of this enthusiasm.

The prototype Lotus Mk 6 attracted much interest in the paddock at the MG Car Club meeting at Silverstone on 5 July 1952. Michael Allen finished second in the novice's handicap after a tussle with Peter Gammon's MG TC; Gammon was later to fit his overbored, alloy Laystall-headed 1497cc MG engine and close ratio gearbox in a Mk 6 of his own. This was UPE 9, probably the most famous Mk 6 of all, and for the last twenty-nine years residing in the tender custody of Caterham's Graham Nearn, who restored it – with Len Pritchard beating out the body panels – and kept everything as was, except for fitting Spax dampers and 145SR × 15 radials. UPE 9 ran on skinny Elektron cast magnesium alloy split-rim wheels, far more sophisticated than Chapman could provide.

Chapman drove the prototype Mk 6 in the Silverstone relay event, in which they placed second after Chapman made contact with the oil drums lining Woodcote bend. The car promised much, but was subsequently wrecked en route to the *Daily Mail* 100 miles International race meeting at Boreham aerodrome, through no fault of driver Nigel Allen. In the wake of the Mk 6's demise, with no tangible assets apart from his day job, Chapman elected to go for broke and set up the Lotus Engineering Company. In February 1953, Hazel Williams as co-director brought the princely sum of £25 – a month's wages then – into the business and this formed the basis of the Lotus Engineering Company's working capital. The Chapmans were sole directors, but Mk 6 production began in earnest at this point. Chapman built eight Mk 6s virtually single-handed, taking a fortnight to construct one car, working mostly at night. The aluminium panelling that clad the chassis was fabricated by another small firm called

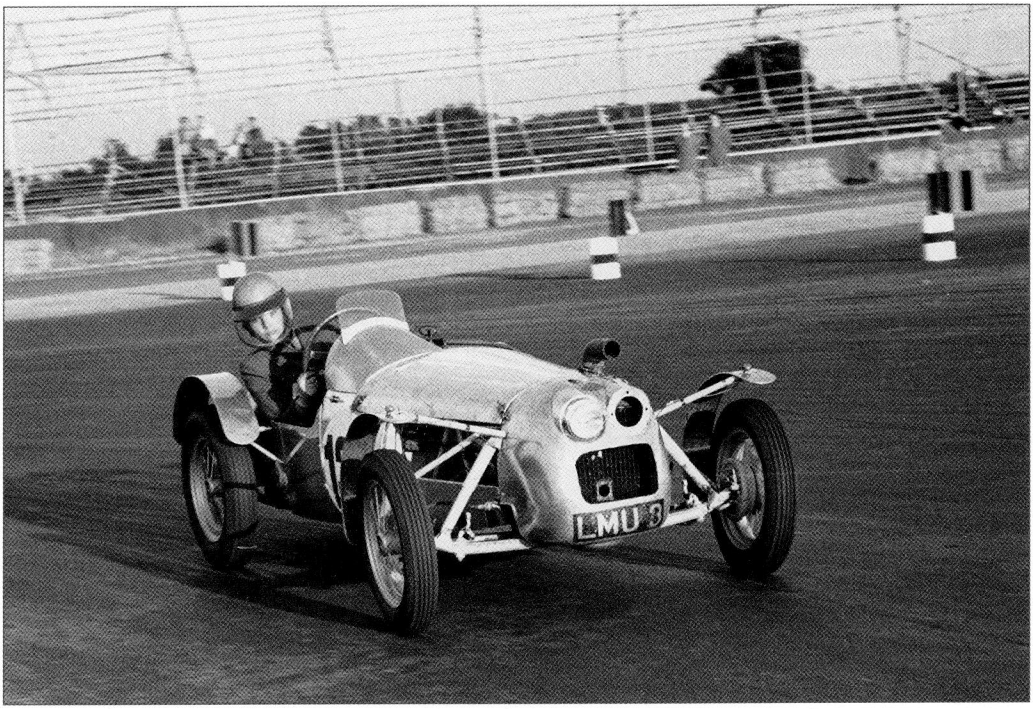

Hazel Williams – the future Mrs Chapman – drives a Mk 3 to victory in the five-lap handicap race at Silverstone in July 1951. One headlight is absent and there appears to be a ram-air induction pipe.

Williams and Pritchard. The company had moved to Tottenham Lane, and produced the panels by day, while at weekends, Chapman was more often than not at a race-track. The company's fortunes were assured at Crystal Palace on 19 September 1953, when Chapman held off the leading 1500cc contenders in a Ford 10-engined Mk 6, and all of a sudden there were customers for the cars. During the next four years, over 100 Mk 6s were sold, with some going to the States, Canada and Australia.

The Mk 6 chassis was built by Dave Kelsey and Johnny Teychenne's Progress Chassis Company, set up specially and operating from garage workshops at Edmonton. Their methods were typical of a small engineering company setting up in the make-do-and-mend climate of post-war austerity. The original chassis jig was made out of an angle-iron bed-frame, and suspension and other bracketry fashioned from whatever light-gauge steel was around. Williams and Pritchard was soon to be based in the same complex and the car was fettled by Chapman and Mike Costin. Very often the layout was assembled before its design was committed to paper on the drawing board, but that was true of evolving Lotus racing cars in the 1960s too.

The Lotus Mk 6 chassis broke new ground in Britain, because it was the first proper triangulated spaceframe, as opposed to traditional sports cars that had bodies based on a ladder chassis that was stout but crude by comparison. The Mk 6 chassis was composed of lengths of 1 inch and 1⅛ inch diameter circular and square-section tubing welded together in lengths and diagonals and braced at crucial stress points. The Mk 6 chassis was not only light, at just 55lb, but coped well with bending and torsional stresses generated at racing speeds. All loads in a spaceframe are in tension

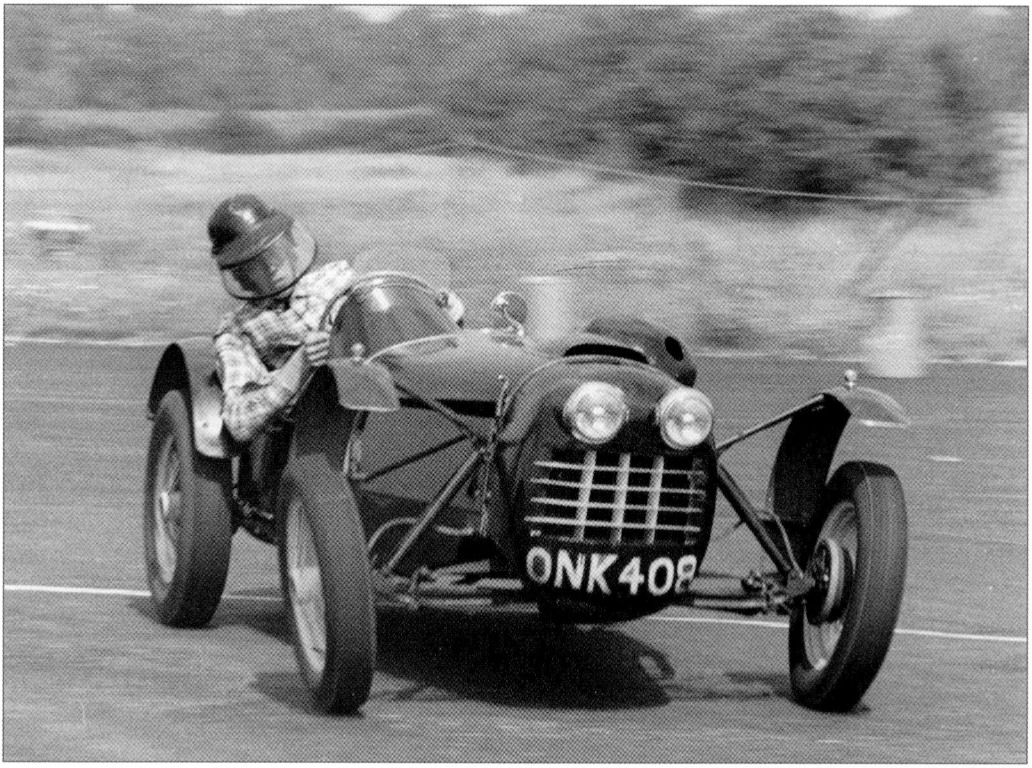

Adam Currie's Mk 3B displays negative camber angles at Beckett's corner, Silverstone, in 1953. Three of these cars had grilles modified from the original Mk 3, and ran with much-modified Austin 7 engines and A-frame chassis, panelled in aluminium. Chapman devised siamesed head ports to improve performance. Currie, meanwhile, went on to marry Chapman's cousin.

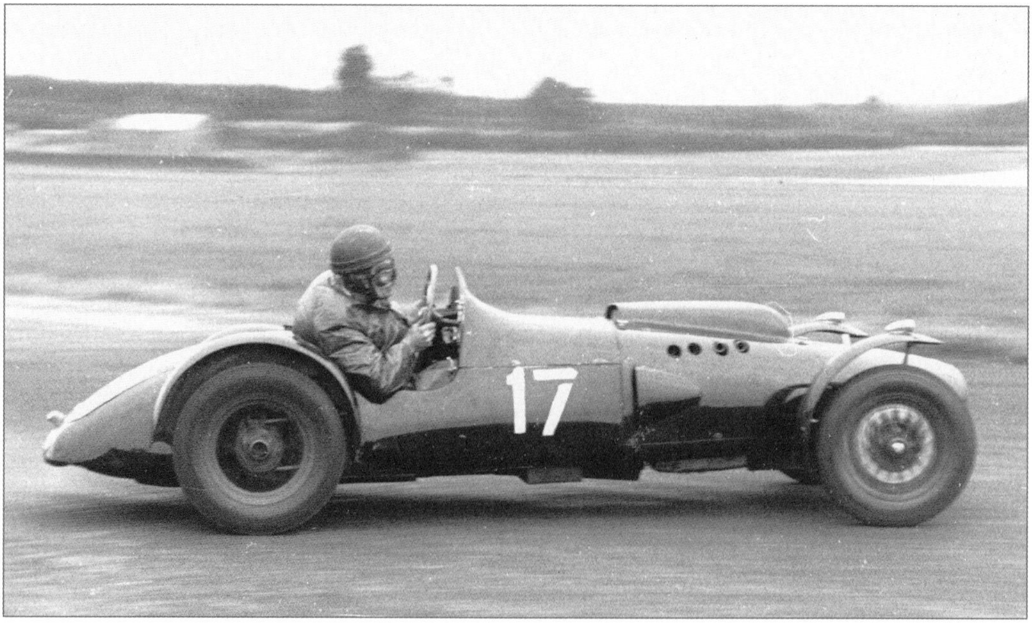

When Chapman was starting to work on the Mk 6 in 1952, the Clairmonte brothers commissioned him to make a one-off front-engined Formula 2 car. Although Lotus designed the chassis, suspension and body as the Mk 7, Chapman virtually disowned the project when the designated ERA engine blew up. The Clairmontes raced their Special with a Lea-Francis engine.

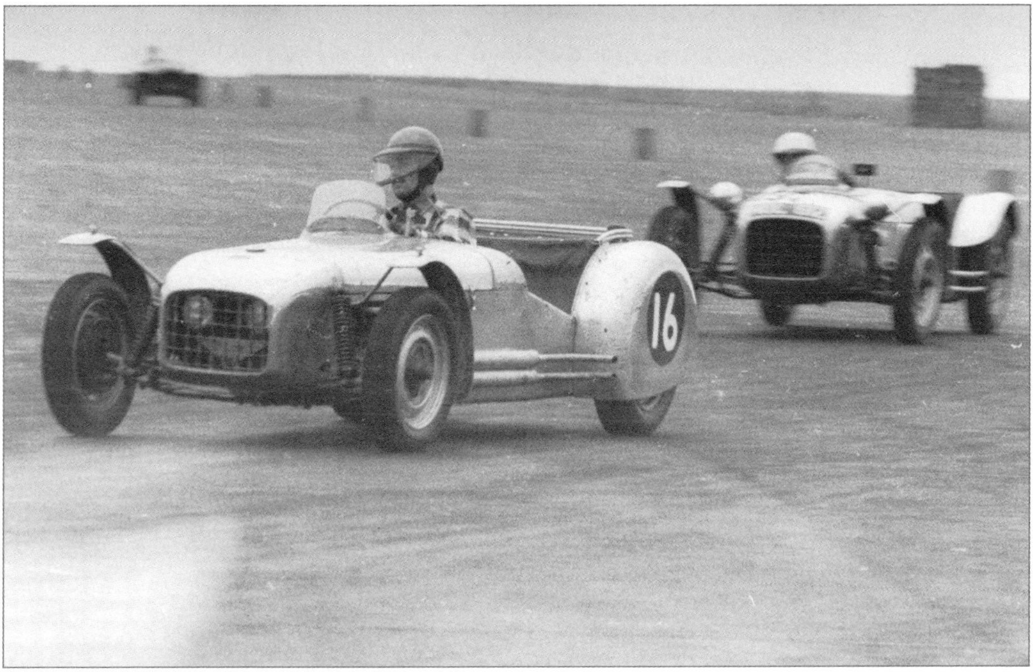

Chapman in the prototype Mk 6 rounds Woodcote at the end of the Silverstone Club straight in 1953 ahead of P.A. Desoutter in another Mk 6. Chapman's car still has the hood hoops in place, while the second car has modified rear wings and external headlamps – turned back to front.

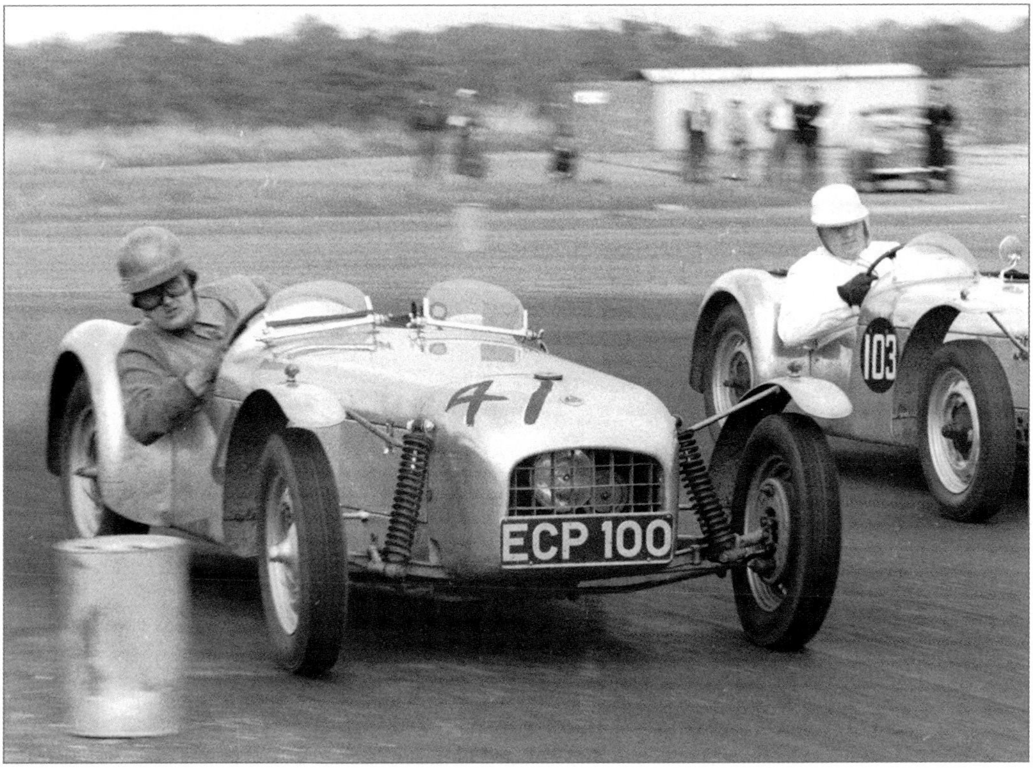

It's September 1954 and the Mk 6s of Mike MacDowell and F.V. Lambert dual for position during the SUNBAC meeting at Silverstone. It's interesting to note the varying treatment of aero screens and headlights.

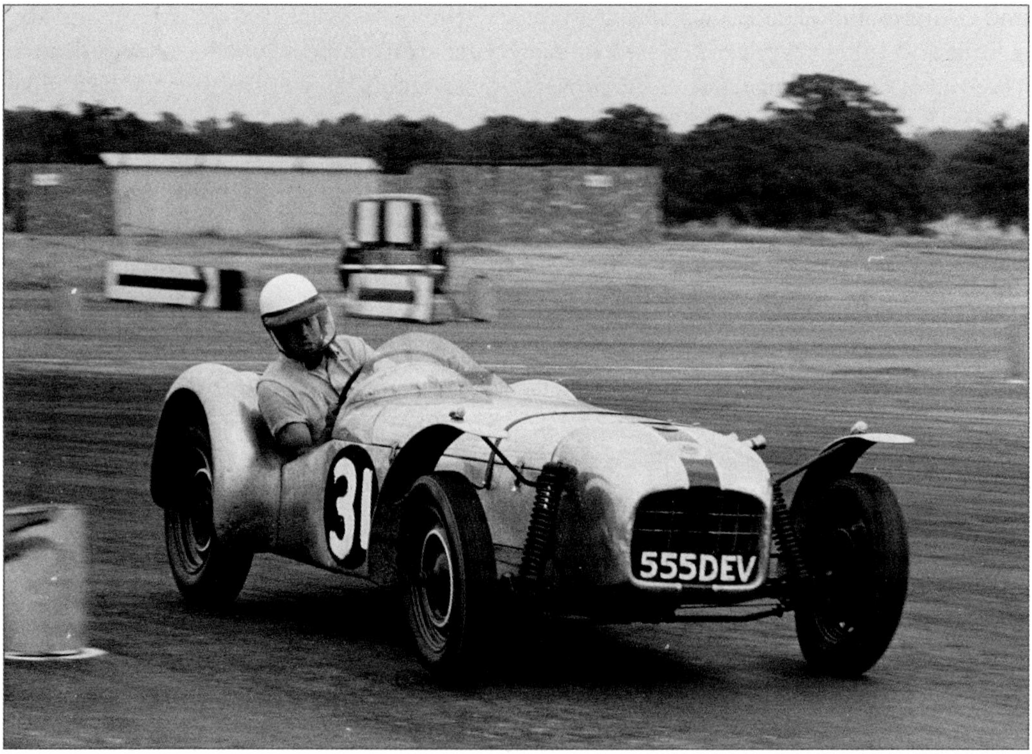

The future Team Lotus works driver Alan Stacey in his Mk 6 at Silverstone on 10 September 1955. He drove five races that day, posting two wins, a third, a fourth and a one from last.

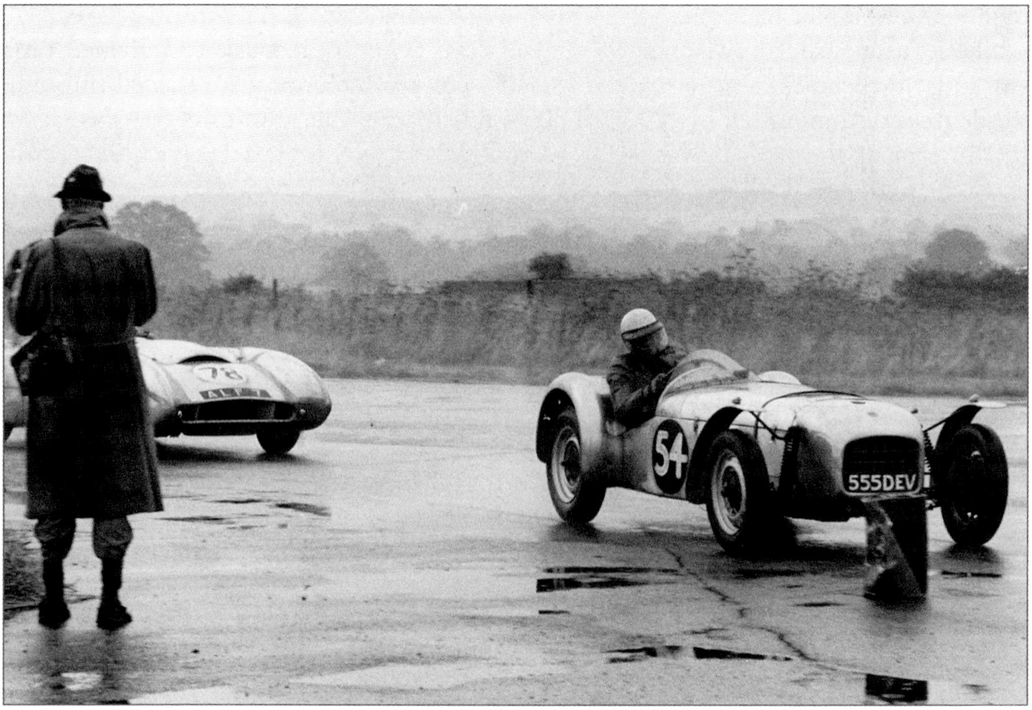

A wild and windswept West Essex Car Club meet at Snetterton on 13 August 1955. The 5 gallon oil drum serves as an apex marker for Alan Stacey's Mk 6 and John Coombs' Connaught-engined Mk 8. Forty-five years later, this Mk 6 was rebuilt by Ian Bentall and driven by Stacey's nephew Alan.

and compression. The floor and side panels of the Lotus were in aluminium, and the scuttle and undertray were riveted to the chassis rails to provide additional stiffness. The Mk 6 set the stage for every subsequent Lotus chassis up to the revolutionary Lotus 25 monocoque of 1962.

The stock Ford engine chosen by Chapman, and also coveted by Sydney Allard for one of his cars, was the Consul's. It would be reliable, economical and potentially long lasting because of its efficient oil supply and large bearings. This contrasted with the Climax engine, which used a lot of oil because it had originally been designed as a wartime fire pump unit that could be run at maximum revs from cold. But because the Consul's 1508cc engine was so new, no dealers had a spare to sell, and despite correspondence between Stan Chapman and Ford Chairman Sir Patrick Hennessy, the manufacturers wouldn't sell one. So the resourceful Colin Chapman assembled an engine out of individual components, sourced by doing the rounds of Ford dealers in the Greater London area. Michael Allen did much of the assembly work, reducing the stroke to bring the cubic capacity down to 1499cc, which brought the car into a different class of racing. Among the tomes available to Chapman in the course of his research was tuning guru Sir Harry Ricardo's book on high-performance engine design, and although it was written in the 1920s, the guidance it provided would not have been lost on Chapman.

The Mk 6 chassis had mounting points for five engines: the 1172cc Ford 10, the 1500cc Consul, the 1100cc Coventry-Climax, and 1500cc MG TC or TD units. Its suspension consisted of the split front swing-axle fabricated from modified Ford parts. The beam axle was sliced in half and plates welded over the ends. A pair of bushes fitted here pivoted on a bracket fixed to the lower front chassis cross-member, and a pair of radius arms was made by splitting the V-shaped Ford tie rod, while separate track rods incorporated a slave arm. With track rod divided and working on a central slave arm, the steering box and column were turned through 90 degrees and mounted by the base plate; a track rod then ran from the slave arm to the arm of the steering box. The regular Mk 6 used a Ford 10 back axle with shortened prop shaft and torque tube. It was located laterally by a Panhard rod, and suspension front and rear was completed by coil springs and Woodhead Monroe dampers. Brakes were regular Ford units, and consisted of cable operated 10 × 1¼ inch Girling drums, which were adequate for cars powered by the 1172cc motor.

Start of the Legend

At the beginning of 1954, Colin Chapman formed Team Lotus to distinguish his racing activities from the business enterprise of Lotus Engineering Ltd, although neither he nor Mike Costin gave up their day jobs to concentrate on Lotus until the end of the year. The Mk 8 that appeared in 1954 was a sports racing car and was a major step forward for the fledgling company in terms of chassis design and body styling. It was built up on a fully triangulated space frame chassis, made of 1¼ inch 20- and 18-gauge tubing, with two sections front and rear of a central bulkhead. The fore section was triangular in plan and the rear part was triangular in side elevation. One of Chapman's band of helpers at the time was Gilbert 'Mac' Macintosh, who worked in the stress analysis department at De Havilland aircraft. He made the calculations that dictated the stress points on the Mk 8 chassis, and it

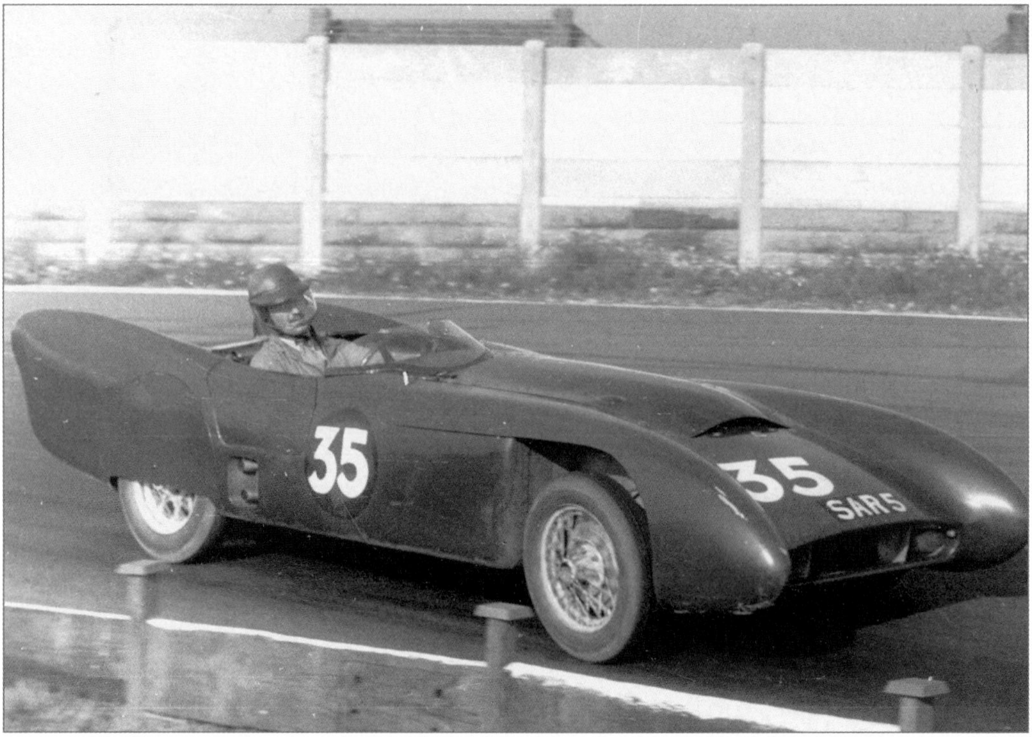

Styled by Frank Costin, the Mk 8 was the first aerodynamic sports racing car built by Lotus, and it was powered by a 1.5-litre MG, Connaught or Coventry-Climax engine. A relaxed-looking Chapman drives the works' car – SAR 5 – at Aintree in 1954. Ten of these cars were built.

was built accordingly. Chapman, on the other hand, tended to design for lightness, subsequently strengthening whatever component failed.

The Chapman front swing-axle and coil spring and damper suspension was located on a triangular frame of sheet steel, the base of which was mounted across the top chassis rails, with the upper chassis members running through the top to absorb the front suspension loads. A pyramid of four steel tubes formed the engine mountings. The downside of this layout was that it took some 12 hours to remove the engine, and twice that to replace it. At the rear, a designated mounting at the back of the chassis located a de Dion axle, transverse coil spring and Armstrong piston-type dampers.

Initially, the Mk 8 was powered by a 1467cc MG-based engine, with certain Morris components, and a Laystall-Lucas alloy cylinder head. It was fed by twin SU carburettors, and drove through a four-speed MG gearbox. Lockheed hydraulic brakes with Alfin drums, outboard at the front and inboard at the rear, slowed it down. Power output was said to be 85bhp at 6200rpm.

The aerodynamic body came as a something of a surprise after the spindly, if effective, creations seen previously. After the design had been thrashed out on the tablecloth of a West End restaurant, the definitive rotund panelling, featuring a low nose, faired-in rear wheels and prominent twin tail fins, was drawn by Frank Costin (Mike's brother). Frank was an aerodynamicist and stresses man for the De Havilland Aircraft Company. Lotus's by now customary body shop, Williams and Pritchard, expertly crafted the idea into a reality. The drag produced by underbody componentry was countered by a full-length undertray, with ducting for the rear brakes. The passenger seat could be covered by a metal tonneau for racing, and a small perspex screen around the cockpit provided a decent wind break. Instead of being prominent at the front of the wings, the headlamps folded down into the engine bay. In the days before pop-up headlights, they had to be raised by hand so they stood up through apertures in the engine cover. On the prototype SAR 5 that was campaigned by Chapman himself, the front section was the only removable part of the bodywork. The rest was riveted firmly in place, although on further models access was less restricted.

The Mk 8 debuted at the British Empire Trophy meeting at Oulton Park in April 1954. It wasn't an auspicious first appearance. In fact the car nearly didn't make it at all, since it was crashed en route to the circuit. But after the kind of feverish activity that would come to characterise Team Lotus's presence at a race circuit, Colin Chapman contrived to start from the back of the grid. The Mk 8 expired in its heat with a blown head gasket. Chapman was on a steep learning curve, and after a luckless Easter Monday at Goodwood, he posted the car's first success in the Sports Car race at Silverstone in May, winning the 1500cc class. The car was also driven to fourth place by Erwin Bauer in the Eifelrennen at the Nürburgring. Following a crash in the wet at Aintree in May, Chapman won the 1500cc class in the Sports Car race at the Whitsun Goodwood meeting. This was followed by a first at Crystal Palace, and a second place for Peter Gammon's Mk 6 at Brands Hatch. There was victory in the 1500cc Sports Car race supporting the Grand Prix at Silverstone, vanquishing even Hans Herrmann's works' four-cam Porsche 550. This achievement

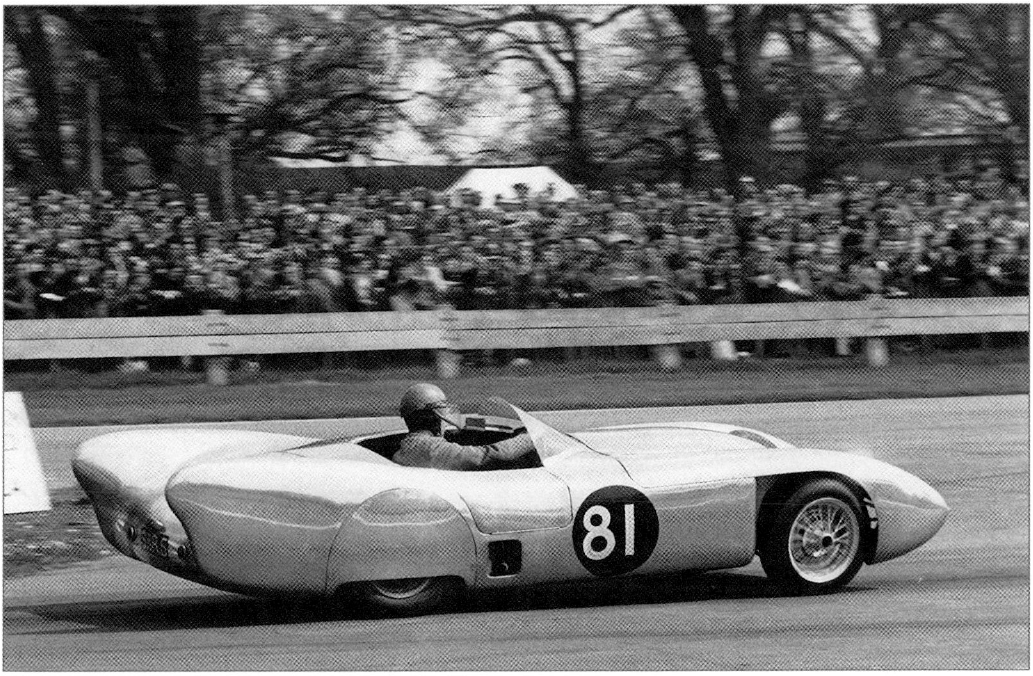

Looking shark-like in its bare aluminium finish and showing off the fabulous lines of its tail fins, Chapman's works SAR 5 takes to the track at Goodwood on Easter Monday 1954 for the five-lap Chichester Cup race. The Mk 8's MG engine promptly went onto three cylinders.

also satisfied one of Chapman's long-held ambitions – to beat one of the up-and-coming German sports cars.

Chapman took another 1500 class win at the Fairwood track in Wales, but holed the crankcase in the Unlimited category event. Such was the demanding calendar of racing every weekend that he began to examine ways of improving access to the car's internals. Even extracting the engine proved exceedingly frustrating for mechanics Mike Costin, John Standen, Peter Mayes and Tony McKusker.

With the engine rebuilt, next stop was the Nürburgring. Two privately owned Mk 8s were also present for Nigel Allen and Danny Margulies, but the Lotus trio retired, Chapman's with a snapped de Dion tube. The hectic schedule was maintained, however, and Chapman drove two further events the same weekend – at Brands Hatch, where he came fourth in his heat, then whizzed up to Crystal Palace, only to retire on the second lap due to carburettor settings that were more in keeping with German fuel. Then it was back to Brands Hatch, where he drove the Mk 8 of Nigel Allen in the Sports Car final, only to retire again with fuel feed trouble.

Meanwhile, Frank Costin and some of his De Havilland colleagues carried out aerodynamic tests on SAR 5, covering it with tufts of cotton wool and observing how they behaved in the airstream at 110mph on a disused Cheshire airfield. Costin found that the tufts weren't ruffled until the car reached 80mph, at which point the airflow appeared to go in a forward direction. His observations were projected into future Lotus bodies.

Colin's forays to race circuits tailed off at this point due to the realities of running a business building, selling Mk 6s and Mk 8s, and his impending wedding to Hazel Williams. But he logged a third at Castle Combe behind Roy Salvadori's Maserati and

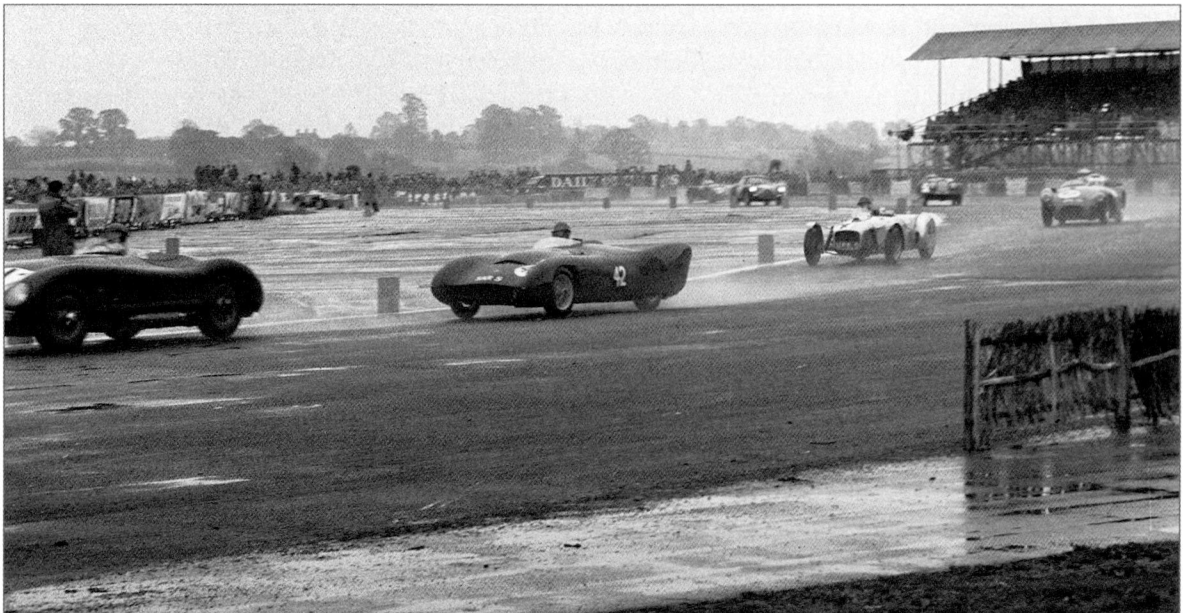

The first victory for the works Mk 8 was at a damp Silverstone at the International Trophy meeting on 15 May 1954. Chapman won the 1100–1500cc class ahead of John Coombs and Peter Gammon at an average speed of 76.42mph and was fifteenth overall, just three-quarters of a lap behind Gonzales' 4.9-litre Ferrari.

Archie Scott-Brown's Lister-Bristol, while at Ards in Northern Ireland for the Tourist Trophy he crashed out. Further placings included sixth overall and 1500cc class winner at Crystal Palace, and second to McAlpine's Connaught at Aintree. He also had a run in with the Lister of arch-rival Archie Scott-Brown at Snetterton.

At the end of the season, SAR 5 was sold to Lotus racer Austin Nurse for the 1955 season, and subsequently went to Roy and Jean Bloxham. After an errant excursion into the weeds at Mallory Park, SAR 5 was rebuilt minus its fins. In an *Autosport* magazine road test in November 1954, John Bolster recorded a top speed of 121.5mph, 0–60mph in 8.0 seconds, 0–100mph in 23.8 seconds, and the standing quarter-mile in 15.5 seconds. All this performance and 30mpg as well.

Further MG-engined Mk 8s were commissioned by Brian Naylor (JBN 1), Alan Brown (VPD 97), and Tip Cunane – who mostly went hillclimbing – while John Coombs installed his Connaught engine in the Mk 8 Lotus. Following the example of Kieft's 1954 Le Mans entry, Dickie Steed fitted an 1100cc Coventry-Climax FWA fire-pump engine in his Mk 8 (HUD 139). In spite of teething troubles, this was the start of a long association between the two manufacturers.

On New Year's Day 1955, Chapman and Mike Costin began their full-time careers at Tottenham Lane. While Costin ran the offshoot Racing Engines Ltd, Nobby Clarke was production manager of the car side. There was already a degree of overlap in Lotus type numbering – the Seven had yet to appear, and the Mk 9 was conceived as a smaller engined version of the Mk 8. Chapman was already draughting the more substantial chassis of the Mk 10, which it was intended would accommodate the bigger Bristol engine advocated by Mike Anthony.

The essence of the Mk 9 was that its multi-tubular triangulated chassis was more compact than the Mk 8's, thanks to smaller-section tubing, which was thus lighter yet stronger than its predecessor. There was also a revised de Dion rear end set up with beefier hubs. The bodywork was subtly reworked by Frank Costin so that it was rather shorter overall but with higher tail-fins. The frontal treatment was altered to provide better airflow to the radiator and somewhat shrouded front brakes. Crucially, accessibility was vastly better, so that the top half of the front bodywork could be removed for engine and suspension access as well as the instrument panel. A panel could be removed from the back end to give access to the final drive and the inboard brakes. Cockpit access was addressed, with hinge-down doors provided on both sides of the car, while the perspex windscreen was a lower wraparound shape. Fuel tanks ranged in capacity from 7 to 11 or 19 gallons, installed at the rear of the MG-engined car and, to improve weight distribution, the tank was located alongside the passenger seat in the lighter Climax-engined cars, with the battery at the rear. Auxiliary tanks could be accommodated within the passenger side of the car for long-distance competition. Headlight installation remained as for the Mk 8.

Examples of the Mk 9 were sold to overseas buyers at £1,150 with Coventry-Climax FWA 1098cc power. This was a tax-free price, elevated considerably on the home market by the addition of purchase tax. The loophole was to sell the cars in kit form so that purchase tax could be avoided. This kind of production also required less space in the assembly area than constructing complete cars would have done. Team Lotus started off building a pair of Mk 9s for its own use. They were an MG-powered car (9 EHX) that Chapman frequently drove, and a Coventry-Climax-engined car (XPE 6) seen

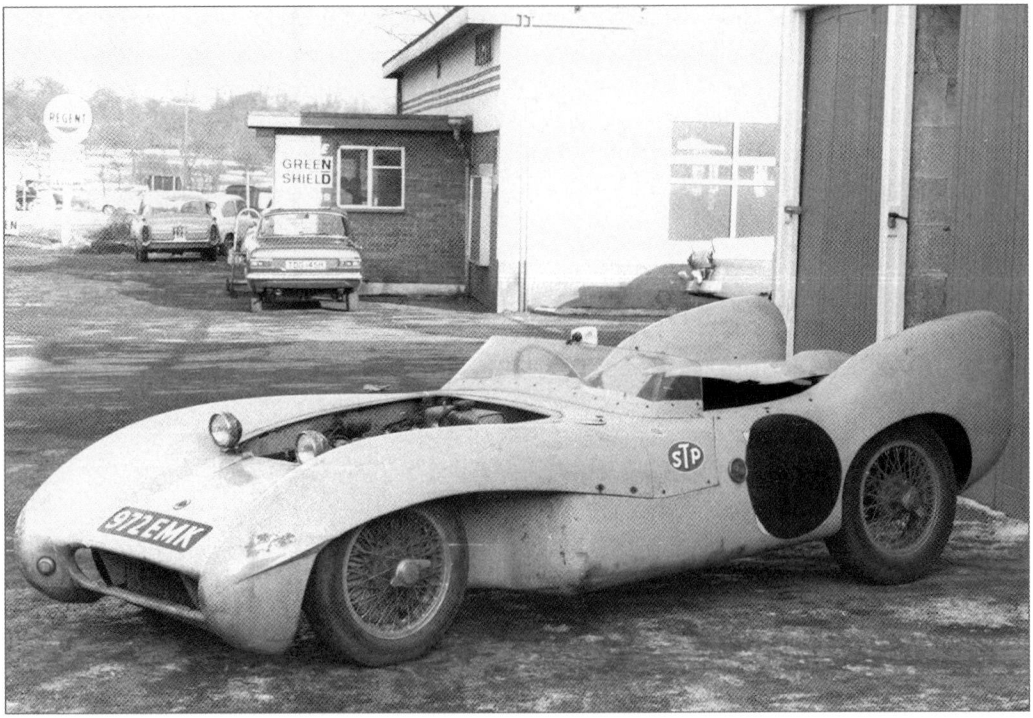

A Mk 9, looking slightly sorry for itself more than a decade after its introduction, demonstrating its fold-down driving lights – the headlights were mounted in the grille aperture – and tail fins. It was basically an improved edition of the Mk 8 and intended for smaller capacity classes. Twenty-three units were made during 1955, using 1098cc and 1460cc Coventry-Climax or 1487cc MG power.

in the hands of Peter Jopp. In both cases MG J2 crash gearboxes were fitted, although private buyers qualified for an MG gearbox with synchromesh on the upper ratios.

Two of the first customer cars were dispatched to Florida in March 1955 for the annual Sebring 12-Hours, running with Coventry-Climax engines and 1954-spec 9 inch brakes. For a while all went well and the Mk 9s led the 1100cc class. But one crashed and damaged its sump, while the other was disqualified for being push-started in the pits.

The beginning of the 1955 UK season started off with Chapman's back-of-the-grid dash through the field at Oulton Park, only to retire with overheating. He hit a concrete post at Goodwood, and broke a crankshaft at Silverstone. There was, however, the satisfaction of having set fastest lap at Goodwood. With a rebuilt and now dry-sumped MG engine, success came at Snetterton and Goodwood over the Whit weekend, where Chapman was victorious in the 1500cc class, and Jopp won at Brands Hatch.

Chapman had always cherished the prospect of tackling the Le Mans 24-Hours marathon with his own cars, and Team Lotus was present at the 8.35-mile Sarthe circuit for the first time in 1955. The 1100cc Climax-engined works car was to be driven by Chapman and the experienced Ron Flockhart. Other British manufacturers running in this class were Cooper, Kieft and Arnott. For this particular race, the works Mk 9 was panelled in heavier 22-gauge aluminium cladding, and its headlamps were housed in the front wings under perspex covers. Girling disc brakes were fitted to cope with the stresses of heavy braking.

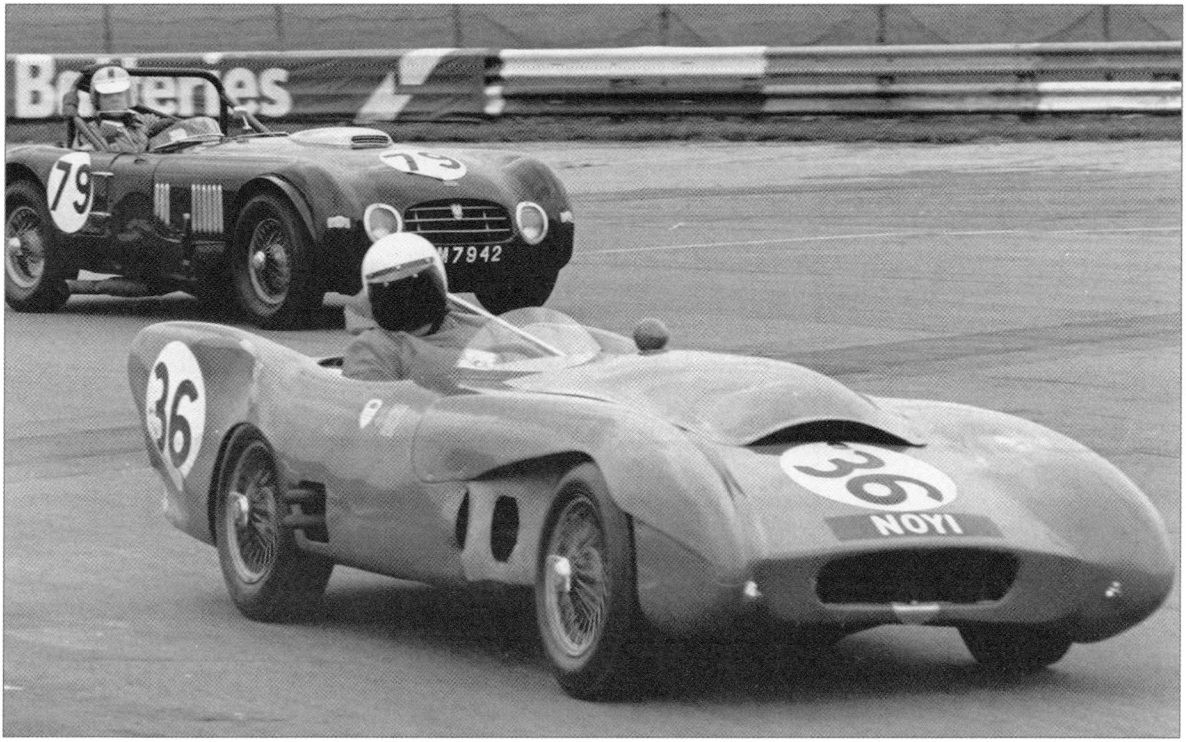

Lotus built seven units of the Mk 10 during 1955. It was a strengthened version of the Mk 8 and was designed to take bigger engines, such as the 1971cc Bristol or 1960cc Connaught units. Here is R. Waldron's example during a BRDC event at Silverstone in 1977.

As became customary, the Team was billeted some 18 miles south of Le Mans at Mayet in the Auberge de St Nicholas under the patronage of M et Mme Mica. The car was prepared in the hotel's garage, and after scrutineering hassles involving the relocation of the retractable spotlights into the nose, the evening practice sessions took place. The problem of oil leaking onto the clutch via the rear main bearings was sorted during an all-nighter in the Cunningham garage. Early in the race, with Chapman leading the class, clutch slip was cured with the contents of a fire extinguisher, and subsequently by Flockhart casting sand from Tertre Rouge sandbank into the bellhousing. Around 10pm Chapman slid off into the sand at Arnage, and because of the presence of a stranded Jaguar, he reversed back onto the track without permission from the marshals. This was an infringement of the rules and the Lotus ended up being disqualified, mainly because the authorities were justifiably shaken by the appalling accident involving Levegh's Mercedes-Benz and could ill afford a further disaster.

Back on home territory there was better luck, with Chapman tying with Bueb's Cooper T39 for second place at Brands Hatch in July, duelling with the big-engined cars in the sports car event supporting the British Grand Prix meeting at Aintree and winning the 1500cc class. There was a victory at Crystal Palace on August Bank Holiday Saturday, followed by a third at Brands Hatch on the Monday. The Goodwood 9-Hours was an interesting venture. Chapman and Jopp were partnered in the 1500 and Flockhart and Cliff Allison in the 1100, with privately entered Mk 9s for John Coombs/John Young and R.A. Page/Paul Emery. Chapman's car shed its flywheel while leading the 1500 class, Flockhart was hit by a wayward DB3S, Coombs' car lost a wheel, and the Page/Emery car came eleventh overall and fourth in the 1500 class. Both MG and Climax works cars retired in the *Daily Herald* International Trophy at Oulton Park – Cliff Allison crashed the smaller engined car and Chapman suffered the oil-on-clutch syndrome with the MG-engined Mk 9. By now the Mk 10s were on the scene, and Peter Scott Russell and Mike Anthony came first and second in the under-2000cc class to uphold Lotus's reputation. A week later, Chapman won the 51-mile 1500cc sports car race at Aintree, setting a new lap record, and followed home by Brooks' Connaught and Edward Lewis in his Mk 9.

The Tourist Trophy race at Dundrod found Chapman/Allison in spectacular form, leading the 2000cc class by an astonishing 9 minutes – until an oil pipe broke. Once repaired, the works Mk 9 recovered to finish eleventh overall and second in class. Around forty Mk 9s were built, and Lotus Engineering Ltd was now welcomed into the fold as a member of the SMMT – the Society of Motor Manufacturers and Traders – thus qualifying as an exhibitor at the Earls Court Motor Show.

Privately entered Mk 9s running with Ford engines were successful in 750 Motor Club events in 1956, and Peter Lumsden was quick with his Climax-powered car. He was awarded the Brooklands Trophy for cumulative performances during the year at BARC Members' meetings at Goodwood.

Following the promising showing at Le Mans, customers were offered race replica Climax-powered Mk 9s called the 'Le Mans', while Ford 1172cc engined cars were marketed as 'Club' versions. These were rather more basic, running Ford Popular gearboxes, back axles and drum brakes, although discs were optional.

BRISTOL MOTORS

The concept of the Mk 10 had been under consideration since 1954, when Mike Anthony lobbied Chapman to produce a version of the Mk 8 that could accommodate the 2.0-litre Bristol motor. The Mk 10 was thus based on the Mk 8 chassis with a larger engine bay, but incorporated the lower driveline and rear axle layout of the Mk 9, which culminated in a Salisbury hypoid bevel final drive. The fuel tank, fuel pumps and battery were located behind the rear axle for better weight distribution because of the heavier engine. Visually, the Mk 10 differed from its siblings in having no rear wheel spats, and a somewhat disfiguring hump on the bonnet. Dunlop disc brakes were fitted front and rear as standard, but only seven chassis units were made, mainly because there was more interest in the Climax-powered models. Among the non-Team owners of Mk 10 chassis were Mike Anthony (PCD 13), Cliff Davis (NOY 1), George Nixon, Bill Short (RCR 520), Peter Scott-Russell (JBW 648), Mike Young, and Dr Vaughan Havard. Teen idol and silver screen rebel James Dean ordered a Mk 10 chassis into which he planned to fit an Offenhauser engine, but he was killed before it could be delivered.

In the 2000cc category, the man to beat in 1955 was Archie Scott-Brown in the better handling Lister-Bristol, and while other Lister drivers could be bested by Mk 10 protagonists, Archie was virtually invincible. Possibly the Mk 10's finest hour was in the Goodwood 9-Hours of 1955, when Cliff Davis and Reg Bicknell finished second in the 2000cc class, although Mike Anthony and Peter Scott Russell notched up some victories and high placings in club events. Both continued to run their cars in 1956, but Scott-Russell's car was incinerated at Silverstone. Meanwhile, Cliff Davis was still racing his Mk 10 in 1957

CHAPTER TWO

INTERNATIONAL RESPECTABILITY

For 1956 Team Lotus concentrated on the Eleven in sports car racing. Visualising a long unbroken lineage of Lotus models, Chapman decided to do away with the clumsy 'Mark' abbreviation, and cars would now be known simply by their type number. Thus, the Eleven and subsequent Seven names are usually spelled out and later cars are referred to simply by their type numbers.

Although it was a development of the Mk 9, the Eleven had a new spaceframe chassis made constructed in 1 inch and ¾ inch 18- and 20-gauge tubing. A lower pivot point for the swing-axle front suspension lessened the Mk 9's tendency to understeer. A lighter de Dion back axle was located by twin radius arms on either side, and the Eleven also got a new rack-and-pinion steering system, Girling disc brakes front and rear, and a lighter radiator. To achieve a better weight distribution, the 9.5-gallon fuel tank was relocated from the rear end to the car's passenger side. An additional 11-gallon tank could be fitted on the right-hand side for endurance racing.

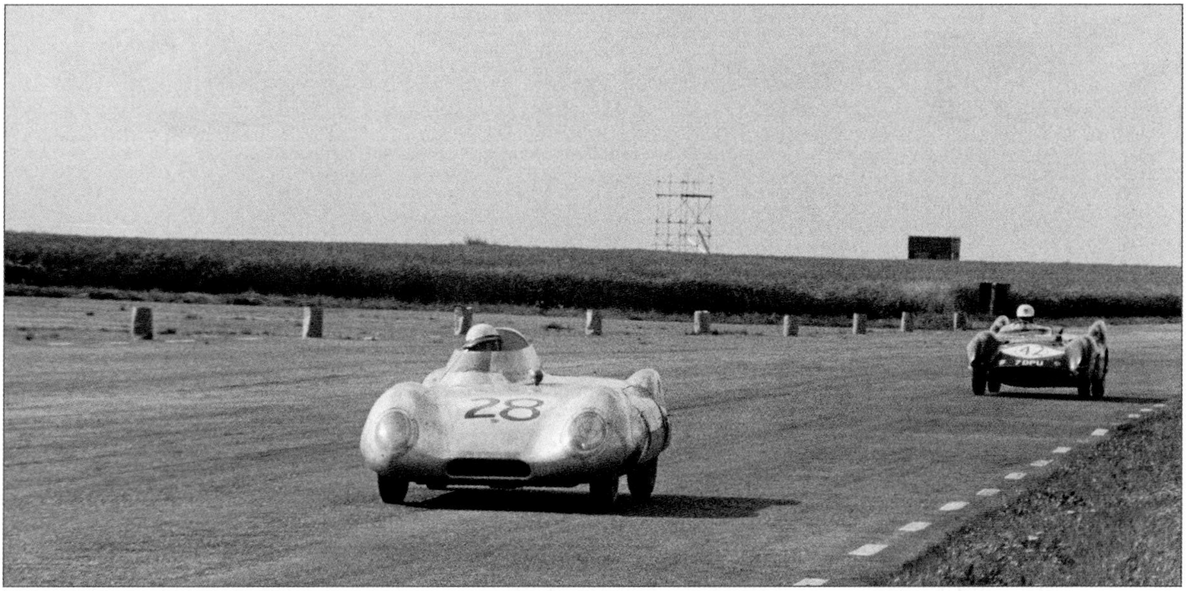

The Eleven was much the most numerous of Lotus's aerodynamic sports racing cars produced during the second half of the 1950s, with 250 cars made. Its spaceframe chassis could accept 1098cc and 1460cc Climax engines, although a variety of other power units were fitted by customers, including 1100 Stanguellini, 750cc and 1457cc Climax twin-cams, 1484cc Maserati and 2.5-litre Lancia. Here, Alan Stacey's 1098cc Eleven leads Peter Ashdown's Mk 9 at Silverstone in 1956.

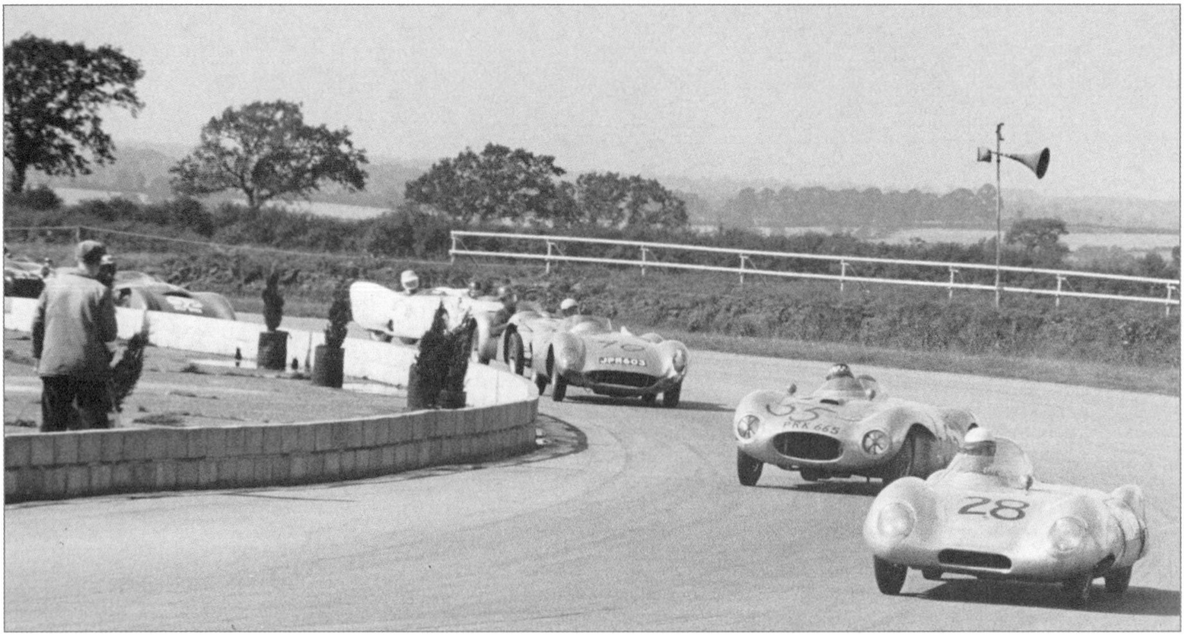

Alan Stacey's Eleven leads an assortment of sports racers, including Cooper T38 and T39 and other Lotuses, at Copse corner, Silverstone, in 1956. Incidentally, the Lotus Mks Seven and Eleven have their names spelled out, rather than expressed numerically.

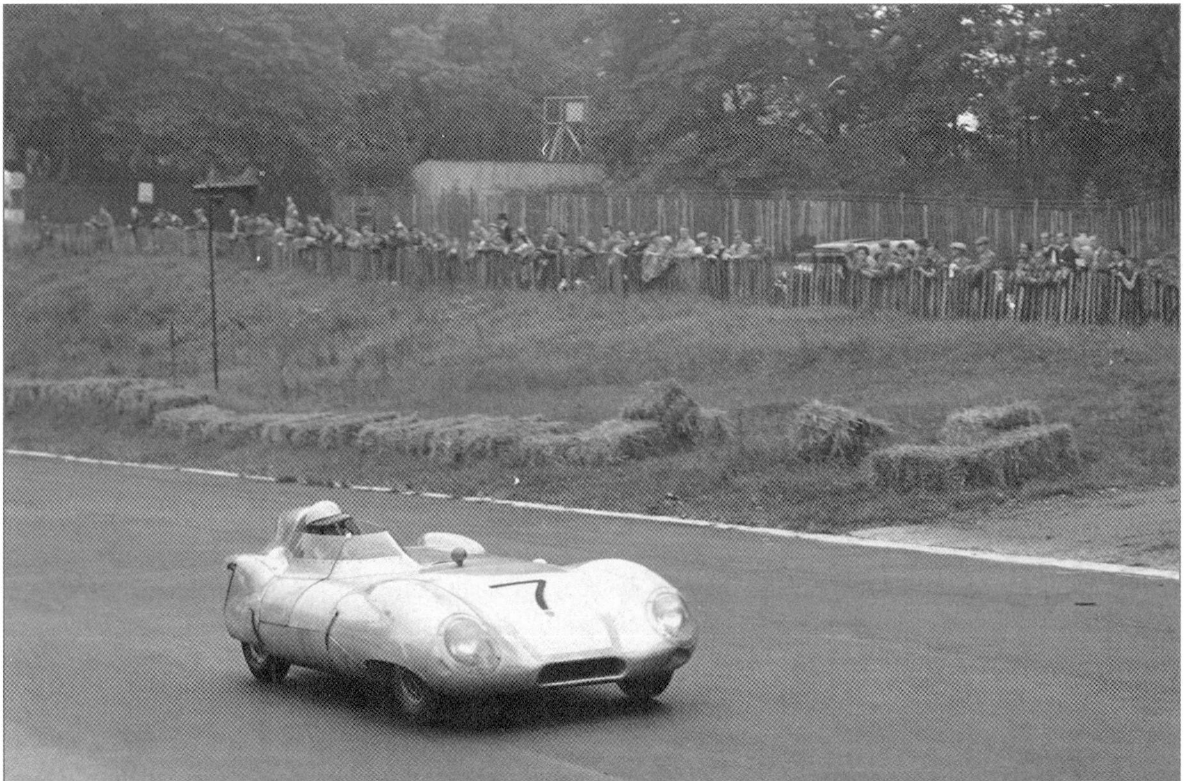

It's the BARC meeting at Crystal Palace on Monday 6 August 1956, where Alan Stacey's Series 1 Eleven came fifth in the final of the sports car race behind a Tojeiro Jaguar, a DB3 Aston, a C-type Jag and Cliff Allison's Eleven.

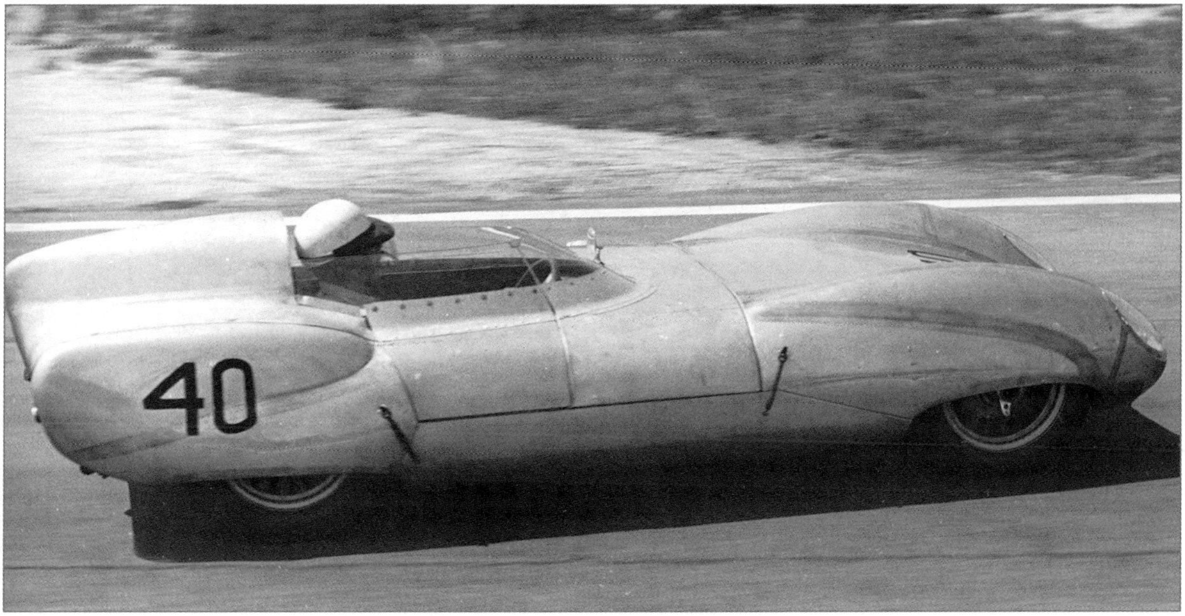

The Lotus Eleven Series 1 of Alan Stacey pictured at the BARC Goodwood National Open Race meeting on Monday 21 May 1956 where he was seventh overall and fourth in the 1100cc class. The race featured an epic dice between the Elevens of Chapman and Mike Hawthorn, in which the two leaders swapped places several times. The race was decided in favour of Chapman after Hawthorn collected Colin's spinning Eleven and had to pit for emergency repairs. The streaks up the front wheelarches are actually curves in the aluminium panelling.

There was no question that the Eleven was a huge improvement aesthetically over its predecessors. The overall shape was smoother, more refined and coherent as a design. The rotund front wings bulged over the 15 inch wire wheels to give a more streamlined shape, and instead of fins, there were higher rear wings. In fact, front and rear wheels were enclosed to the knock-off spinners, making for a wide turning circle. Accessibility to the running gear was now first rate, since the front and rear body sections hinged on the ends of the chassis frame. Bottom-hinged doors were fitted on both sides, and even the interior was trimmed. The headlights were visible in their perspex-covered nacelles in the wing fronts. In common with sports racing cars of the time – notably the Jaguar D-type – most Elevens were produced with faired headrests that tapered away at the rear of the car. For racing, there was a metal tonneau to cover the cockpit on the passenger side, and a wraparound perspex screen for the driver. International events such as Le Mans demanded a full-width screen, and in this configuration the headrest was absent.

In Series 1 form, Elevens were available in three differing specifications. The out-and-out racing car was powered by the Coventry-Climax FWA 1098cc engine, or in some cases – including Team Lotus – the 142bhp Climax FWB 1460cc engine was used. Coventry-Climax units were canted over at 10 degrees to the left so the induction manifold could be horizontal. Transmission was initially via an Austin A30 casing fitted with Lotus close-ratio gears, superseded later by the MGA gearbox, and a de Dion rear axle was fitted. Girling 9½ inch disc brakes were outboard at the front and inboard at the rear. Price wise, the basic Le Mans 75 model cost £1,337, the Stage 2 FWA-engined Le Mans 85 would be £1,387, and the FWB engine cost an extra £250.

The Club version was also specified with the Climax engine and Lotus gearbox, with a live rear axle located by parallel trailing arms, plus drum brakes. With all mod cons including wipers, a full-width screen and hood, the Club model cost £1,083. Offered in similar trim to the Club, the Sport version was generally powered by the 1172cc Ford side-valve engine allied to a Ford three-speed gearbox, and priced at £872. Although intended principally for road use, the Sport model's chief claim to fame was its success in the hands of Graham Hill in the 1956 Autosport Production Sports Car Championship. The following year Ian Walker took the same title in this car, equipped with a Willment cylinder head.

Now Lotus privateers as well as the Team had got the measure of the Cooper-Climax Manx-Tails and successes during 1956 were legion. Among the most successful of the private entrants was Alan Stacey, who impressed Chapman sufficiently to gain works status. Like Scott-Brown, Alan suffered from a physical disability, in his case caused by the partial loss of one leg in an accident as a youth. Despite an artificial lower limb, he was still able to outperform the majority of his rivals. Stacey's mechanic Bill Basson only had one hand, and the partnership's success was much admired. Team Lotus Elevens were usually piloted by Colin Chapman and Reg Bicknell, while other notable private entrants and their drivers included John Coombs, Brian Naylor, Edward Greenall, Pete Ashdown, Innes Ireland, Graham Hill, David Piper, the Equipe Endeavour car of Tommy Sopwith, and Ivor Bueb's Ecurie Demi-Litre Eleven that was frequently

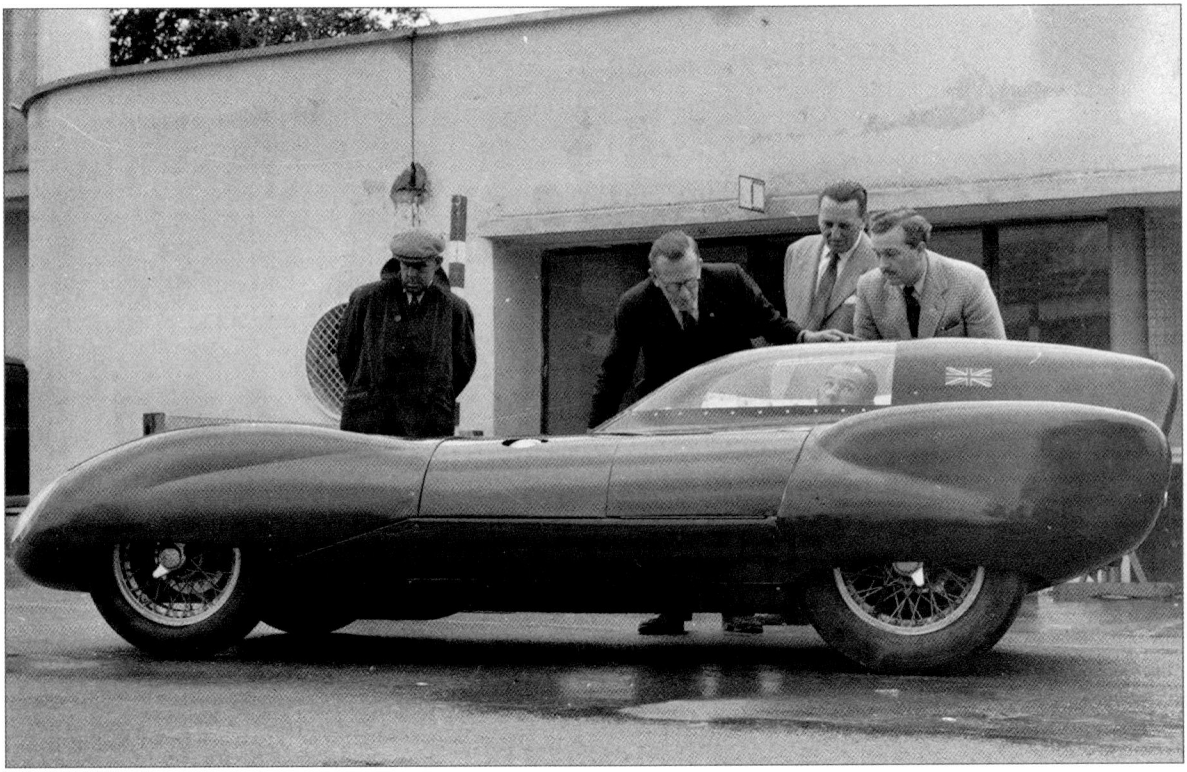

In September 1956 a specially prepared Eleven fitted with a bubble-top over the cockpit was taken to Monza to attempt the International class G speed record. Stirling Moss tries the car for size. He went on to break the 50km and 50 mile records at 135.5mph (218kph) and 132.7mph (214kph) respectively.

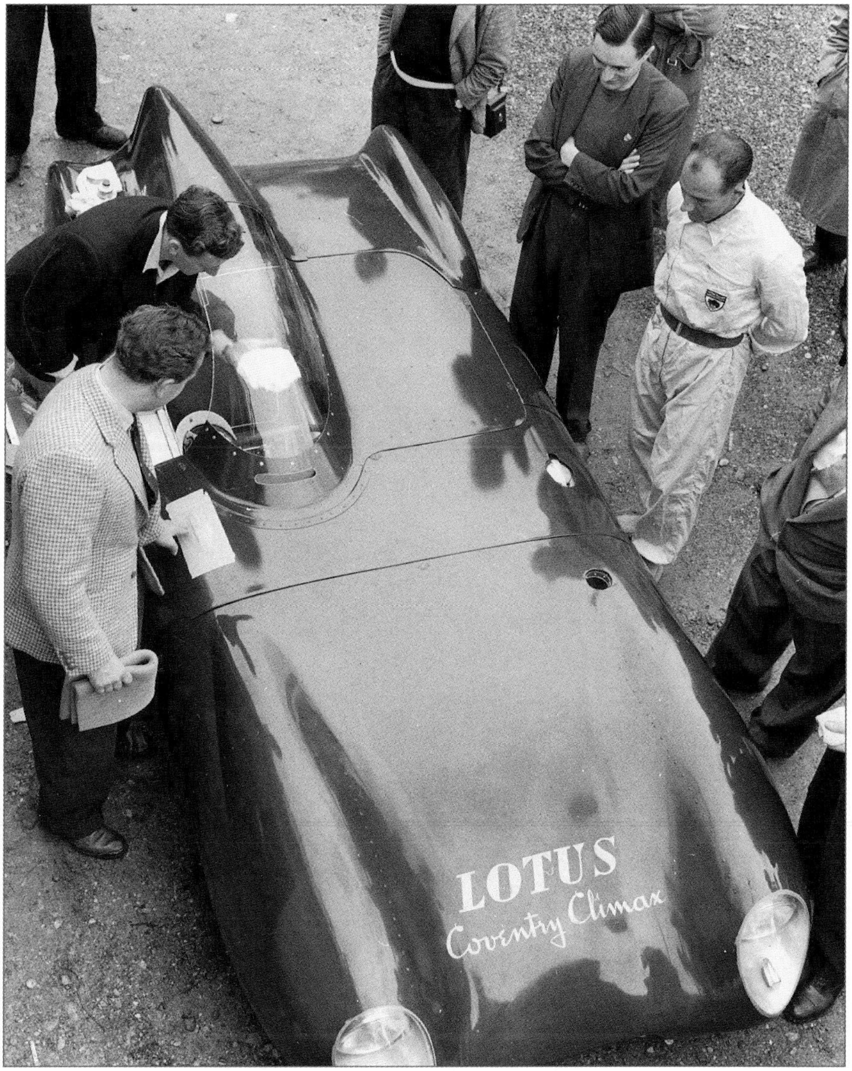

Moss and Chapman survey the Eleven in the pit lane at Monza prior to Stirling's record-breaking runs in September 1956. Further attempts were thwarted when a rear subframe fractured. Notice the neat detailing of the cockpit fairing and bubble top.

driven by Mike Hawthorn. This car had wells specially sunk in the undertray to accommodate the latter's heels, on account of his larger than average feet.

In March 1956, the first two Elevens built were dispatched to the States for the Sebring 12-Hours. One entered by Briggs Cunningham for himself and Chapman to drive had the FWB engine but was crashed in practice. Chapman co-drove the other car with an 1100cc engine, but a burned-out starter motor caused its retirement. Following a couple of defeats at the hands of its Cooper rivals, Team Lotus had a success at the Coupe Delamare de Bouteville, a 75-minute race serving as a warm up to the Rouen Sports Car Grand Prix. Chapman was first overall and Cliff Allison second, also taking 1100cc class honours, while Harry Schell brought the Sopwith FWB-engined car home third.

At Goodwood on Whit Monday there was a dramatic duel between the Elevens of Chapman and Hawthorn, as they gyrated in unison at Madgwick corner, the Team Lotus boss going on to win as Hawthorn pitted for a look at his damaged car. At

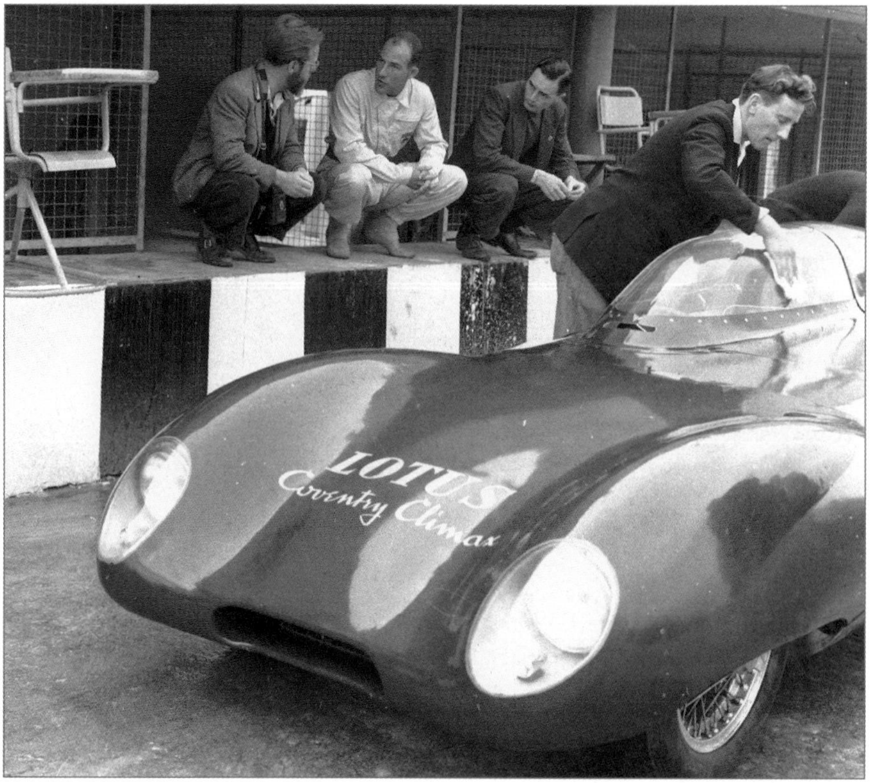

While Stirling talks to Denis Jenkinson on the Monza pit wall, Lotus's John Standen wipes the windscreen of the Eleven record breaker. Early in October 1956 the car returned to Monza where Mac Fraser created six new records and lifted the 50km record to 138mph (222kph), which was quite an achievement for an 1100cc-engined car. His fastest lap averaged 143mph (230kph).

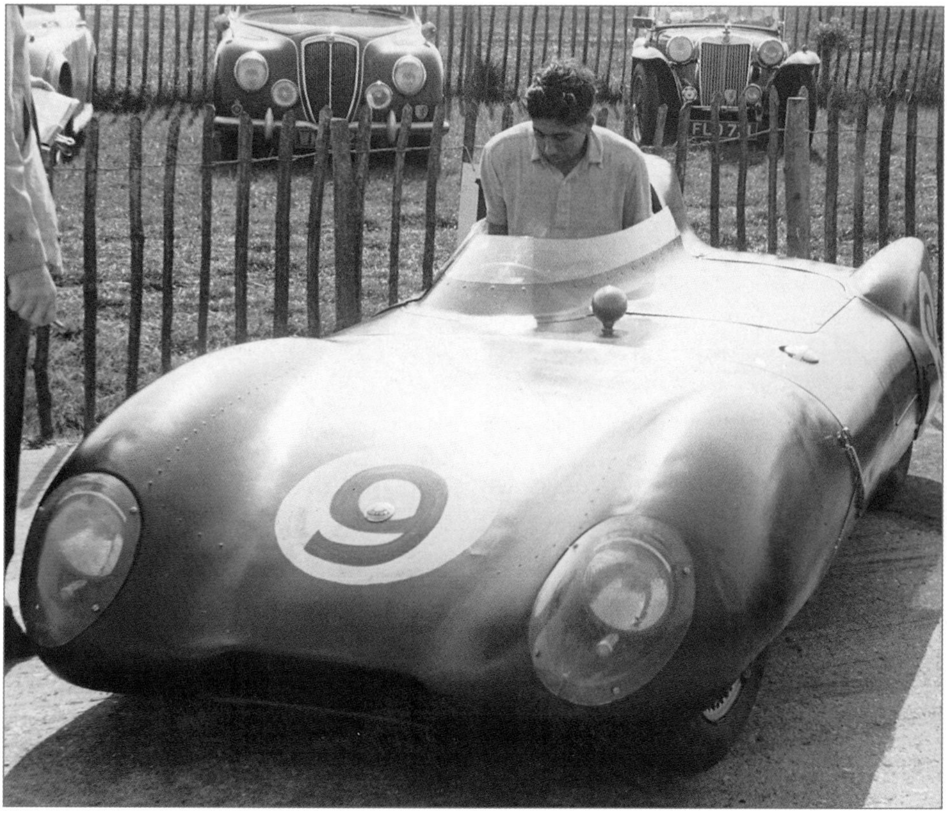

Alan Stacey climbs aboard his Eleven in the Goodwood paddock during the BARC meeting on 22 September 1956. He won the 1500cc sports car race by 11 seconds.

Oporto in June, Reg Bicknell came third in the City Cup street race, while the Ecurie Demi-Litre Eleven driven by Bueb/Mac Fraser led the Reims 12-Hours until first and second gears disappeared. Unfortunately, this car was wrecked at Oulton Park in August when Hawthorn hit the bank and rolled during the Sporting Life Trophy race, and the race was won by Stirling Moss's Willment Cooper.

As Chapman had averred the previous year, Team Lotus was back at Le Mans in late July 1956 with a trio of Elevens bent on achieving a result. Two were 1100cc cars for Reg Bicknell/Peter Jopp and Cliff Allison/Keith Hall, and the other for Chapman/Mac Fraser ran with the 1500 Climax unit. In order to comply with new Le Mans regulations, their spaceframes were wider at the centre to permit mandatory full-width open cockpits and full-width aerodynamic screens to be fitted. The cars' track width remained the same. Team headquarters was once again at the Auberge de St Nicholas at Mayet, but race preparation was carried out in a rented garage elsewhere in the town. The cars' immaculate presentation impressed the scrutineers – and other teams, including Porsche – and the French began to speculate on Lotus's chances of scooping the Indices of Performance and Thermal Efficiency, territory hitherto sacred to the native Panhard équipes.

Chapman was running sixteenth overall during the early laps, and the smaller-engined cars equally well up in the running order, in spite of deteriorating weather. But their pace meant they were having to contend with faster cars approaching from behind and to watch out for rear lights of slower ones ahead – all this in a wall of spray and getting increasingly soaked into the bargain. In the mists of Sunday morning, Cliff Allison hit a stray dog, damaging the car too much to continue. Running second in class at noon, the Chapman/Fraser car was eliminated by the failure of a big-end bolt. However, the Bicknel1/Jopp 1100 carried on to finish seventh overall and first in the 750–1100cc category, placing fourth in the Index of Performance. Its average speed for the 24 hours was 87.97mph, and it had been timed at 119.44mph over the flying kilometre on the Mulsanne Straight, which spoke volumes for the Eleven's aerodynamics. Needless to say there were celebrations that night back at the Mayet hotel. Monsieur Mica had ministered to the Team at the circuit for the duration of the race as well.

A special Eleven was built for Stirling Moss to attempt the International class G record at Monza on 3 September 1956. The engine spec was raised to Stage 2 with Weber carbs fitted, the headrest tail section was welded in place instead of being held by Dzus fasteners, and a clear perspex bubble served as the cockpit canopy. Stirling, who had just won the F1 Grand Prix, set 50km and 50-mile records at 135.54mph and 132.77mph respectively, but the rough condition of the high-speed banking wrecked the car's rear subframe and the attempt on the main goal was halted. When Team Lotus returned to Monza at the end of the month, Mac Fraser improved on the times set by Moss, and left the flying lap at an average of 143mph, which was quite some achievement for an unsupercharged 1100cc car.

Later in the season, more privately entered Elevens appeared on the scene, driven by Gordon Jones, L.I. Bramley, G.R. Baird, H. Deschamps, and Peter Ross. The cars were also eligible for F2 races, and at Oulton Park in September there were no fewer than eleven Lotuses and fourteen Coopers on the grid. Attractive prize money was the key draw as the organisers sought to promote the new 1.5-litre formula.

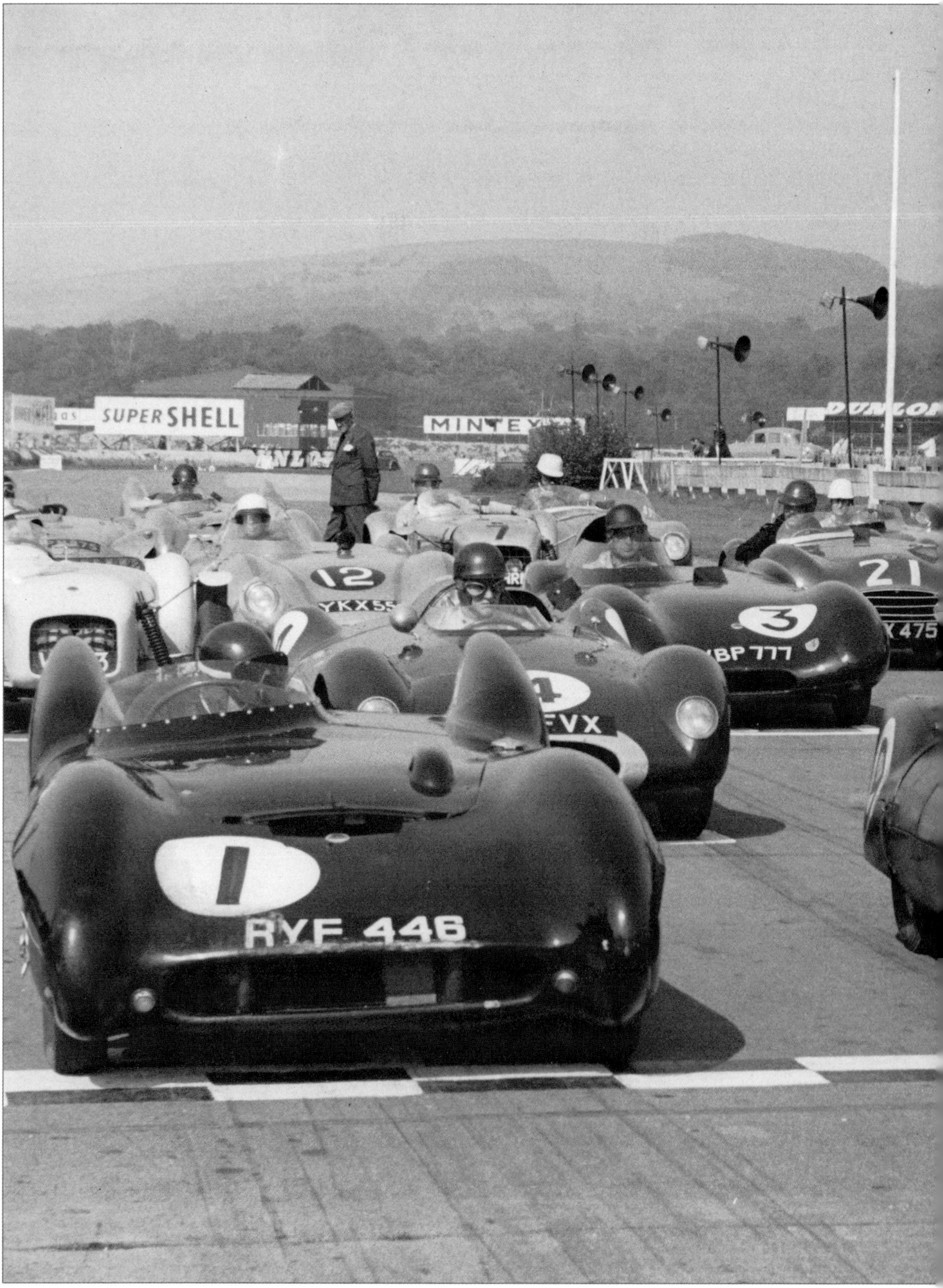
SUPER SHELL
RYF 446

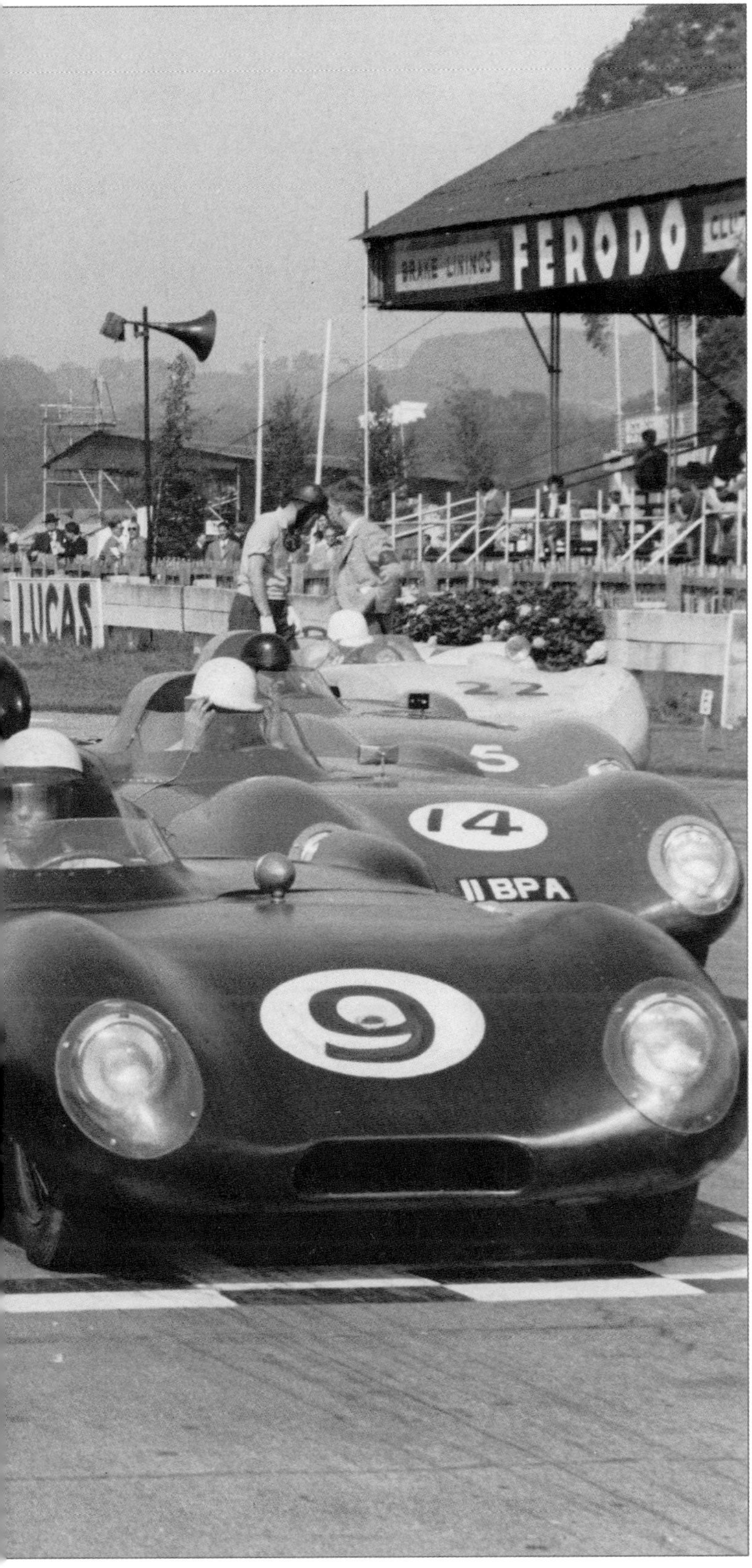

On the grid at Goodwood for the up-to-1200cc sports car race on 22 September 1956 are Peter Lumsden (who won the Motor Sport Brooklands Memorial Trophy) in his Mk 9, eventual winner Alan Stacey in a Series 2 Eleven, with Keith Greene's Cooper Manx-Tail (better known nowadays as Bobtail) and Gordon Jones' Eleven on the second row.

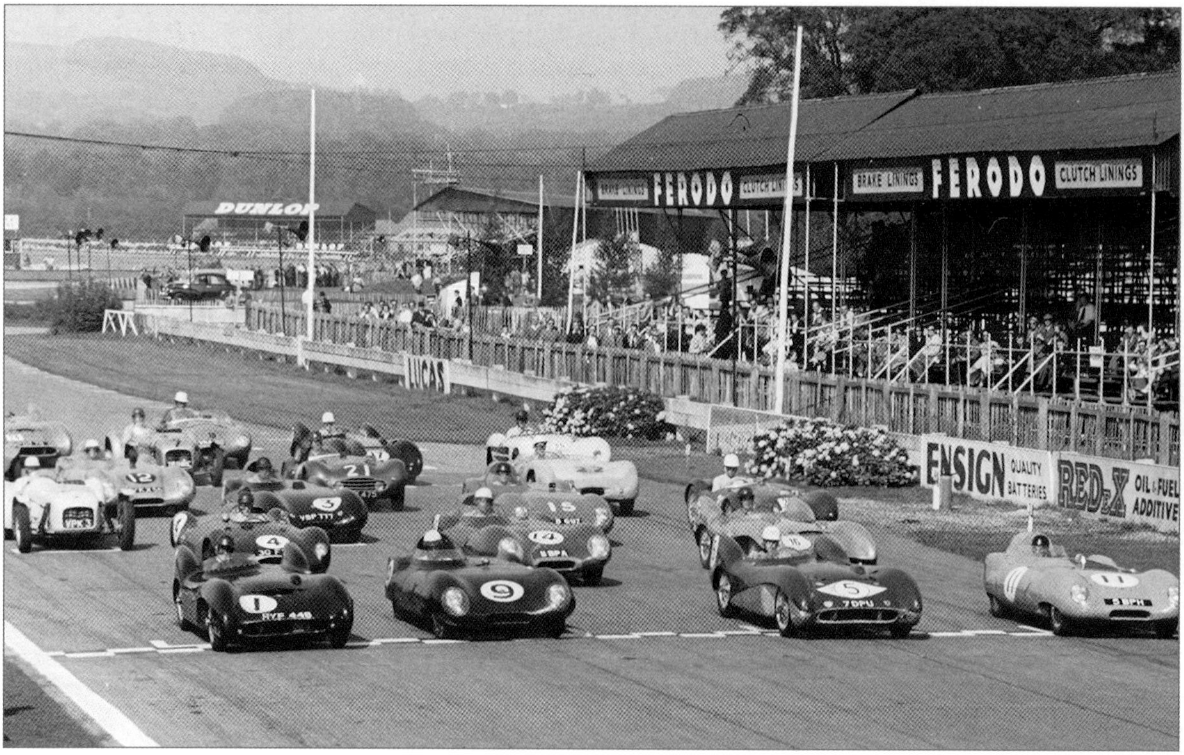

Here's that grid in full for the 1200cc Scratch sports car race at Goodwood on 22 September 1956, with Lumsden (Mk 9), Stacey (Eleven), Ashdown (Mk 9), and Innes Ireland (Eleven) on the front row. On the second row are Greene (Cooper T39), Jones (Eleven), and Bryan Hewitt (Mk9), with assorted Lotuse, Coopers and a Tojeiro on the rows behind. Stacey beat Lumsden by 11 seconds.

Flockhart and Chapman in Elevens followed home the Coopers of Salvadori and Brooks. Salvadori's T41 was a proper single-seater model.

For the second year, Lotus took a stand at the Earls Court Show, and in addition to a pair of Le Mans-spec Elevens, a new single seater F2 car appeared – the 12. Present at the show was Eleven protagonist and Californian Lotus agent Jay Chamberlain, and things augured well for more sales to the States. The Burbank-based dealer was also reaping rewards in racing, using his own modified camshafts in the Climax engines.

For 1957, Lotus came out with an improved version of the Eleven known as the Series 2, featuring slightly larger diameter chassis tubing, a front suspension set-up much like the Mk 12 F2 car, with double wishbones and coil springs. To handle the greater power of the new 1475cc Coventry-Climax FPF twin-cam, a strengthened de Dion rear axle was fitted. The main visual difference between Series 1 and Series 2 Elevens was that the lower panels between front and rear wheelarches were normally unpainted aluminium on the Series 2 cars. The range was expanded slightly to four models. The Sports 45 was the entry level model, fitted with the 1172cc Ford side-valve engine, drum brakes and live rear axle. Next up was the Club 75, equipped with 1100cc Climax power, drum brakes and live rear axle, followed by the Le Mans 85 with stage 2 tuned Climax 1100 engine, while the range-topping Le Mans 150 ran the FPF Climax twin-cam.

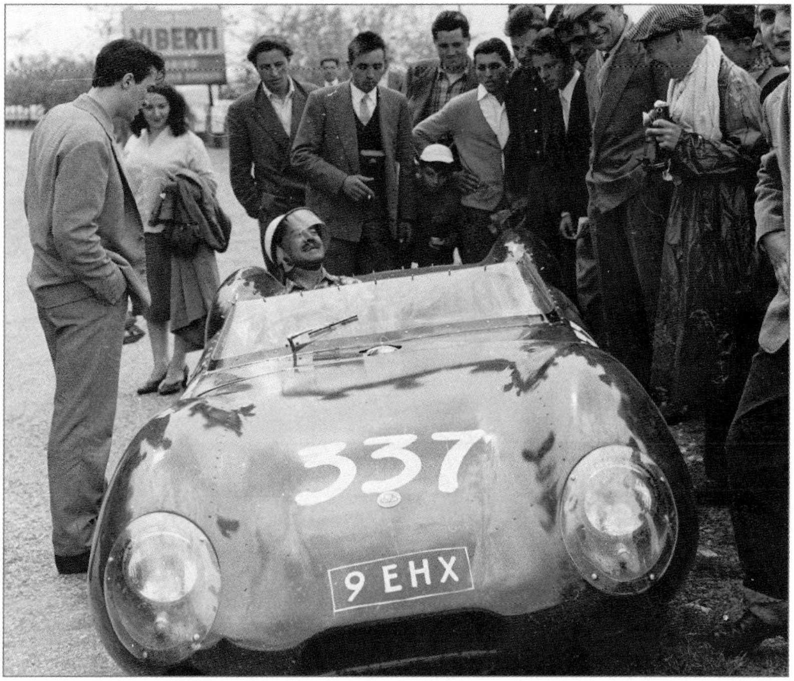

A crowd gathers around the Eleven of *Autosport*'s founding editor Gregor Grant during his highly eventful single-handed run in the 1957 Mille Miglia. A leaking fuel tank led to him racing with a jerrycan lashed to the bonnet in the later stages, but although he finished the course, he'd run out of time because of the roadside lash-up. The number 337 indicates his start time from Brescia.

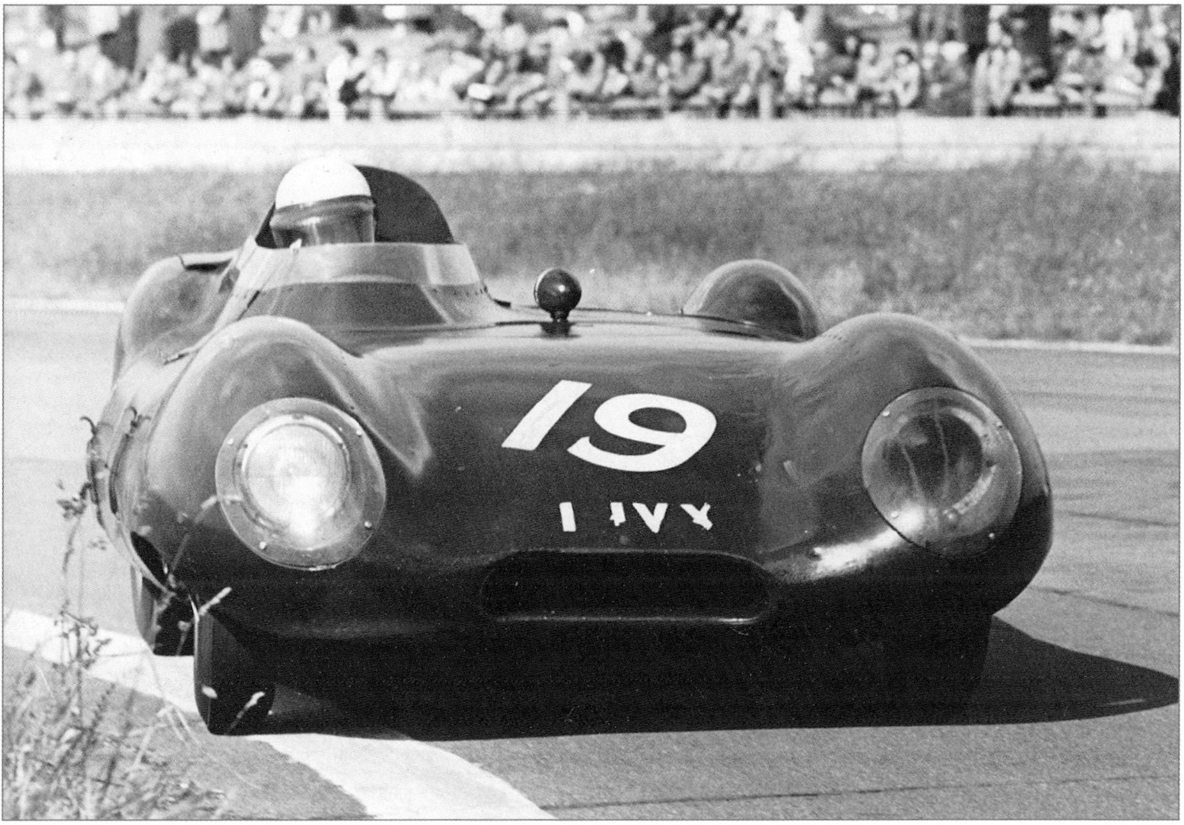

An excellent head-on shot of Stacey's Series 2 Eleven at Goodwood on 10 June 1957, when he won the up-to-1200cc class. There never was a Mk 13 Lotus model. That should have been the Series 2 Eleven, but for superstitious reasons it never received the designation.

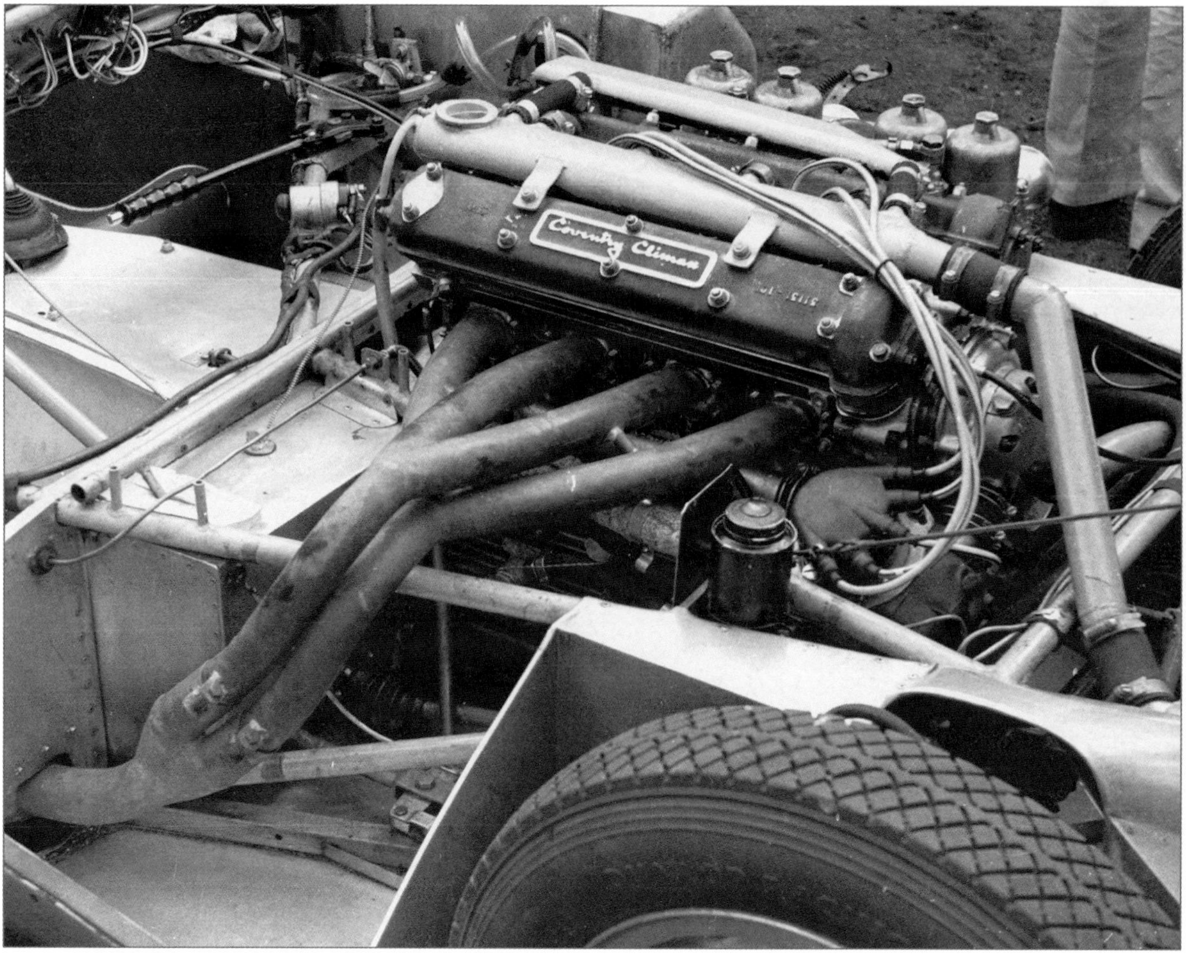

The Coventry-Climax engine revealed in John Coombs' Eleven in the Rouen paddock in 1957. Note the spaceframe tubes and aluminium bulkhead, the angle at which the engine is canted over and four-branch manifold, the contemporary side-pull handbrake lever and tread pattern of the Dunlop racing tyres.

At this point, Lotuses were virtually in complete command of the 1100 and 1500cc categories in British sports car racing. Club Lotus Secretary and historian Ian H. Smith calculated that the marque had notched up no fewer than 132 wins, 121 second places and 91 thirds during the preceding season. For 1957 there were to be two 'official' teams – the works effort consisting of Chapman, Fraser and Allison, and the unofficial works-backed privateers' team of Stacey, Ashdown and Keith Hall. Chapman agreed to build a one-off chassis for Brian Naylor to run his 2.0-litre Maserati engine on the condition it was entered as a Team Lotus car.

Chapman's first foray of the season was to the Sebring 12-Hours, where the 1100 Eleven was co-driven by himself, Joe Sheppard and Dick Duggan into eleventh overall and third in the Index of Performance. He then embarked on a tour of US dealers with Chamberlain. There were numerous successes for the works and works-supported cars and the better-known privateers, both at home and abroad. Mac Fraser came seventh overall at Spa, a lap ahead of the Porsches in his class and in front of a pair of Aston DB3Ss and a D-type Jag. Tony Hogg won the inaugural

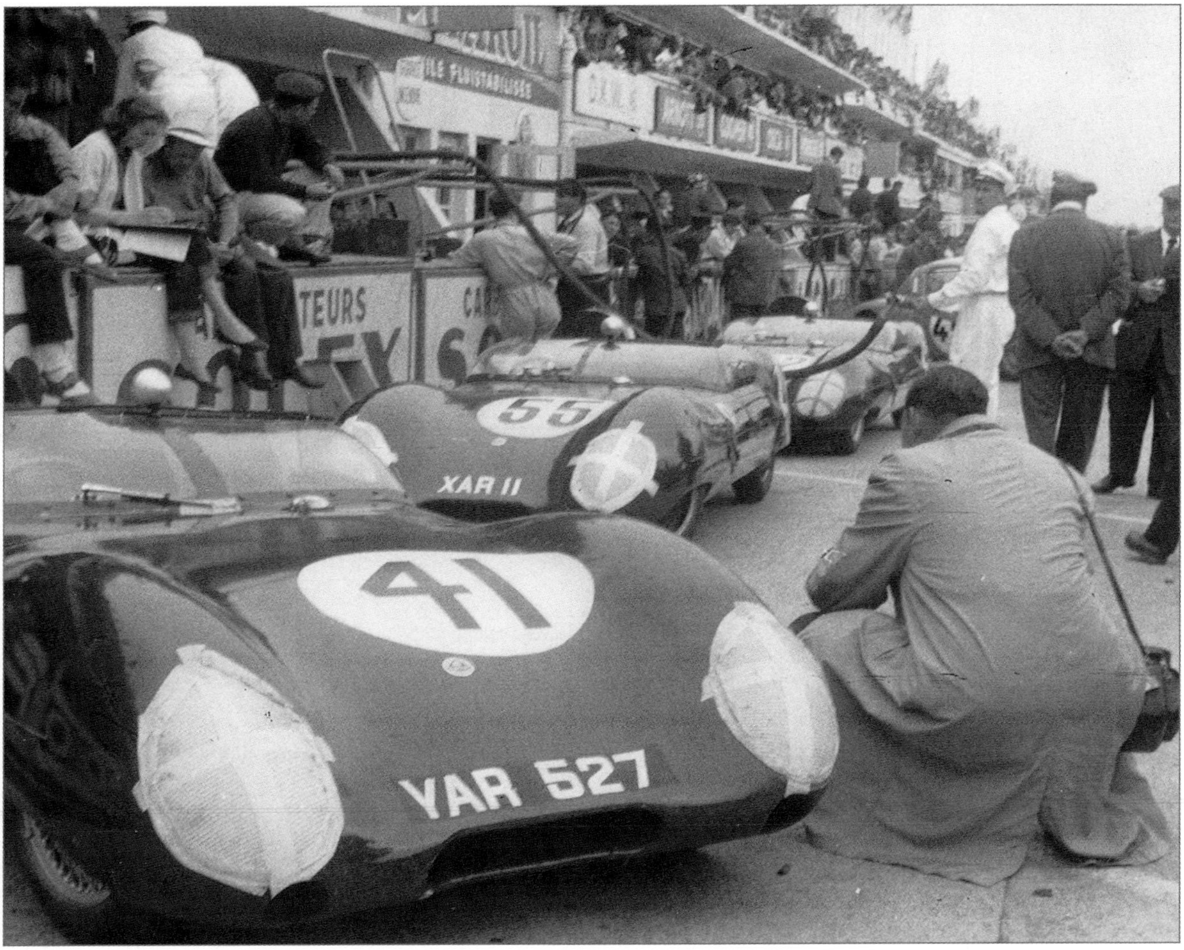

The works-supported Lotus Eleven of Andre Héchard/Roger Masson (41), the Team Lotus entry of Cliff Allison/Keith Hall (55) and the other works-backed privateer Eleven of Bob Walshaw/John Dalton (42) are fuelled up prior to the start of the 1957 Le Mans 24-Hours. Headlights are protected until the covers come off at night time. 1100cc class winners were Mac Fraser/Jay Chamberlain in the other works entered Eleven (62), while the 750cc FWC Climax-powered car of Allison/Hall was fourteenth and won the Index of Performance.

Austrian sports car event at Aspern in Dan Margulies' car, and *Autosport* editor Gregor Grant had a spirited attempt on the Mille Miglia, retiring in the closing stages with a fuel leak. There's a thoroughly entertaining personal account of Grant's exploits at the back of the Ian H. Smith book *The Story of Lotus*.

Team Lotus built five new Elevens especially for the assault on the Le Mans 24-Hours that year. There were three 1500s, an 1100, and a 750. All ran with wire wheels, except for the 750, which was fitted with magnesium alloys. Changes in the regulations called for a full-width screen, two regular seats, and a flexible tonneau cover. To improve the aerodynamic potential, the windscreen's top lip was flipped over in line with the rear bodywork. This was now elevated to correspond with the height of the rear wings, and the flexible tonneau covered the void of the passenger seat. In a bid to push the aerodynamic envelope a bit further, Colin intended to use inflatable tonneau covers that would streamline the entire passenger compartment, but there was no time to make these. They were required to produce the hoods for each car at scrutineering, however, and whether the hoods would actually have worked in practice was a moot point.

In practice, the 1500cc car of Fraser/Chamberlain dropped a valve, so the pair took over the 1100 of Le Mans novices Ashdown/Stacey, and finished the race ninth overall, winning the 1100cc class and coming second in the Index of Performance. The privately entered Elevens of Dalton/Walshaw and Héchard/Masson finished thirteenth and sixteenth overall, classified second and fourth in the 1100cc class. More remarkably, the 744cc Climax FWC-engined car of Allison/Hall was fourteenth overall, winning its class and the Index of Performance. All four Elevens contrived to cross the finishing line together at 4.00p.m. and, justifiably, Team Lotus was in jubilant mood. It had trounced the class favourites – the DB Panhards, the Stanguellinis, even the Coopers and Porsches – in the Index stakes. After all drivers and mechanics had been paid off, Team Lotus still retained £3,500 worth of francs in prize money, and there was still the gala dinner to attend.

The Team moved north to Rouen for the Coupe Delamare de Bouteville, which was won by Flockhart in the Coombs 1500cc Eleven; Chamberlain was second driving the Team Lotus car. Chapman was second in the 1100cc category in the car that had won the class at Le Mans, while Allison took the 750 class win. It was another excellent result, but the jubilation was short lived. Running in the Formula 2 event at Reims, Mac Fraser was killed when he crashed at the high-speed Garenne Curve.

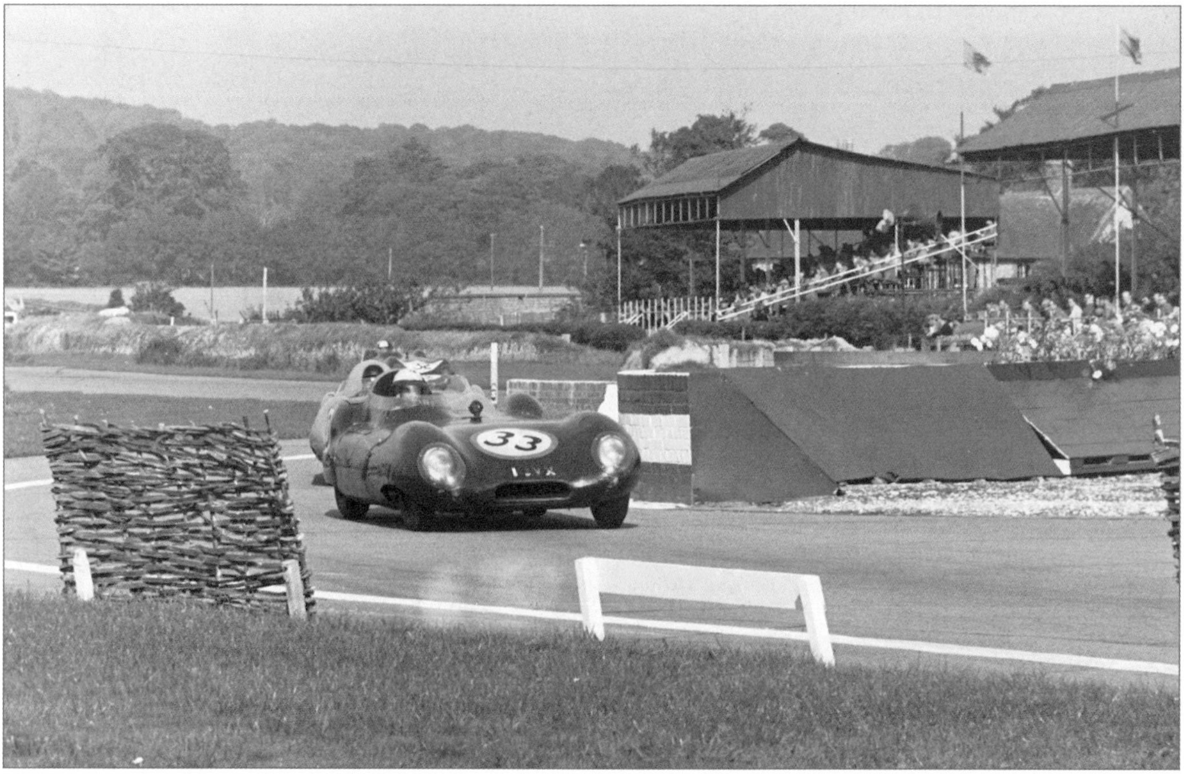

One of the best known vantage points of any circuit, Goodwood's chicane, with Stacey's Eleven hurrying by during the BARC National open meeting, 28 September 1957. This model was available in various specifications, including Club, Le Mans and Sports, with specifications and trim-levels according to use.

The career of the Eleven continued in 1958, when Alan Stacey was the most successful driver in his Series 2 model with a host of successes and class wins at both club and international level. Elevens driven by Mike Taylor, Innes Ireland and Keith Greene also performed well. Team Lotus concentrated on the Mk 15 sports racer and Mk 16 single seater, and only fielded Elevens at Le Mans. These were an 1100 car for Ireland/Taylor, and a 750cc car for Stacey/Dickson, fitted with the new Coventry-Climax FWMA 745cc engine. In the interests of reliability and lightness they had live rear axles to reduce transmission power loss, drum rear brakes and magnesium-alloy wheels. The Stacey/Dickson car blew its engine in practice so an older 744cc version was substituted for the race, and the other 750 was that of privateers Masson/Héchard. Another privately entered 1100 Eleven ran for Bill Frost/Bob Hicks.

At this point, let me introduce Jim Endruweit. He was Chief Mechanic from 1960 to 1968, and Lotus Cars Service Manager thereafter until being recalled to Team Lotus to run the Texaco Stars Formula 2 Team in 1973 – which was when I first met him. Jim joined Team Lotus in 1957 having served as a mechanic in the Fleet Air Arm. He did a six-week voluntary stint while waiting to be demobbed – during which time he was paid by the MoD, and was put to work by Mike Costin assembling the idiosyncratic Lotus gearbox known as the 'queerbox'. Jim Endruweit's first foray was to the 1958 Le Mans 24-Hours. As before, the team based itself at M. Mica's Auberge de St Nicholas at Mayet, and prepared the cars in a nearby garage. On the

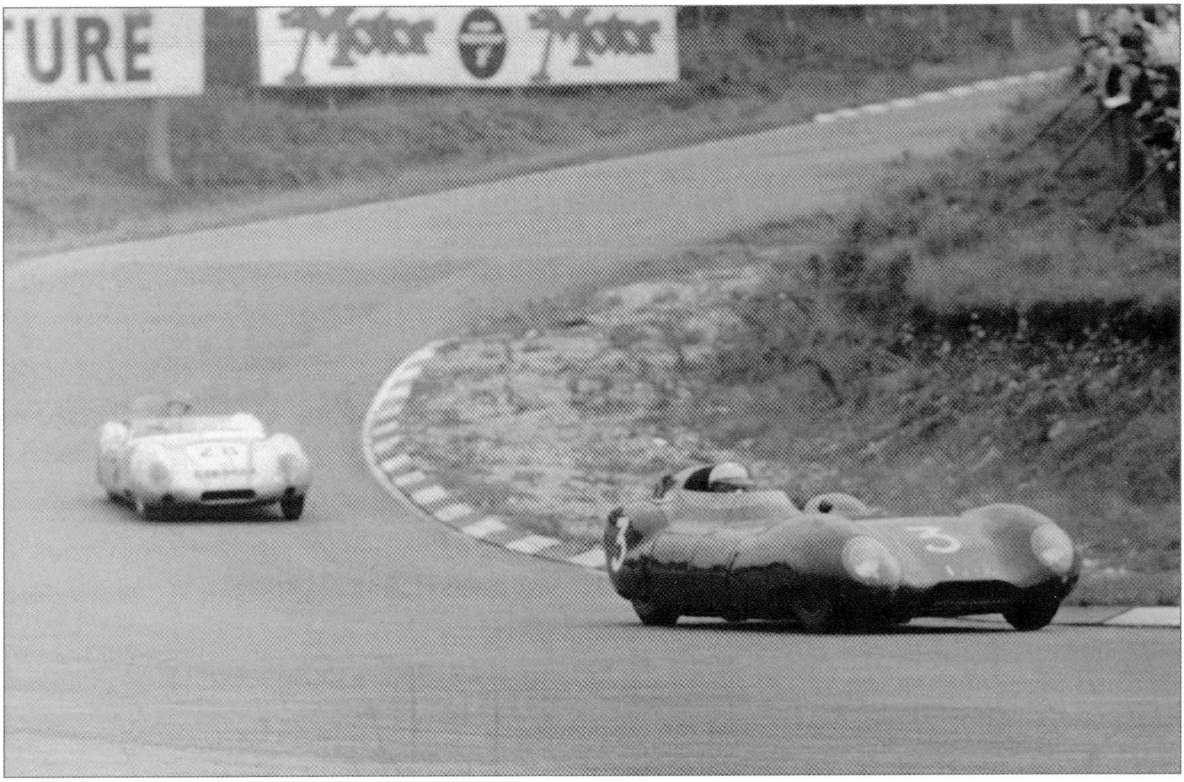

Alan Stacey's Series 2 Eleven leads Keith Hall's Eleven at Brands Hatch during the BRSCC Sporting Record Trophy on 6 October 1957. Hall's car is running with the full width windscreen and two-seater cockpit, which lacks the fairing behind the driver's head.

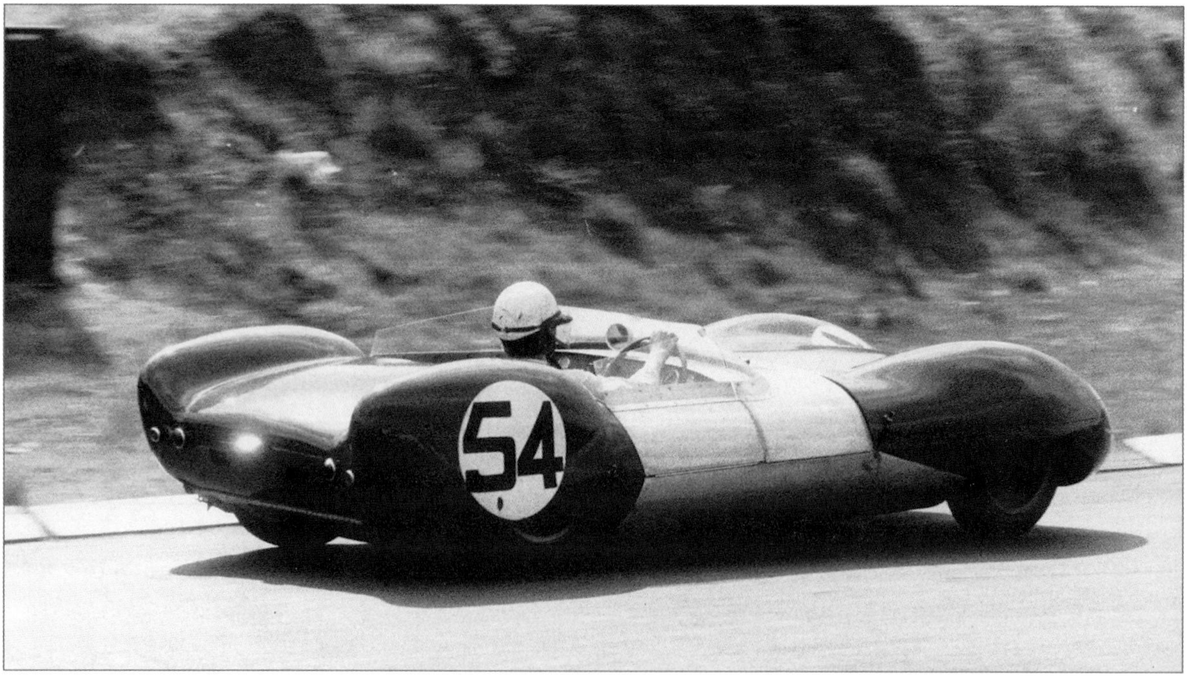

The Series 2 Eleven of Alan Stacey at Brands Hatch on 18 May 1958, showing the car in its two-seater cockpit configuration. The unpainted aluminium centre section illustrates where the panel gaps are. The race was won by Stacey ahead of Innes Ireland's similar car, and Stacey left the 1100cc lap record at 60 seconds.

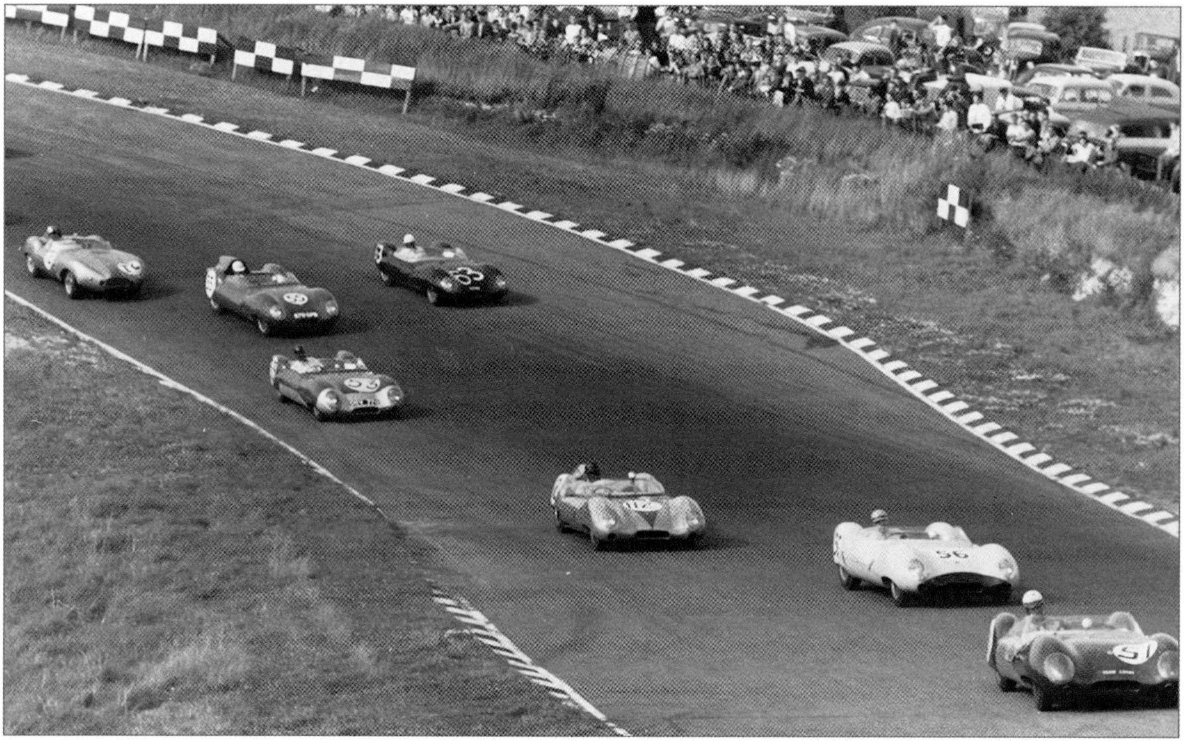

The pack swarms down Paddock Hill bend at the start of the Rochester Trophy event, with six Elevens heading a Lola. Stacey leads Keith Greene and Mike Taylor. Stacey was by far the most successful Eleven protagonist in the 1100cc class in 1958, and Taylor won the Brooklands Memorial Trophy for consistent performances in his first season, with Greene second.

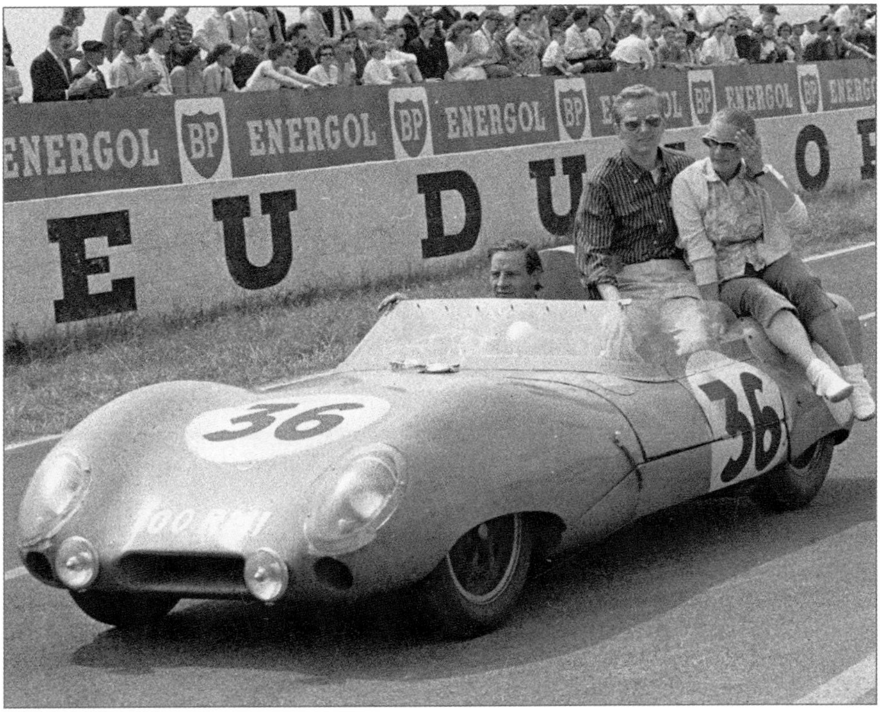

The Eleven of Innes Ireland/Mike Taylor won its class at the 1958 Reims Grand Prix meeting. The Team Lotus 1100cc car was running a stage 3 tuned engine and MGA gearbox.

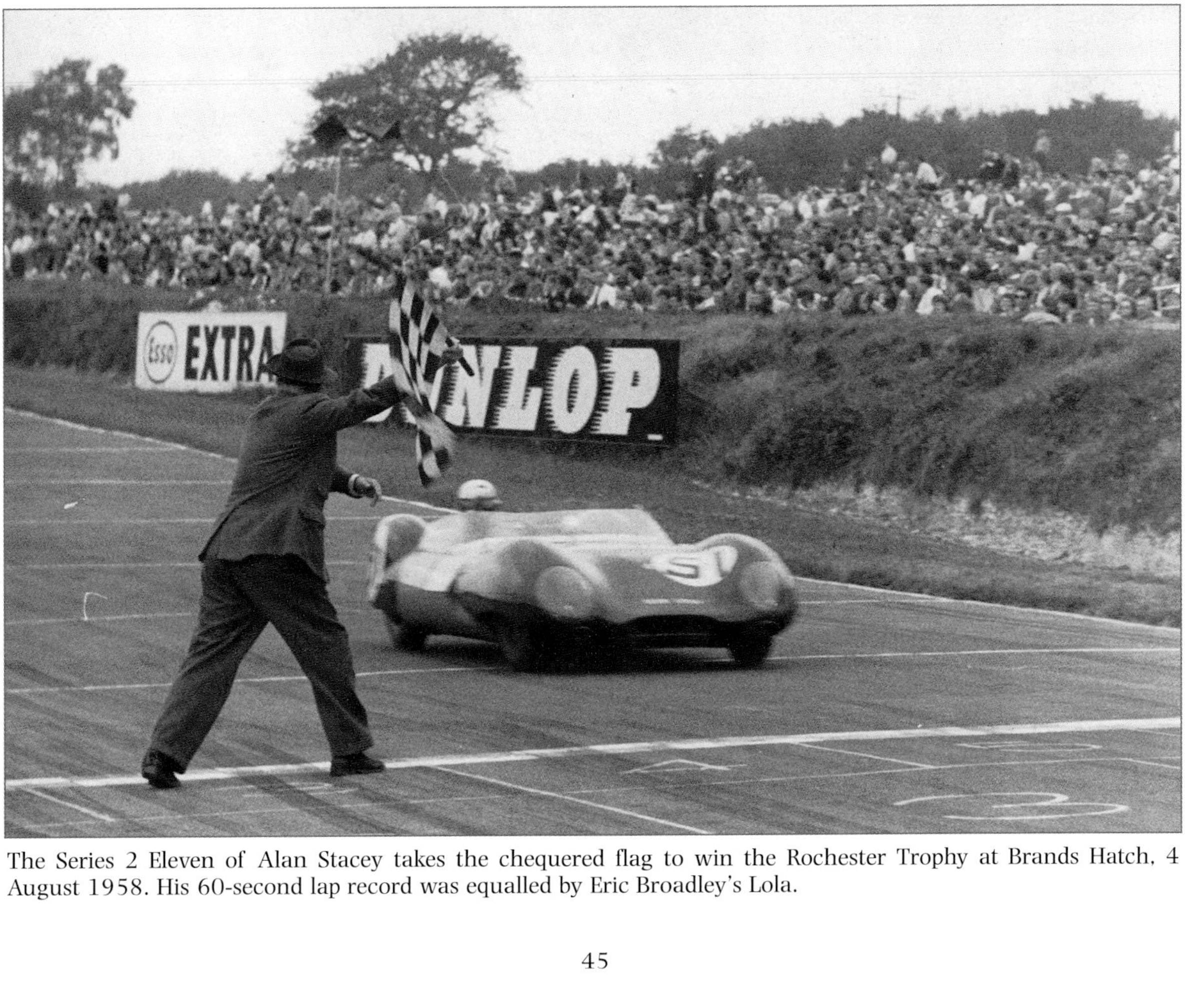

The Series 2 Eleven of Alan Stacey takes the chequered flag to win the Rochester Trophy at Brands Hatch, 4 August 1958. His 60-second lap record was equalled by Eric Broadley's Lola.

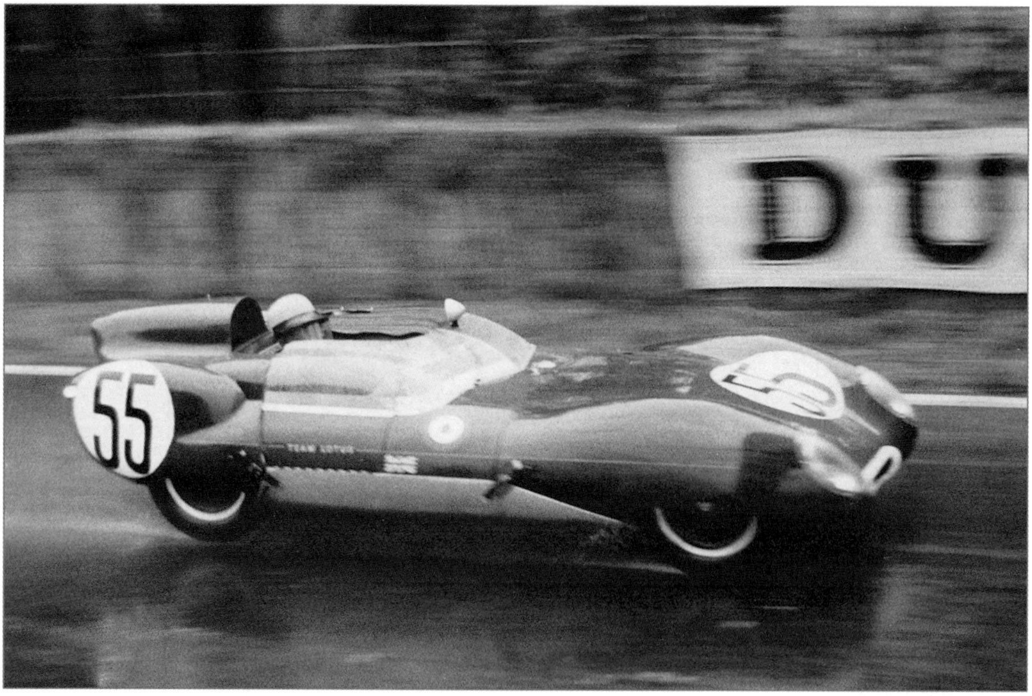

For the 1958 Le Mans race, Team Lotus also entered an Eleven fitted with a 750cc Climax twin-cam – adapted from the previous year's 1100cc unit after the designated engine failed in practice. Down on power according to Stacey, it was the only Lotus to go the distance, finishing twentieth – and last.

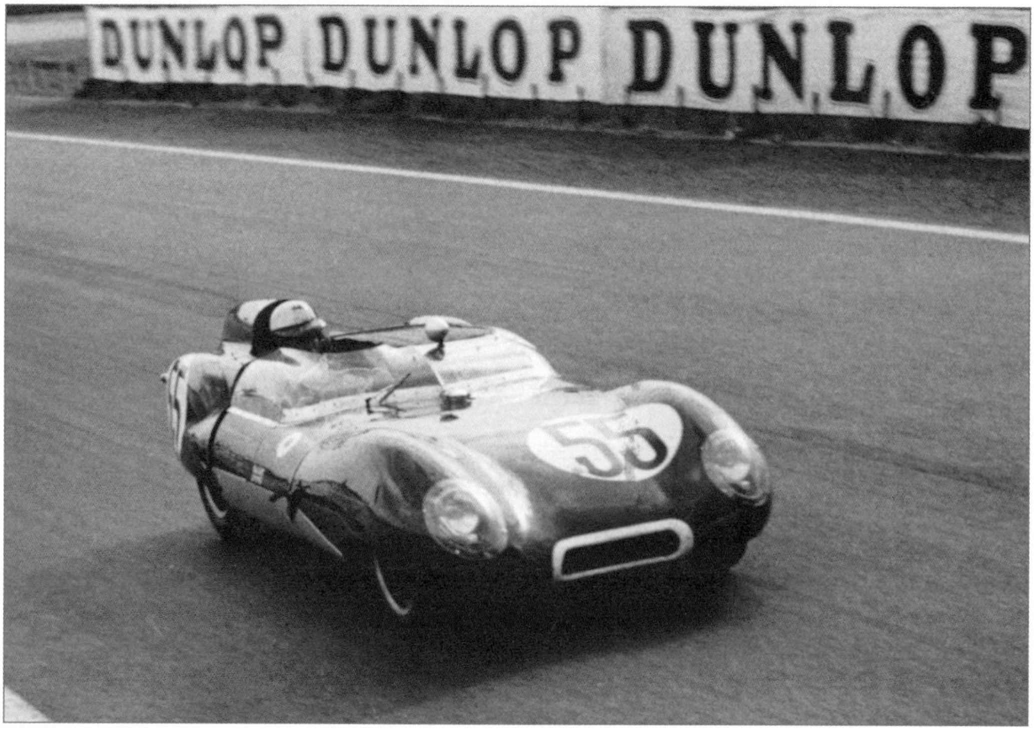

The 750cc Eleven of Tom Dickson/Alan Stacey survived the rigours of Le Mans in 1958, only to be put into the sandbank exiting Tertre Rouge early on the Sunday morning. It took Dickson two hours to dig it out, using a headlamp cover, hence the late finish.

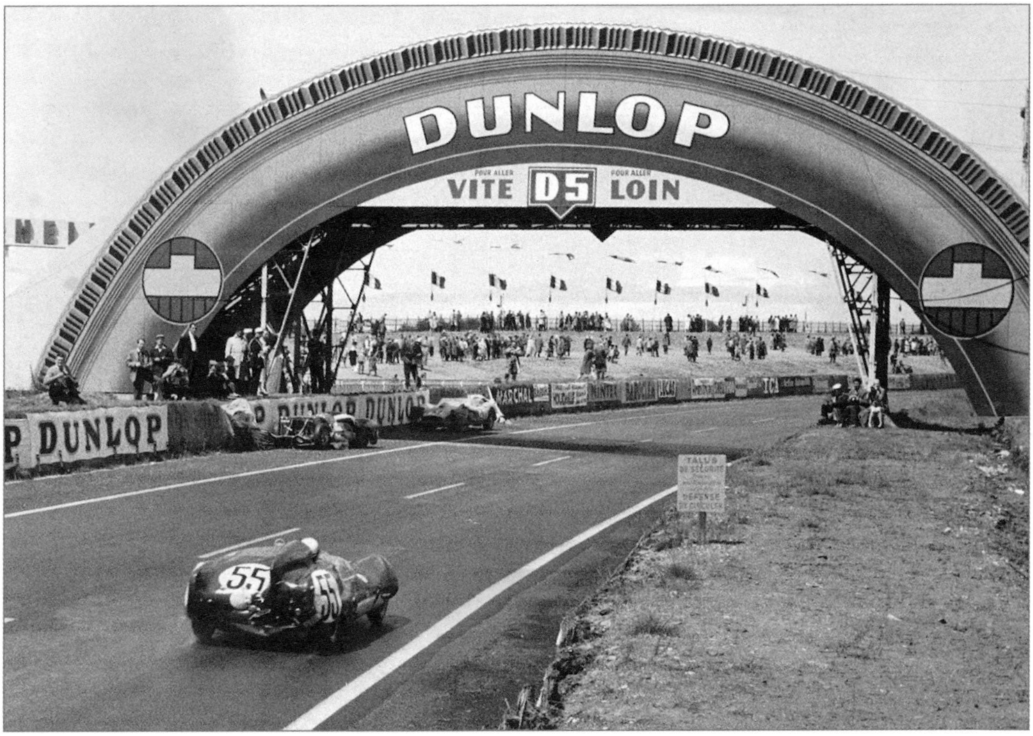

Passing the carnage of an earlier accident beneath the Dunlop Bridge, the Stacey/Dickson 750 Eleven bears the scars of its altercation with the sandbank in the early hours of Sunday morning.

first day they were there, recalls Jim, the mechanics were left to their own devices, building and fettling the cars that had arrived pretty much in kit form. At lunchtime, they repaired to the hotel for a midday meal. At this point, Mike Costin returned, found no one at the garage and bawled them out, banning any further lunchtime excursions. The following day, a flamboyant M. Mica brought a table out to the garage, complete with tablecloth, and the mechanics lunched there. It didn't go down well, and that was the last time they enjoyed a civilised lunch.

The cars were driven to the circuit for scrutineering and practice by Team personnel, and Jim remembers having to drive an Eleven far quicker than he'd have liked in order to keep up with Costin. The pace of the proceedings was such that Benzedrine tablets were issued to mechanics prior to the race, and at intervals during the night as the effects wore off.

It was not a good Le Mans for Lotus. The new 15s and the 1100cc Eleven failed during the race, until only the 750cc Mk 15 of Stacey and Dickson was running. In the middle of the night, Jim Endruweit had dozed off behind the pit counter, only to wake during a lull in the cacophony when all the cars still running were around the back of the circuit. Assuming he'd overslept, Jim leapt onto the pit counter, ignoring the rules about how many personnel were allowed there at any one time. This was the moment that Dickson arrived at the pit with a sorely deranged car that he'd stuffed at Tertre Rouge and spent two hours digging out of a sandbank. Jim shouted instructions and the Eleven's mutilated rear end was secured. The car rushed off on its way to twentieth – and last place.

Lucky Seven

Together with Mike Costin, Chapman had intended to uprate the Mk 6 back in 1954. Its successor, a Bristol-engined sports racer built as a one-off for the Clairmonte brothers was known at the time as the Mk 7, but Chapman withdrew that designation because it was eventually delivered as just a chassis unit. So the Seven wasn't introduced until the time of the 1957 Earls Court Show. The fact that it's still produced today, as is the Caterham, the chassis of which is still made by Arch Motors at Huntingdon, speaks volumes for Chapman's original conception. The Seven continues to espouse his ideals for an affordable sports racer, and is equally at home on the road and on the track.

The Seven's specification was based on a car ordered in 1957 by Mk 6 racer Edward Lewis, which had Mk 12 Formula 2 front suspension and the Eleven's de Dion rear end and inboard disc brakes, with 1100cc Climax power. Because this was thought to be a tad too radical for the regular enthusiast, the cheaper live axle, drum-braked, 1172 Ford-engined car went into production. While the chassis was re-engineered accordingly, the new Grand Touring car, the Elite, stood alone in the Earls Court limelight. Potential Seven buyers had to take a trip out to Tottenham Lane.

Not many Mk 6s were produced after 1955 as Lotus concentrated on the more sophisticated Mks 8, 9, 10 and Eleven, and the launch of the Seven in 1957 was greeted with mild surprise. Predictably, there was method in Chapman's timing, because he was also looking for a money-spinner to tide things over while the Elite progressed through its gestation period. The Seven was closely based on the Mk 6 concept and the Eleven chassis, so all the elements were in place to go ahead

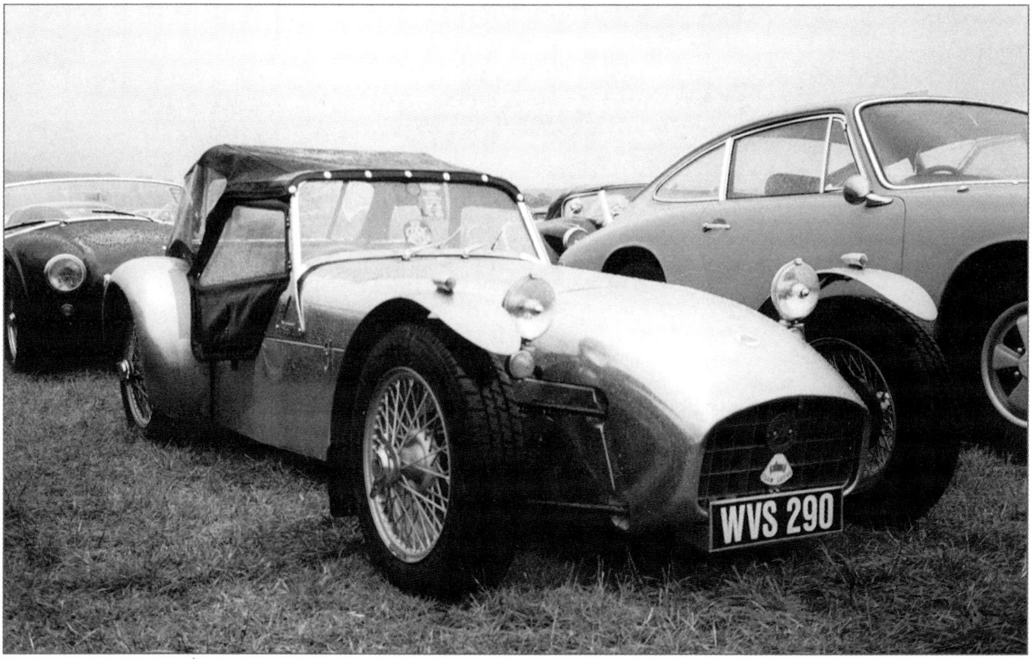

The Seven was launched at the Earls Court Show in 1957. It was a road-going component car intended to be a replacement for the Mk 6, and was initially powered by the Ford 10 engine. I snapped this fine Series 1 model in the Goodwood car park in 1999.

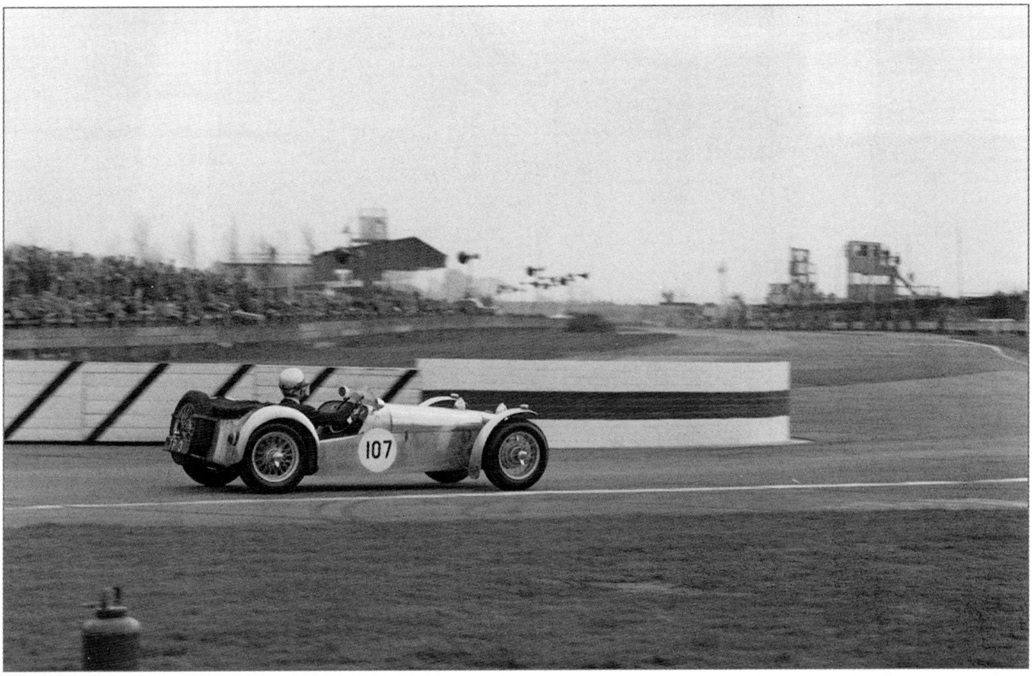

A Series 1 Seven rounds the Goodwood Chicane. The driver is not overly concerned about weight saving since the spare wheel is still fitted.

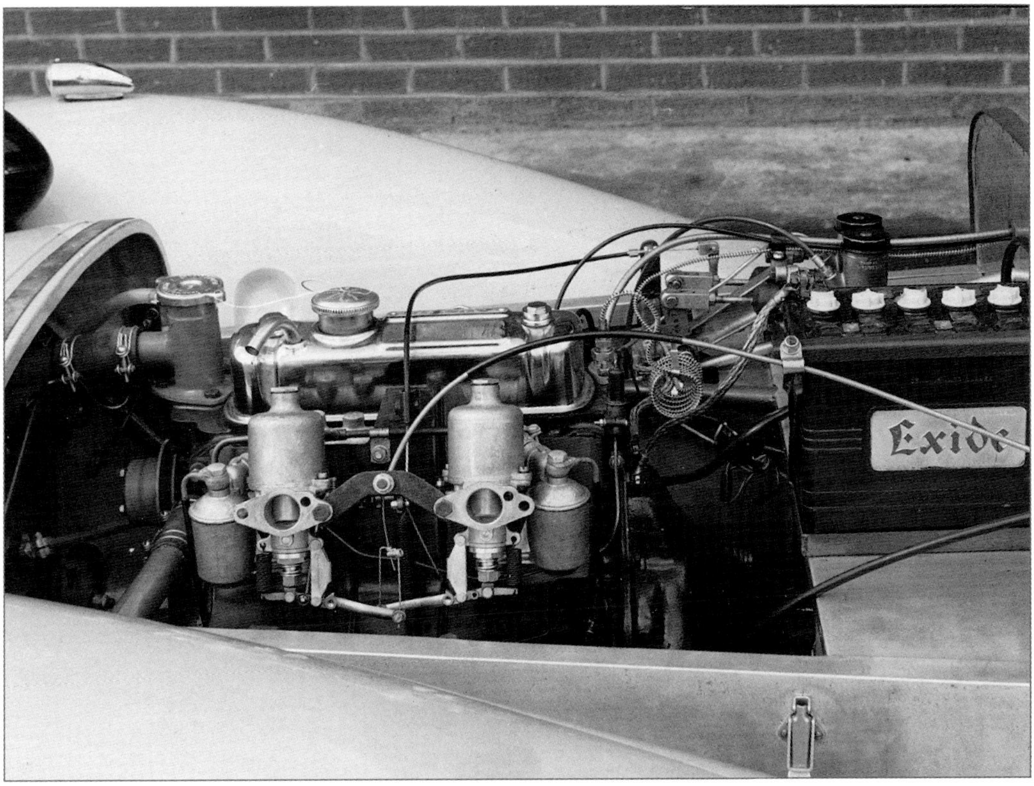

The Seven A – for America – came out in October 1959, powered by the 950cc BMC A-series engine with twin SU carbs, which were used in the Austin Healey Sprite. It had the advantage of an established BMC dealer network in North America, which made for easier servicing.

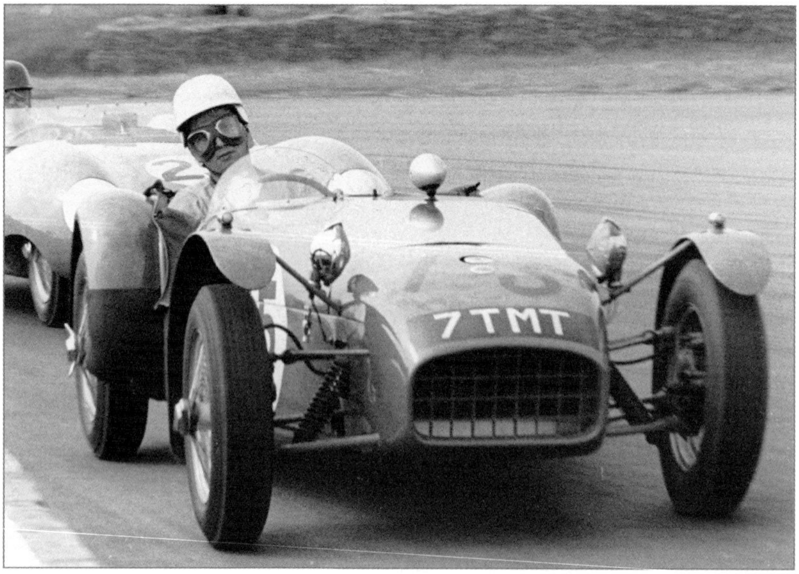

Peter Warr corners his Series 1 Seven at Silverstone ahead of an Eleven at the NSCC meeting in July 1960. Note the headlights turned in to minimise wind resistance. Peter joined Lotus Components in 1958 as Sales Assistant to Colin Bennett, and after a career in model slot-car racing, he went on to become Team Lotus's Team Manager in two stints, from 1969 until 1977 and, after a brief spell at Walter Wolf Racing, from 1982–9.

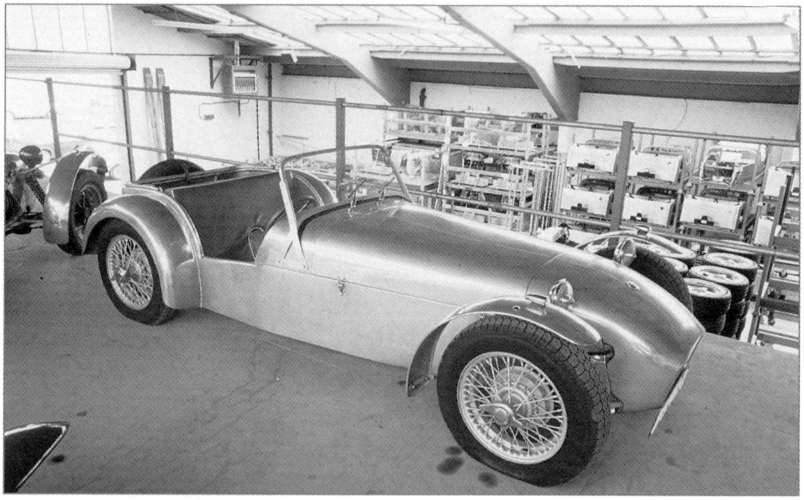

This Seven belonging to Caterham Cars demonstrates the external differences between the Series 1 and Series 2. The S1 has aluminium cycle-wing front mudguards and slightly drooping nosecone. The S2 would adopt the fibreglass clamshell front wings of the Seven A US export model – although cyclewings were always optional fitment.

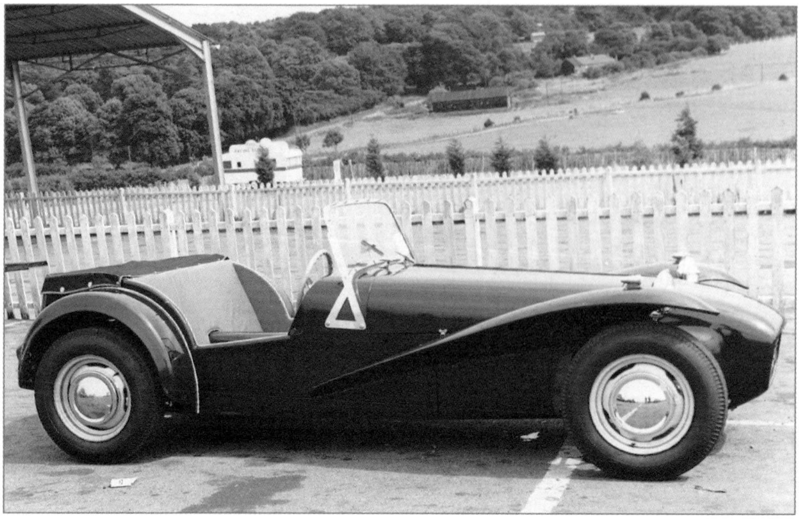

The Series 2 Seven came out in June 1960, and this car is pictured at Brands Hatch. It featured flared clamshell front wings, a straighter nosecone, and was offered with the Ford 1500 pre-crossflow engine and speedo mounted ahead of the passenger. Amazingly enough, the company couldn't sell them and unsold kits were wrapped in polythene and stored at an airfield near Cheshunt. A price cut announced at the Racing Car Show in 1961 got sales moving again.

straight away. It differed from the Mk 6 in having double wishbone front suspension, borrowed from the Lotus 12 F2 car and also featuring on the Elite. At the rear was a mass-production live axle located by twin parallel trailing arms and a diagonal strut brace with coil springs and dampers. Much ground had been covered in the years between the development of the Mk 6 and introduction of the Seven, and the Seven's chassis frame was a rationalisation of the Eleven's, but clearly lacking the sports racer's all-enveloping bodywork. The chassis of the Series 1 Seven was more akin to the Eleven than the Mk 6, although smaller section 18-gauge 1 inch or ¾ inch tubing was used. There were fewer triangulations as a weight-saving rationalisation, which was made up for by the riveting of all the aluminium panels to the frame. These now included the transmission tunnel and side panels, instead of just the scuttle and undertray as in the 6. Where components were attached by bolts or rivets, the Seven chassis used square-section tubing.

The Climax-engined prototype Seven first appeared at the Brighton Speed trial in September 1957, and such was its potential that Edward Lewis almost went off the promenade and into the sea. The following week he won the 1100cc class at the Prescott Hill Climb. The specification of the standard car was rather less exotic, however. The first Sevens, priced at £587, were powered by the ubiquitous 40bhp Ford 1172cc unit, allied to a single dry-plate clutch and three-speed Ford gearbox. Steering was by the Burman worm-and-nut system, and road wheels were 15 inch bolt-on steel discs, shod with 5.20 × 15 tyres, whereas the Mk 6 often had bigger tyres at the rear. Brakes were hydraulic twin-leading shoe drums in cast iron, 9 × 1¾ inch at the front and 8 × 1¼ inch at the rear. There was a horizontally mounted handbrake under the passenger-side scuttle, acting on the rear brakes, and both front and rear were served by a combined hydraulic fluid reservoir and master cylinder. For racing, tiny perspex fly screens could be substituted for the windscreen, and the height of the car was just 2½ inches from ground to scuttle top. Wheelbase was 7 feet 4 inches and track front and rear was 3 feet 11 inches.

The 1172cc 'autothermic' side-valve engine with its steel inserts frequently suffered from broken rings, so the trick was to fit a set of Hepworth pistons and a Willment overhead-valve conversion. Cord piston rings that adjusted according to bore wear were a cheaper alternative. The firm of Aquaplane at Oulton Broad, Suffolk, specialised in hydroplane engines, and their alloy heads and flywheels were sensational.

In broad terms, Chapman was influenced by Ford component design. Items like the MacPherson strut and integral roll bar were cheap to productionise, and there was a BMC Nash Metropolitan back axle, as used in the Elite. This BMC axle had separate nose pieces so there was a choice of ratios. The axle was located by a side A-frame bracket, attached by a fixing welded to the differential case. Thick triangular bars were eventually welded to the axle to counteract flexing. As with the Mk 6, the earlier E93A torque tube had to be shortened. Likewise, the valve guides in this engine had to be modified before the mushroom-based valves could be removed, and the 100E unit used subsequently was far easier to work on.

Because of the Seven's narrower track, only the top arms that formed a wishbone with the anti-roll bar, trunions and uprights fitted, and the lower wishbones were specially fabricated. These provided the mounting points for the coil springs and

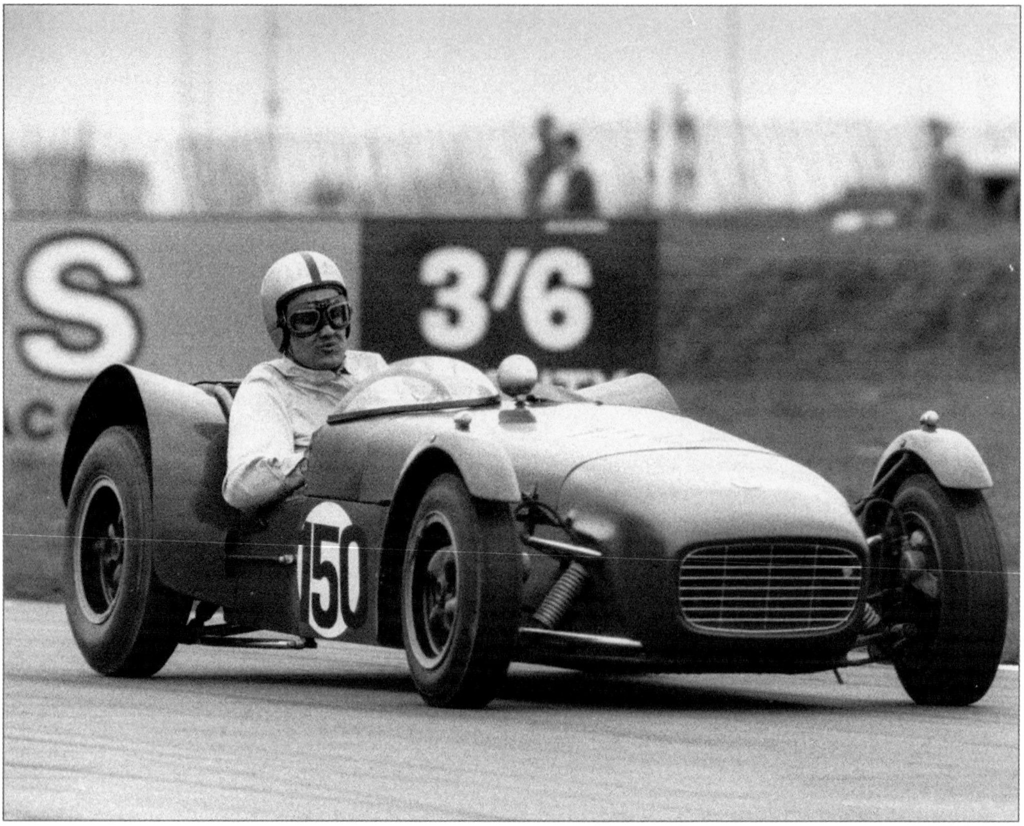

A one-off Series 2 Seven (actually based on an S1 chassis) known as the 7/20 was built for customer Keith Hamblin by Lotus project engineers Hugh Haskell and Don Gadd using Eleven suspension and fitted with a 997cc Cosworth Formula Junior engine. It was raced with much success by Wendy Hamblin, Natalie Goodwin, Clive Lacey, Colin Chapman – notably in the 1962 Birkett 6-Hours Relay – and David Porter, seen here at Snetterton in April 1963.

Armstrong damper units and uprights, which were linked to the upper arm by a ball joint; all chassis pickup points used rubber bushes. Initially, steering gear was of the Burman worm-and-nut variety, brought up-to-date with an Elite-style rack-and-pinion system soon after production started. This gave two and a half turns lock to lock, and the basic car – the Seven F – was endowed with an unsexy Wilmot Breedon ivory plastic steering wheel. The Seven C or Super Seven fared better in this respect with a leather-rim version.

An early Lotus Seven racer was Brian Luff, who worked at Lotus from 1958 to 1970, during which time he progressed from Project Engineer to Chief Engineer, thence to Manager of the Vehicle Engineering Division. On the strength of his previous Thorneycroft training, he was taken on as assistant to Technical Director Mike Costin. 'There was never any intensive effort on development of the Seven until the Series 4, and that was a retrograde step in engineering terms,' said Luff. 'The car evolved mainly due to parts becoming scarce or obsolete, or when a choice of engines was possible. There was always a supply and cost problem, but the Chief Buyer, John Standen, always managed to get components cheaper.' Revealingly, Luff said 'Chapman's main interest was the racing cars; he wasn't interested in the Sevens.'

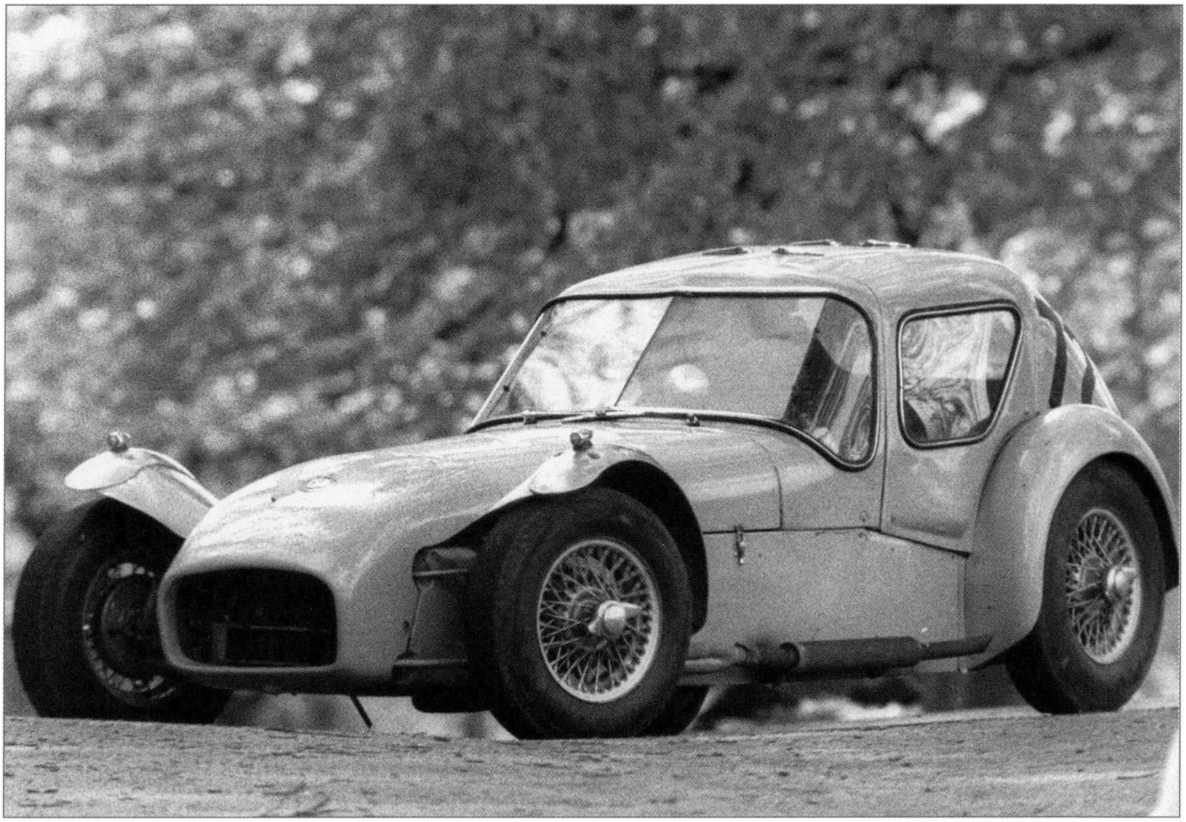

The Lotus Seven could be converted into a GT car by fitting a Fibrepair hardtop, complete with gullwing doors. This is M.J. Hollebone's S2 Seven A pictured in May 1964. Another was raced in GT events in the early '60s by Tony Goodwin, where the lightweight Seven could be highly effective against heavyweight opposition like Jaguar XK-120s.

At first, the Series 2 Seven's live rear axle was located by twin trailing arms attached to the chassis inside the wheel arch and a diagonal member located the differential to the rear of the chassis. Coil springs and dampers echoed the frontal arrangement, and all joints were rubber-bushed. The majority of Sevens made before 1968 used Triumph Mayflower and Standard 10 back axles, although both these models were defunct by 1959. In its original incarnation, the axle had merely been required to serve in saloon cars, but in the Seven it transmitted power from race-tuned 1172cc Ford and 75bhp Coventry-Climax engines, leading to blown oil seals, and the solution was to brace it with a sturdy A-frame from a bushed mounting welded to the differential casing, combined with a radius arm on either side. Cast-iron drum brakes were fitted front and rear, 8 inch diameter and 1¼ inches wide, with twin leading shoes at the front, and a handbrake acting on the rear. Road wheels were 15 inch pressed steel on skinny 4 inch rims, with a spare mounted on the back panel and bound on with a leather strap.

Like its predecessor, the Seven was the car for club racing enthusiasts to take to the circuits with, and the cycle-wing, cigar-tube configuration was rather more practical for the amateur driver, being cheaper to repair if shunted. Until 1962, the Seven's panelling was made by Williams and Pritchard. The chassis was

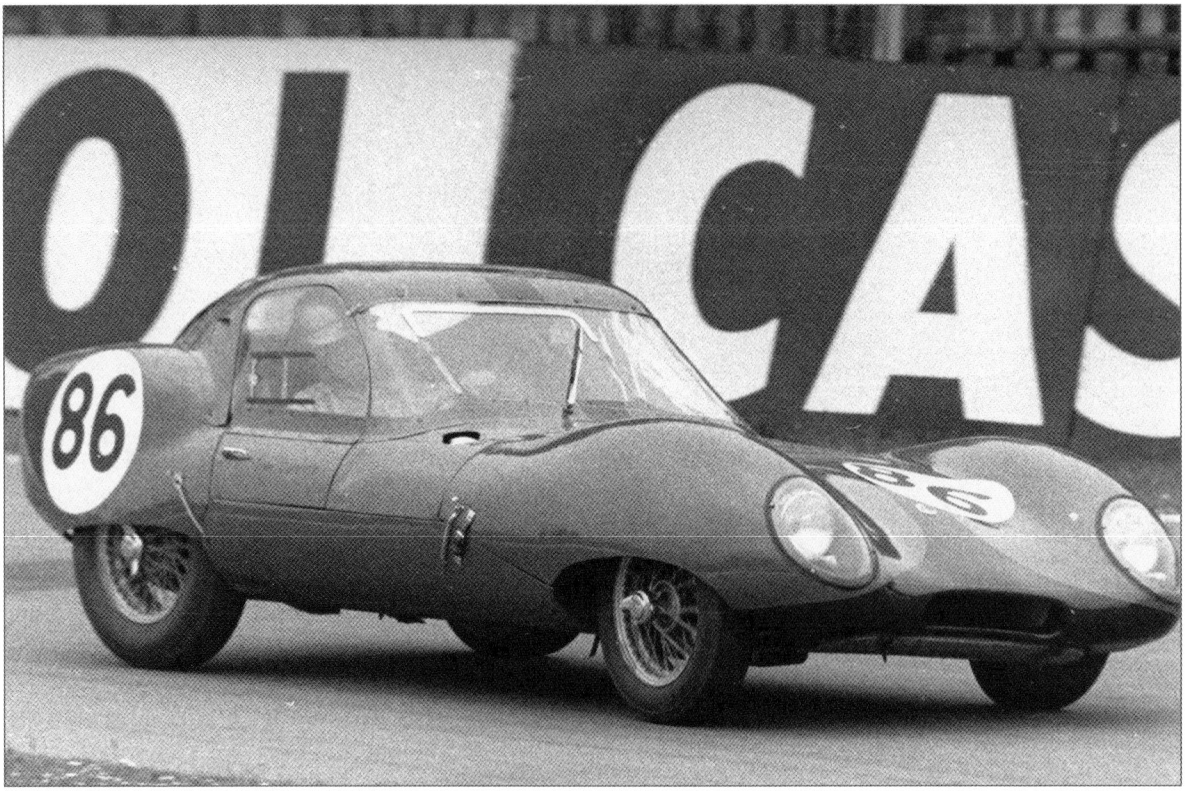

The Eleven was never built as a GT car, but J.P. Dickinson's car was fitted with a hardtop that appeared to suit it quite well. It's pictured here at Crystal Palace in June 1965.

manufactured by Progress until 1965, from which time it was made by Arch Motors. The most obvious differences between Mks 6 and Seven are the sloping nose of the latter, and the squared-off rear end that superseded the Mk 6's rounded rump. The wings or mudguards were now separate entities, although the front mudguards continued to be the pointed cycle-wing type, carrying torpedo sidelights, and the by now familiar egg-crate grille was also carried over. In the cockpit, the instrument binnacle of the Mk 6 was simplified to a flat dash panel. This, like the nose cone, was secured with Dzus fasteners, while four over-centre stays held the bonnet down. Drive a Caterham today and provided it's not over-tyred, you'll capture much of the flavour of the original Seven.

OPEN WHEELS AND SINGLE SEATS

During the early 1950s, one of Colin Chapman's main ambitions was to achieve success in the Le Mans 24-Hours. The event carried more status then than it does now, which was one reason why Lotus concentrated on sports car racing and not single seaters. Besides, sports cars meant sales to other drivers. However, the new 1500cc Formula 2 of 1957 was a tempting prospect as it provided a vehicle for Team Lotus to put into practice many of the lessons learned with the 1500cc Climax-engined Elevens. The main opposition in F2 was provided by the mid-engined Cooper-Climax T43s, but the Lotus 12 retained the traditional front engine/rear drive layout, with transmission via a transaxle. It was built around a tubular spaceframe chassis, stiffened up with a rigid undertray, with double wishbone front suspension consisting of a wide lower member and an upper transverse tubular link that also acted as an anti-roll bar. Like the Eleven, the two prototype 12s used de Dion rear axles, located by single (rather than double) radius rods and an A-frame, and their BMC gearboxes were in unit with the engine. Progress was retarded by Girling discs up front and inboard-mounted at the rear. It ran on Chapman's new 'wobbly-web' cast magnesium-alloy wheels, which were bolt-on fitment rather than knock-ons, since it wasn't envisaged that pit stops would be routinely necessary. Very soon after the 12's competition debut, Dunlop tyres were replaced by Continentals, which suited the car's light weight.

One of the de Dion cars was raced by Mac Fraser at Brands Hatch and Crystal Palace meetings during the Whitsun weekend. However, subsequent cars were built with the Chapman strut layout. This consisted of three members that located each rear hub: a forward-facing radius arm to locate the hub fore and aft, an upright coil spring and damper unit controlled by the hub casting, and a half-shaft providing lateral location. This system compensated admirably for the change in weight that came about when the rear-mounted fuel tank was empty, and it was also lighter than the de Dion set up.

The first Lotus single-seater was the 12, introduced in 1957 as a contender in Formula 2 events, in which its power-plant was the 1475cc Climax twin-cam. This is Cliff Allison negotiating the Goodwood Chicane in 1958, on his way to winning the F2 section of the Glover Trophy – he was harried by Moss's Cooper-Climax on the way.

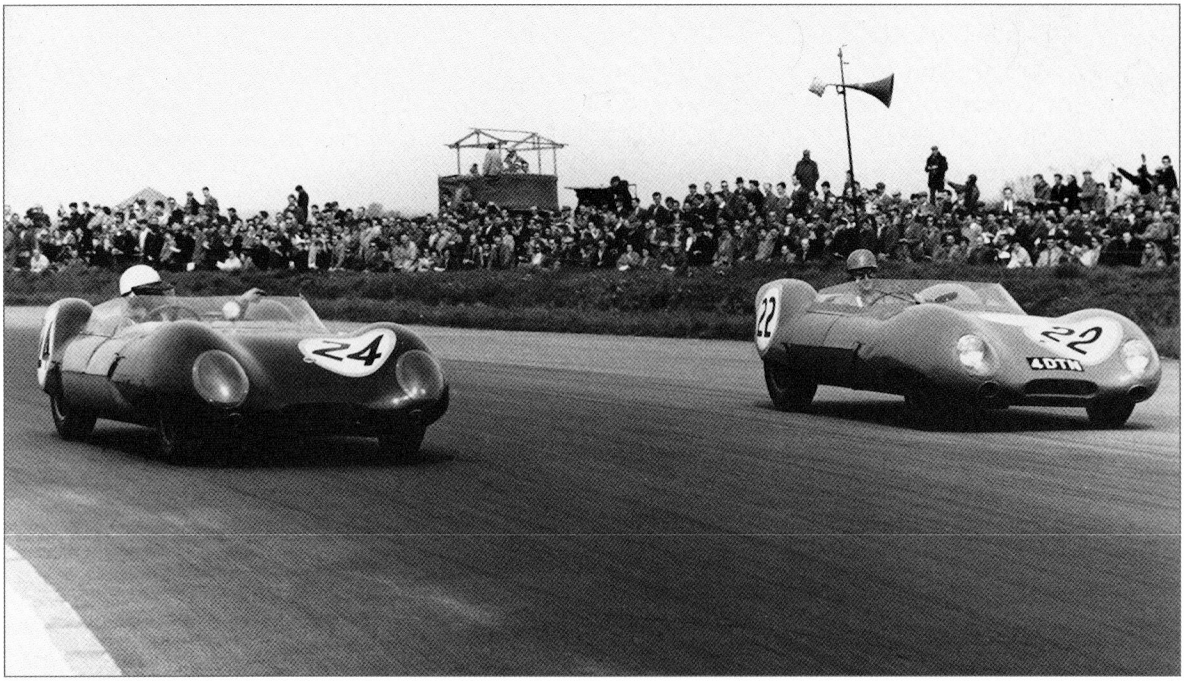

While Graham Hill driving a Lotus 15 in the up-to-1500cc class won the sports car race supporting the *Daily Express* International meeting at Silverstone on 3 May 1958, Alan Stacey and Keith Hall were second and third in their 1100cc Elevens. Here they circulate on the slowing down lap, Stacey having set a new lap record at 1 minute 48.8 seconds.

The new gearbox, designed by Chapman, Harry Mundy and Richard Ansdale, was matched to a hypoid crown wheel and pinion and a ZF limited slip diff. The internal gear ratios were sourced from ZF as well. The gearshift was similar to the modern sequential pattern, with the lever lodging in a series of shallow niches with each shift. Despite the obvious advantages of a straight shift pattern the all-indirect 'box quickly became known as the 'queerbox', and much time and effort was spent on the gearbox lubrication system. Although this was problematic, the gearbox itself worked well. The casings, though, were another matter according to Jim Endruweit.

Like Keith Duckworth, who helped out during student vacations, Jim's first task as a volunteer mechanic in 1957 was to assemble the queerbox. The casings were poorly made, with no apertures for the cross fixings, and Jim recalls having to drill and line up by eye when making holes for the studs. They didn't have the luxury of the conformity of mass production, and perhaps only a couple of dozen units ever existed.

Although Lotus retained its Hornsey premises, the racing team relocated to nearby Edmonton where the tin-clad workshop formerly occupied by Williams and Pritchard had become vacant with their move to Hammersmith, and the 12 was assembled here. Its body panels were created by Williams and Pritchard from Chapman's drawings, and it was as neat a design as any contemporary front-engined single-seater racing cars. Seven cars were built during 1957, including three works' cars.

Although Team Lotus entered the 12 in Formula 1 events in 1958, there was no tangible success. It was powered by the 1.9-litre Coventry-Climax unit, and Allison had the measure of the car rather better than Graham Hill. Running with a 2207cc

Climax engine, Cliff came sixth at Zandvoort in the Dutch Grand Prix and fourth at Spa-Francorchamps. Since the three cars ahead of him at the finish of the Belgian GP were unlikely to have completed another lap, Allison might well have won.

In Formula 2 events, Cooper was still very much to the fore, although Hill and Allison were second and third in their 12s in the Lavant Cup race at Goodwood, and there were several fourth and fifth placings at relatively major meetings. As far as the success of the 12 is concerned, the results speak for themselves, and it is perhaps surprising that Lotus could be doing so well in sports car racing on one hand, and failing to score convincingly in single seaters on the other. It was generally felt that one reason for the inconsistent results was Chapman's obsession with paring components down to the barest minimum to save weight.

Variation on a Theme

The Type 15 sports racing car was made between 1958 and 1960. It was virtually identical to the Eleven Series 2, so similar in fact that Jim Endruweit and I spent ages looking at undated photographs from the late '50s and early '60s to correctly identify some of these cars. The chief differences between the Eleven and the 15 was the latter's lower frontal area, and the windshield, which curved back at the trailing edge on the later model. The 15's Williams and Pritchard-made body also lacked the

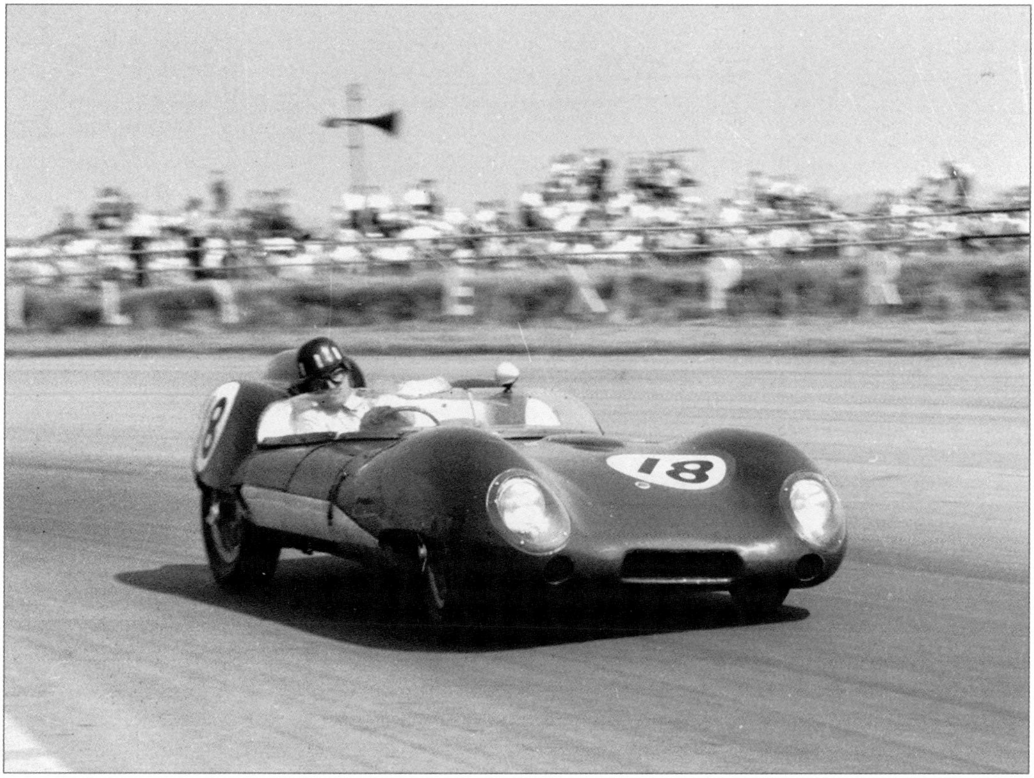

Graham Hill at speed in the 15 at Silverstone in 1958. The 15 was based on the Eleven, and fitted with FPF Coventry-Climax twin-cams ranging from 1457cc and 1960cc to 2208cc and even 2495cc, as well as 1595cc Lotus-Ford twin-cams. This car is running wire wheels rather than wobbly-web magnesium alloy wheels.

crease to be seen on the rear flanks of the Eleven, and the rear wheel arch was more squared off. Its spare wheel was mounted within the driver's headrest.

More fundamentally, the 15's *raison d'être* was to take Lotus into a bigger capacity class at Le Mans. Its chassis was therefore made to accommodate larger-capacity engines, which were mounted in the familiar multi-tubular triangulated spaceframe chassis. This was made up of 1 inch and ¾ inch round and square tubes, and the propshaft tunnel and floor were integral stressed members. As with the S2 Eleven and 12 Formula 2 car, suspension consisted of a double wishbone and coil spring set up at the front, with radius arms, coil spring damper units at the rear, located laterally by the drive-shafts. While customer cars were fitted with 1.5-litre Climax FPF engines, Team Lotus cars were powered by the 2.0-litre motor, subsequently replaced by the full-blown 2.5-litre Climax FPF unit, which promised a maximum speed of 205mph. It was canted over at 60 degrees to the right, and the twin sidedraught Webers breathed through specially made manifolding. The 15 also used the same queerbox and ZF final drive that were to be found in the 12, with reverse gear operated via a separate linkage controlled by a knob behind the passenger seat, so it couldn't be to engaged in error. Although knock-on wire wheels were standard fitment, the works cars ran on Lotus wobbly-web magnesium-alloy wheels – and they could be specified on customer cars as well. It also used Girling disc brakes all round, and inboard at the rear.

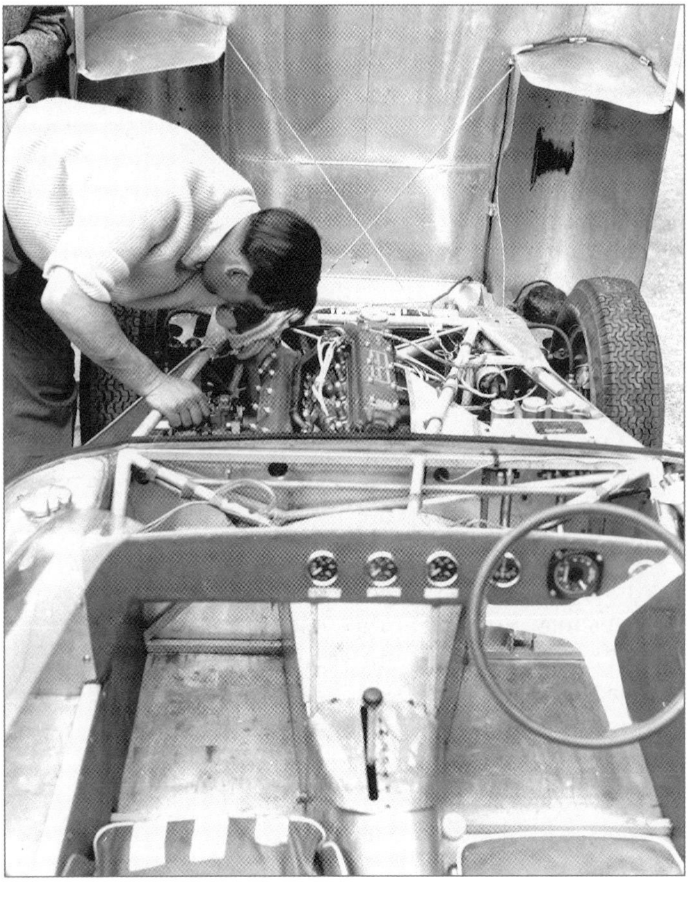

The spartan cockpit of Graham Hill's 15 revealed at the 1958 Le Mans 24-Hours. The engine is canted over at 6 degrees to the left, and transmission is via the queerbox, with a sequential-type shift arrangement. Engine access is good due to the front-hinged bonnet, but the offside wheelarch bears evidence of the tyre rubbing.

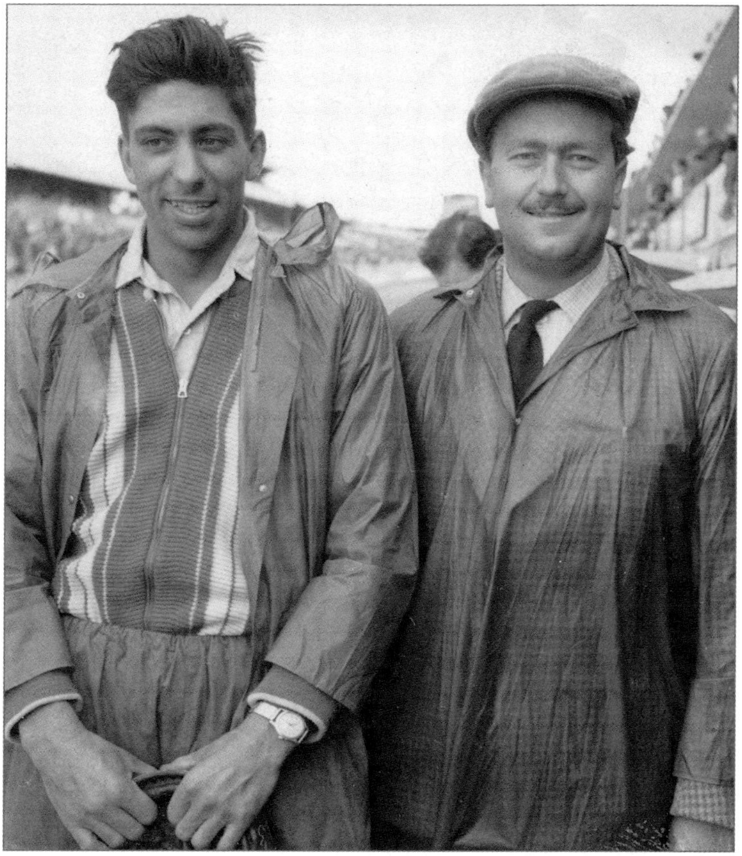

For Le Mans 1958, Team Lotus entered two 15s as well as an 1100cc Eleven for Innes Ireland/Mike Taylor and a 750cc Eleven for Alan Stacey/Tom Dickson. Wearing trendy new packamacs in the Le Mans pitlane are Colin Chapman and Alan Stacey.

The 15 was first raced by Graham Hill in the Unlimited Sports Car race at Goodwood on Easter Monday 1958, but it retired with queerbox trouble. Both Graham and Cliff Allison retired during the British Empire Trophy at Oulton Park. The 15's first success came in the 1500cc Sports Car race at Silverstone in May in the hands of Graham Hill. At this stage the Climax engine was canted over in the opposite direction, veering now at 17 degrees to the left. The low frontal area was thereby compromised, and a bulge appeared in the bonnet to accommodate the more upright engine. The pair of Team Lotus 15s that went to Le Mans that year also sported more aerodynamic tails, and this time Chapman's ingenious inflatable tonneau covers were fitted, running from the top of the windscreen across the passenger side of the cockpit to the rear panel. In practice, Cliff Allison/Graham Hill's 2.0-litre 15 showed much promise, and since it was quicker than all the Ferraris, there was even the prospect of outright victory. But in the race its head gasket blew on the fourth lap before Hill had a drive. The Jay Chamberlain/Pete Lovely 1.5-litre 15 fared only a bit better. It was dogged by a persistent misfire caused by faulty ignition, and when Chamberlain hit the bank by the Dunlop bridge while avoiding a slower car, the 15 was also collected by a Ferrari. The fortunes of the Elevens present at Le Mans have already been told – the only finisher being the Stacey/Dickson 750 car.

Of the privately entered 15s, John Coombs' 2.0-litre car was one of the more successful. Roy Salvadori came second in this in the Sports Car race accompanying

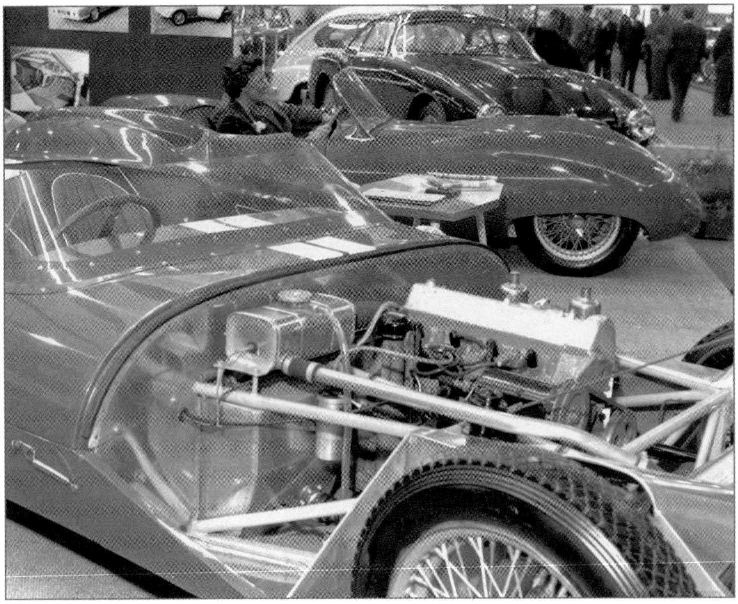

Elevens presented as road-going machines grace the Lotus stand at the 1958 Geneva Show. The high quality of conception and execution is evident in the Williams and Pritchard panel work and Progress Chassis of the Hornsey fabricated model in the foreground. Photographs on the wall show the new Elite.

the 1958 British Grand Prix at Silverstone, while Allison followed him home with a Team Lotus car to win the 1500cc class. While Salvadori won the Oulton Park Gold cup driving the well-prepared Coombs car, Team Lotus resources were stretched to the limit in Formula 2 as well as sports car racing.

In August 1958 a second series 15 was launched, its 1.5-litre Climax engine canted over at 17 degrees to the left, as seen in the Le Mans cars. Its BMC B-series four-speed gearbox was mounted in unit with the engine, and the conventional H-pattern gear-change was used. BMC differential parts were fitted in a chassis-mounted Lotus-made casing. The following year, a Series 3 Type 15 came out, which was available with 1.5-litre and 2.0-litre Climax FPF engines mounted at the same 17 degree angle as the Series 2 cars accompanied by the BMC-based four-speed transmission. The chassis frame was stiffened up, front track narrowed by an inch, and the front upper wishbones were reversed in order to pass behind the wheel centres, facilitating a larger radiator and oil cooler. A works 15 S3 was campaigned regularly on the British club scene by Graham Hill, most notably notching up victories in a club race at Mallory Park, the sports car race at the British Grand Prix at Aintree, Brands Hatch on August Bank Holiday Monday, and the International Brands Hatch meeting at the end of August. But Cooper Monaco T49s were the class of the field in the 2.0- and 2.5-litre category, often driven by top men like Moss, McLaren, Brabham and Salvadori. A 2.5-litre Lotus 15 was entered for Graham Hill/Derek Jolly at the 1959 Le Mans, but Jolly over-revved the engine when it jumped out of gear while running seventh. The Ireland/Stacey 2.0-litre version broke a wishbone and retired with engine failure. Meanwhile, Jay Chamberlain raced a 2.0-litre 15 with much success in the States. Private owners, including Tom Dickson, David Piper and Mike Taylor, continued to race the 15 during 1960, and its last gasp was in the hands of test pilot 'Dizzy' Addicot in 1962, who fitted his car with an aluminium 3.5-litre Buick engine. This motor evolved into the Rover V8, of course, and was used by a few shoe-horning contemporaries like Reg Gubbins and Doc Merfield in 'Fraud' Anglias and Cortinas.

CHAPTER THREE

THE FIRST PRODUCTION CAR

Former Chief Mechanic Jim Endruweit is convinced that Chapman's role model was Enzo Ferrari, and achieving that sort of status was probably his goal – a name synonymous with successful racing cars, and building road-going machines as a lucrative side line. Calculated or not, the beautiful looking Elite coupé would inevitably move him closer to that objective, since it embodied the specification of his racing cars in a relatively civilised bodyshell. The Elite was as technically advanced as it was aesthetically attractive, and with this model, Lotus took its first step on the road to becoming a serious production car manufacturer.

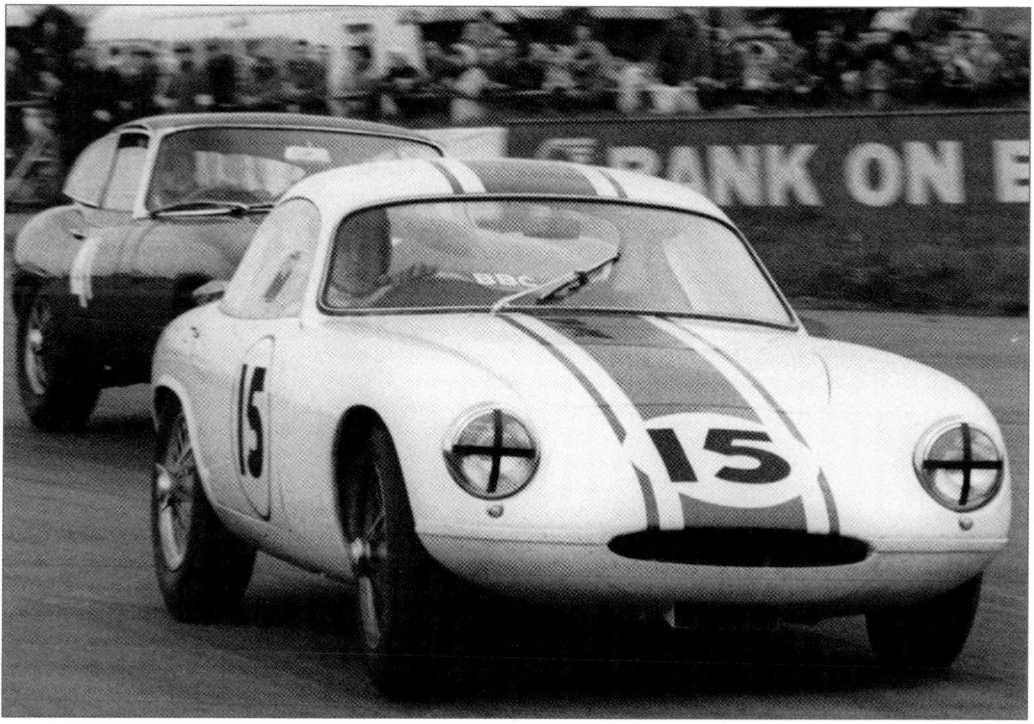

Between 1959 and 1963, the small-capacity GT classes were dominated by the Lotus Elite, and it was Jimmy Clark's skill in the Border Reivers' car against Colin Chapman that first attracted the Lotus boss's attention. Lotus's first production car almost sank the company because its complex fibreglass monocoque was difficult and problematic to manufacture, but on the racetrack it could prove invincible, beating serious opposition like the Alfa Romeo Giulietta SZ and SVZ and Abarth-Simca 1300 GT. This is Trevor Taylor in the Team Elite car ahead of an E-type on his way to a class win at Silverstone in May 1963.

Among Chapman's coterie from the early days was merchant banker Peter Kirwan-Taylor, a wizard accountant and financier as far as Colin was concerned. He was originally a Lotus customer who had fitted his own body to a Mk 6 chassis, and he made the initial drawings for the Elite styling. These were tweaked by Frank Costin, and the flowing lines were transformed into a reality by John Frayling. The latter worked with design engineer Ron Hickman, who eventually took over from Costin. Later on, Frayling worked at Maximar Mouldings at Pulborough, productionising first the Elite and later the Elan body. Frayling would also style the flared wings first seen on the Lotus Seven 'A' America, and at Huntingdon-based Specialised Mouldings he was involved in producing the body for the Lotus 47 Group 4 racer.

The Elite was unique in that its shell was a fibreglass monocoque, constructed from three major moulds and five minor ones. The main moulds created panels for the upper body section comprising roof, wings and boot. Secondly, the base section was formed by the floor pan and wheel wells, which incorporated the metal sub-frame for the front suspension and final drive mounting. And thirdly, a complex moulding formed the engine bay, transmission tunnel, boot interior, windscreen hoop and jacking points. The three mouldings were bonded together with epoxy resin, and the doors, boot and bonnet closures were created in separate moulds. The prototypes had steel and aluminium strategically sandwiched between sheets of fibreglass for strength, but it was found to be an unnecessary measure and not incorporated in production versions.

Chapman lined up a supply of Coventry-Climax's latest version of the 1216cc single-cam 75bhp FWE engines. These were married to the MGA's B-series gearbox, modified with an alloy bell housing to match the Climax engine. Like the Seven, the Elite suspension was derived directly from the 12, with double wishbones and coil springs and dampers up front and the Chapman strut set up at the back. Brakes were Girling 9½ inch discs front and inboard rear, and the car ran on 15 inch Dunlop 48-spoke centre-lock wire wheels. Elite production was increased from 1960 when body construction was transferred to the Bristol Aircraft Company at Filton, specialists in glass fibre injection moulding. Output was just five cars a week, each priced at £1,949. Whereas the Seven didn't make the 1957 Earls Court Show, the Elite was undoubtedly one of the star attractions.

Two other prototypes were farmed out to Ian Walker and John Lawry, who raced them in the interests of the development programme, but the first production cars didn't appear until December 1958. Owners of the earliest Elites included Keith Hall, jazz band leader Chris Barber, Border Reivers team principal Ian Scott-Watson, John Whitmore, and Graham Warner of The Chequered Flag sports car dealership (LOV 1).

Out of Town

In 1959, Lotus was poised to enter the big league. The new plant at Cheshunt was officially opened on 14 October by the local member of parliament, and Jim Endruweit remembers that Team Lotus was the first part of the operation to move in there. The Elite was built in the main block while race preparation and development work took place in the other. This included production of the sports racing cars and the Seven, of which around 100 units had been produced during the model's first two years. By the end of the year, the Elite was all but invincible in production sports car racing, with Graham Warner, Les Leston (DAD 10), Peter Lumsden, Tom Dickson

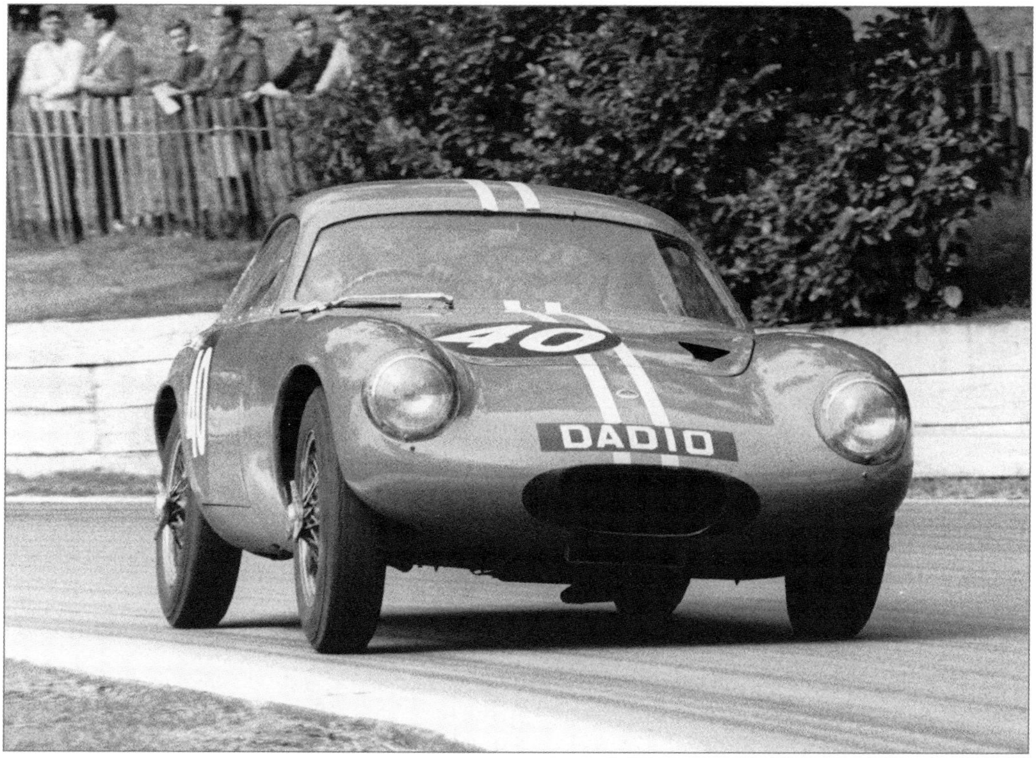

One of the most famous Elites was DAD 10 belonging to motorists' clothing specialist Les Leston. 'Daddio' – a popular contemporary jazz familiarity – was usually to be seen locked in combat with Graham Warner's similar car, the Chequered Flag garage sponsored LOV 1. During the 1961 and 1962 seasons in particular, both were equally like to emerge the winner. DAD 10 was one of a handful of Elites with a rounded off nose that cleaned up its aerodynamics slightly.

(V 50), David Buxton, Edward Lewis, Keith Greene, J.P. Williams, and Jimmy Clark in the highly successful Border Reivers' team car (ESH 700). The latter was owned by Ian Scott-Watson, and Jimmy was his protégé at the time. At the 1958 Boxing Day Brands Hatch meeting, Clark had very nearly beaten both Chapman and Mike Costin in similar Elites. This had much impressed Colin, and was partly what prompted him to bring Jimmy into the Formula Junior team. For his part, Clark had been highly impressed by the roadholding of the Lotus 18, which he first drove at Goodwood immediately after testing the front-engined F1 Aston Martin DB4R that he was contracted to drive in 1960. It was the start of probably the most momentous liaison between driver and team owner in motor sport.

By July 1960, 280 units of the Elite had been made, known retrospectively as Series 1 cars, and these had been followed by a further 749 Series 2 cars by the close of production in late 1963. In order to counteract rear-end steer manifest in some racing versions, Series 2 Elites had bigger springs at the front and triangular wishbones instead of the bent radius arms at the rear. An 85bhp Cosworth FWE-powered model, known as the Series 2 SE (Special Equipment), came out at the 1960 Earls Court Show. It was equipped with the new ZF 84-12 gearbox with a separate aluminium bell housing and synchromesh on all four gears. Five Series 1 cars were fitted with a nose section without bumpers drawn by Frank Costin that was allegedly good for another

10bhp – the Les Leston car was a good example. In addition, a few factory-built competition cars were produced from 1962, including the Super 95, superseded by the Super 100 and the Super 105 the following year. Lotus Components General Manager David Lazenby installed an Elan twin-cam engine in one of the last ones.

The Elite may have been one of the prettiest cars of all time, but there was a continual struggle to achieve reliability and maintain sales against mass-produced sports cars from Triumph and MG. Robin Read was Lotus sales manager from 1959 to 1962, and his book *Colin Chapman's Lotus* throws some interesting light on the period. It's probable that the company lost about £100 on every Elite it made. Over in the US, Jay Chamberlain's franchise floundered for exactly those reasons – the cars were relatively dear and they lacked the stolid reliability of more prosaic sports cars.

As with the Seven, the Elite was available in kit form from October 1961 in a bid to boost sales. Its competition debut came at the 1959 Nürburgring 1000km winning the 1300cc GT class in the hands of Peter Lumsden and Peter Riley. Amazingly, this car came eighth overall at Le Mans the same year, winning the 1500cc class and placing fifth in the Index of Performance, and second in the newly introduced Index of Thermal Efficiency, which was a startlingly good result for a new car. At the following year's Nürburgring 1000km, Elites were first and second in the 1300cc class, with John Wagstaff/Alan Stacey heading Peter Lumsden/Peter Sargent. Team Lotus ran a 1960cc Climax FPF twin-cam engined Elite at Le Mans in 1960 for Innes Ireland. His co-driver Jonathan Sieff was badly injured during practice in a 1.2-litre Elite, and the ill-handling 2.0-litre car was withdrawn. Consolation came in the form of the 1300cc GT class victory of the privately

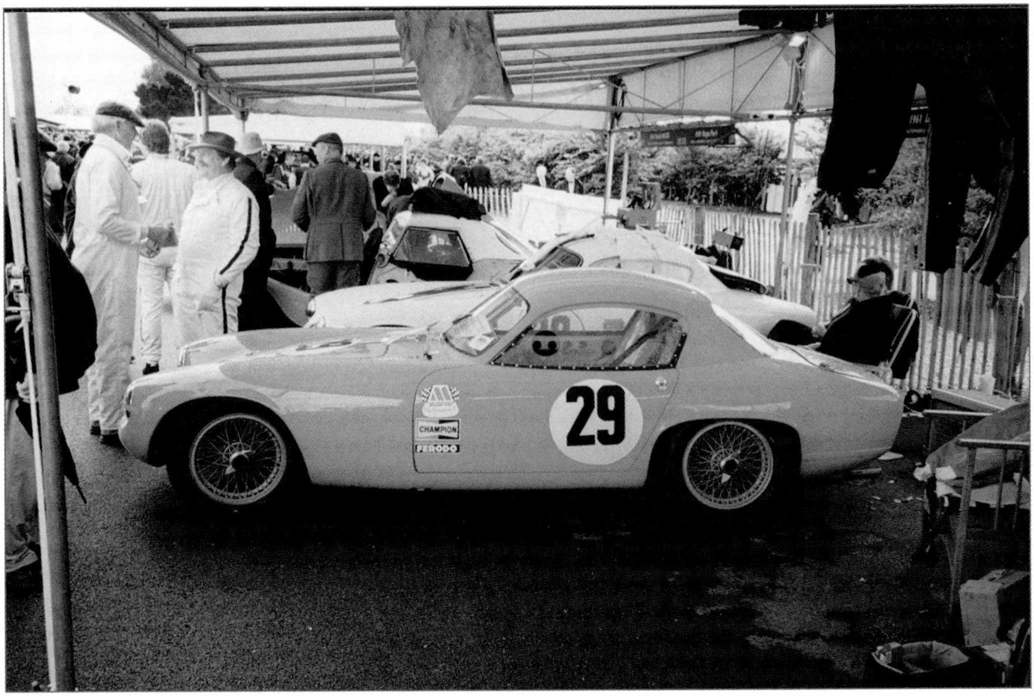

Now owned by Robin Longdon, DAD 10 sits in the paddock shelter at Goodwood prior to an outing at the 1999 Revival meeting. Oliver Glassin's Elite is behind it.

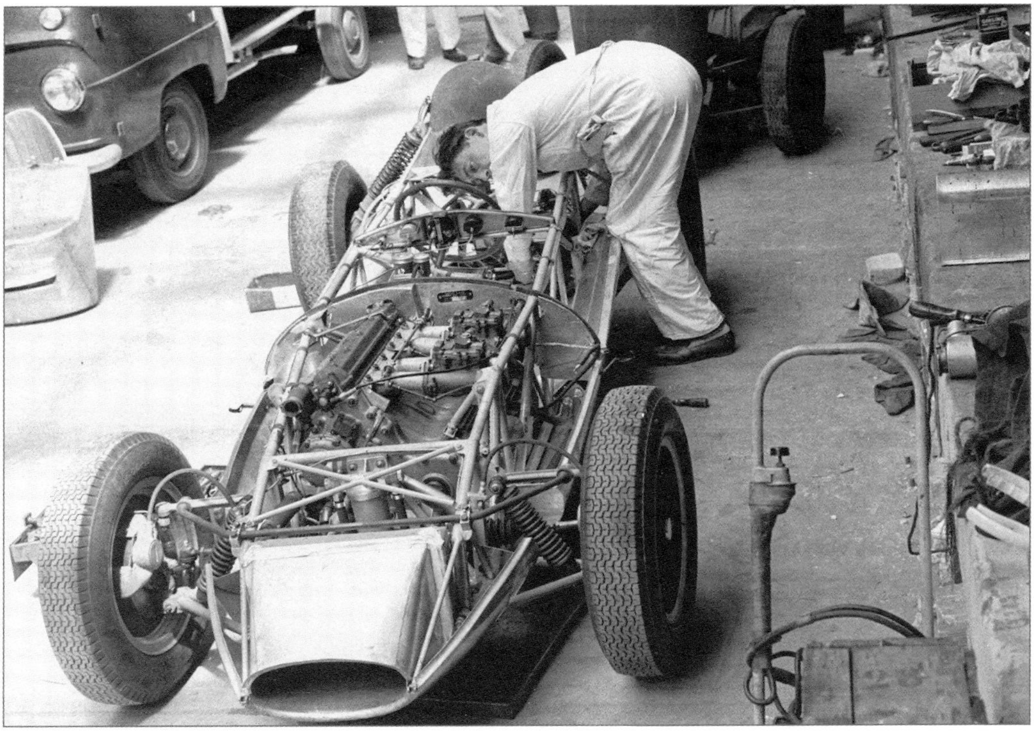

Team Lotus mechanic Jim Endruweit adjusts the linkage for the 'queerbox' on Graham Hill's works 16 on the car's debut at Reims in 1958. The tubular spaceframe chassis is plainly visible and the engine is canted over at a severe angle to obtain a low centre of gravity and optimum body line. In this case, the propshaft ran to the driver's left, and the gearbox transaxle was also offset by 4 degrees. However, 16s subsequently had their engines leaning to 16 degrees the other way, and the driveline was revised accordingly.

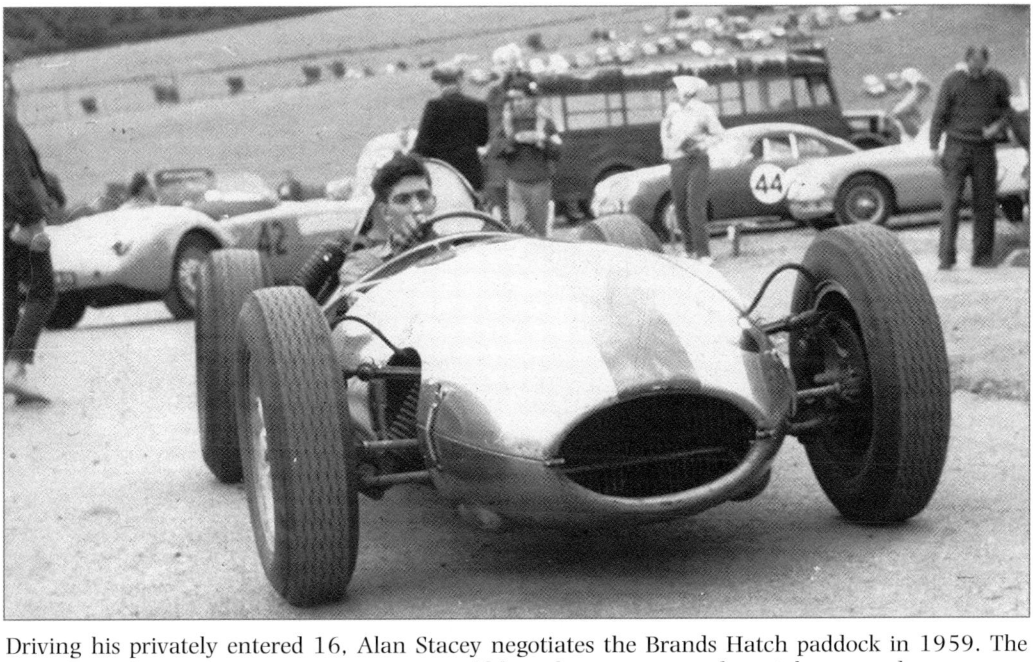

Driving his privately entered 16, Alan Stacey negotiates the Brands Hatch paddock in 1959. The arrangement was that certain privateers could buy their cars at preferential rates and repay some of their winnings to Lotus Components. Note the cockpit section of the 16's bodywork is missing.

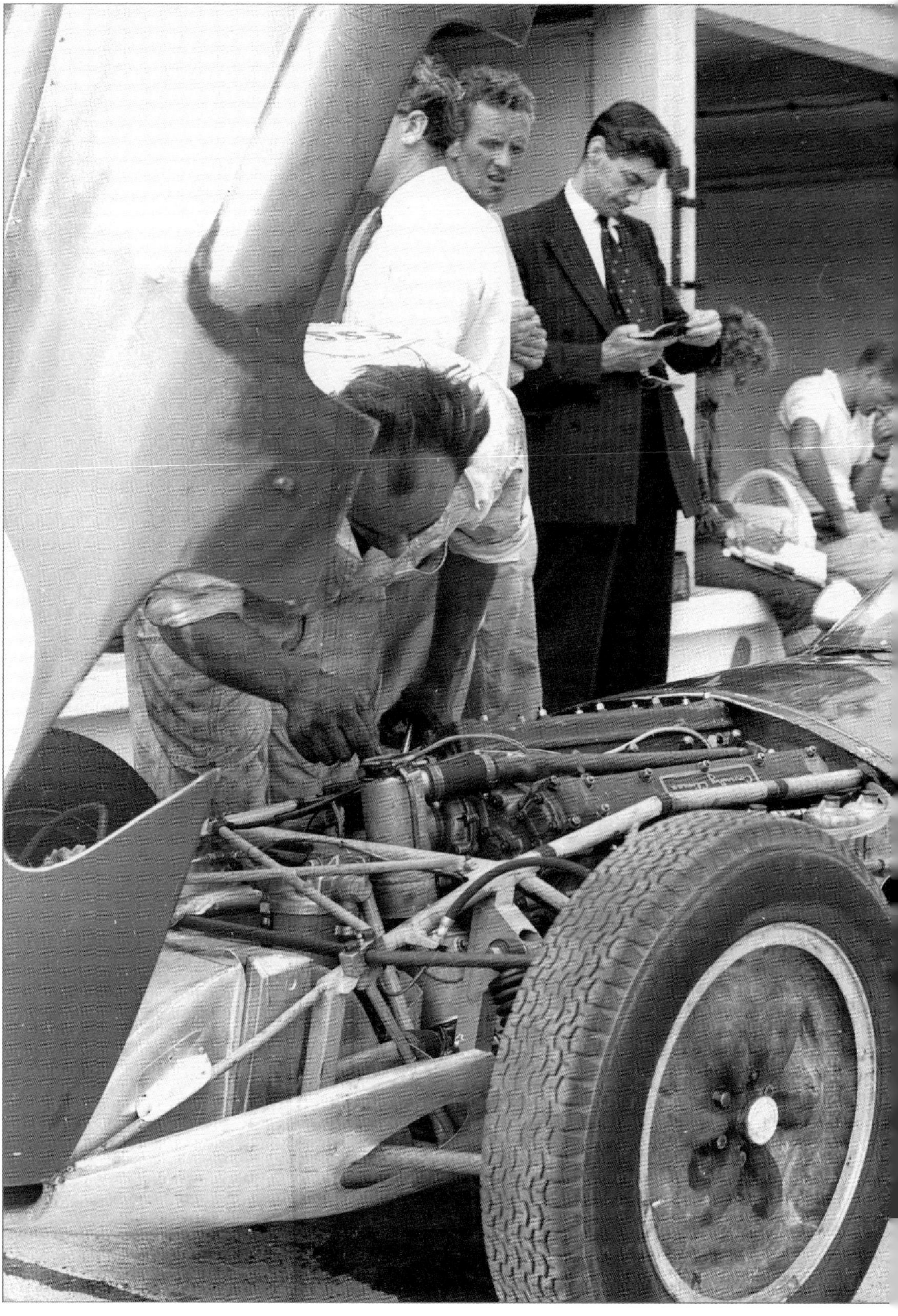

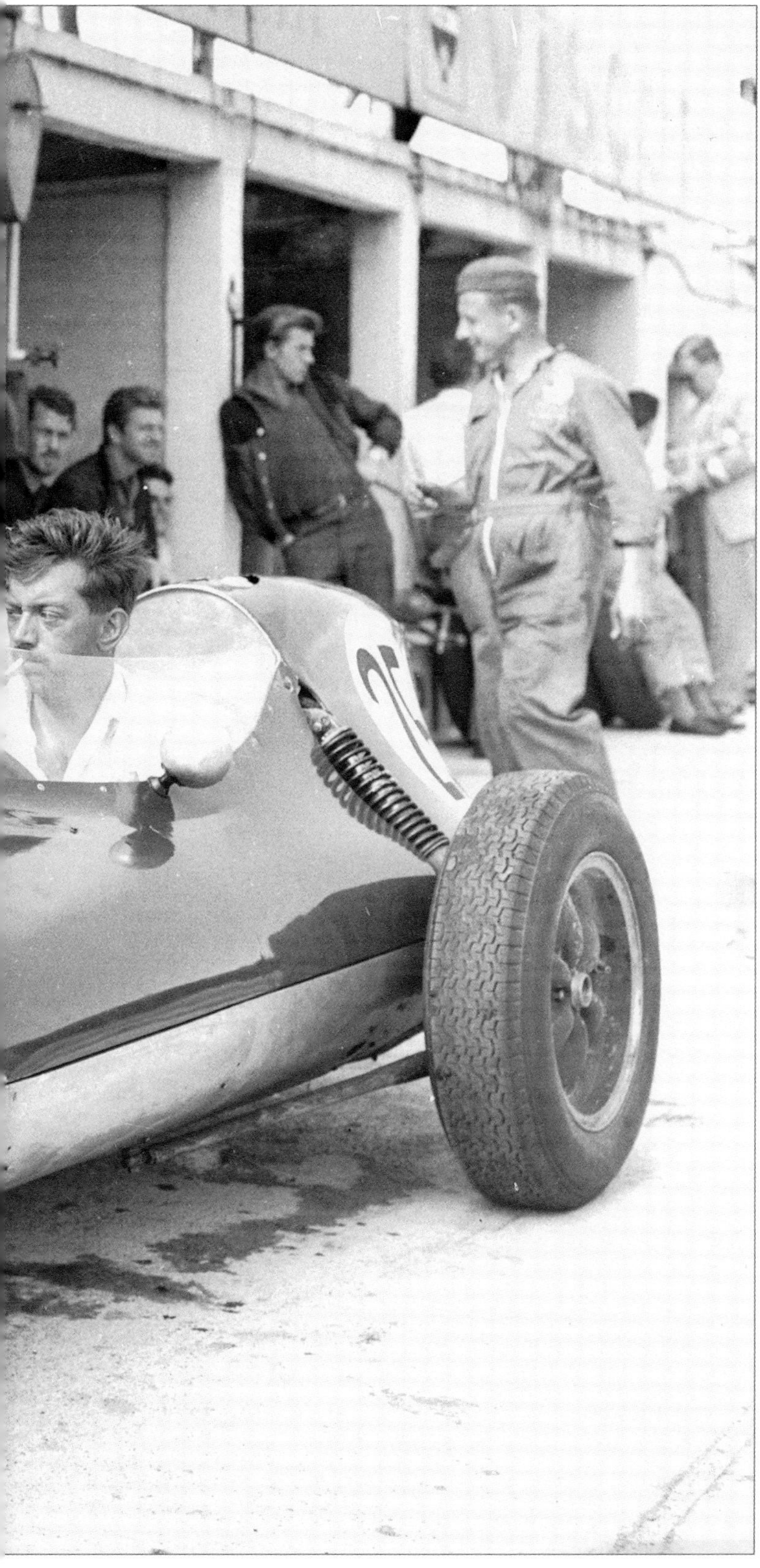

A shortage of Climax engines due to breakages and Allison's crash at Oporto led to his car running at Monza '59 with the only spare unit left – a 1.5-litre F2 engine. Despite a certain amount of overheating, Hill was sixth and Allison seventh. Mechanic 'Basil' sits in the 16's cockpit, while Allison (third left) looks anxiously on. The 16s subsequently got bigger radiators and oil coolers.

entered Elite of Roger Masson/Claude Laurent, followed by the John Wagstaff/Tony Marsh Elite – and paradoxically, the order was reversed in the unfathomable Index of Thermal Efficiency.

During 1961 the works-backed Team Elite ran cars for David Hobbs, which featured the family-produced Hobbs Mechamatic automatic transmission, and Peter Arundell. The latter wrote off his car at the Nürburgring, but Hobbs/Pinckney took the 1600cc sports car class in the 1000kms, while Lumsden/Riley were victorious once more in the 1300cc GT class. The Team Elite car of Bill Allen/Trevor Taylor was successful at Le Mans in the 1300cc class. Another creditable effort came from the UDT/Laystall Elite that was powered by the Climax FWMC 742cc twin-cam engine, an 81bhp screamer with which Cliff Allison/Mike McKee attempted to win the Index of Performance. It was sidelined after ten hours, however, due to overheating and a broken oil pump drive.

The Elite continued to feature prominently in endurance events in 1962, with Wagstaff/Fergusson coming second in the 1300cc GT class at the Nürburgring and Hobbs/Gardner winning the same class at Le Mans and finishing eighth overall. It was a similar story the following year, with the duo of Wagstaff/Baird taking the 1300cc GT class at the Nürburgring, and Wagstaff/Fergusson first in that class at Le Mans. By now, though, the Elite was coming to the end of its competition career at international level, and at the 1964 Le Mans race, Wagstaff/Hunt were twenty-second overall and class winners – the fifth time for the Elite.

Too Sophisticated

The beautiful lines of the Mk 16 that Team Lotus introduced for the 1958 season were never matched by the results that the car deserved. Not only were its Frank Costin-inspired lines among the most attractive, it was also among the most sophisticated of the front-engined Grand Prix cars raced during the 2.5-litre years from 1954 to 1960. It proved overcomplicated and thus unreliable, and it was also up against the Cooper-Climax T41, T43 and T51s that were spearheading the rear-engined revolution.

The Lotus 16 was built up on a multi-tubular spaceframe chassis, benefiting from experience gained with the 12. It was still composed of 1 inch and ¾ inch Reynolds round-edge tubing, mainly 20-gauge, but with 18-gauge tube for highly stressed areas, and wider and stiffer than its predecessor. Detachable tubes supporting the fuel tank and gearbox were adopted subsequently, instead of the original cradle arrangement, and the engine bay was also triangulated by a removable tube. The 16's suspension system was akin to the 12, with longer radius rods at the rear, and its Girling disc brakes were mounted outboard at the front and inboard at the rear.

Among my first Scalextric racers (around 1959) was a Mk 16, and its distinctive Vanwall-esque body always looked just right. Its proportions were better than the tall, bulbous Vanwall and, like Chapman himself, many people still believed that front engine/rear drive was the appropriate configuration for a racing car. In spite of their success, Coopers with their mid-engines were still the exception rather than the rule. The real-life Lotus 16's bodywork was made of 22-gauge aluminium panels by Williams and Pritchard, and it proved to be flimsy enough to distort and split under duress. An absence of ventilation ducts meant that overheating was a problem for engines and drivers alike. Fundamental to its low-slung cigar shape, as

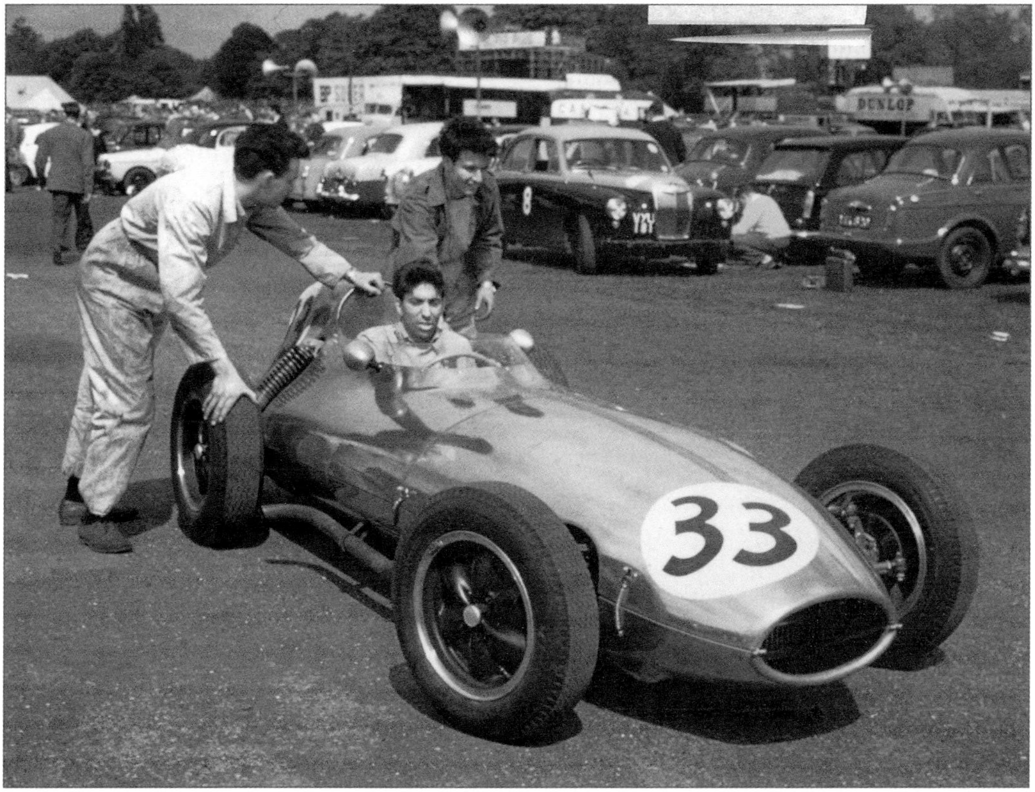

The baby-Vanwall shows off its Frank Costin-inspired, Williams and Pritchard-built bodywork to good effect in plain aluminium as Alan Stacey's 16 is wheeled out through the Oulton Park paddock in 1959. Although Costin didn't actually style the 16, he carried out an exercise on the aerodynamics by having Hill drive it around with tufts of wool taped to the bodywork to assess air flow and areas of turbulence.

well as a lower centre of gravity, was the siting of the Climax FPF engine at 62 degrees from the vertical, further offset by 5.5 degrees from the centre-line and with the transmission shaft running to the driver's left and culminating in the Lotus 'queerbox', with lateral offset between gearbox input and driveshafts. On the car's Formula 1 debut at Reims in 1958, Hill's 2.0-litre car retired with a damaged radiator, while Allison's F2 version overheated due to a stoved-in air intake. The latter car incorporated a revised engine layout, its Climax unit now canted at 17 degrees to the left and orientated longitudinally at 6.5 degrees off-centre to the right.

It wasn't an auspicious beginning in the Grand Prix world. Both 1.9-litre 16s were seen off by overheating in the British Grand Prix, then Allison finished tenth at the Nürburgring, and both cars were eliminated by crashes at Oporto. After numerous dramas – detailed in Ian H. Smith's entertaining *Story of Lotus* – Graham Hill was still running at the finish at Monza and Casablanca, though not classified. Even in Formula 2, the Mk 16 posted retirements in both events it contested.

The 17-degree engine configuration was utilised for the six units built for the 1959 season, and the previous year's cars were sold off to David Piper, Bob Bodle and John Fisher, who ran Bruce Halford in F2 events, while a fourth 16 was built up by Anthony Brooke, probably around the chassis of Allison's wrecked Portuguese GP car. Of the six new Mk 16s, Innes Ireland and Alan Stacey each built up a Formula 2 version from parts provided by Lotus at cost price. The deal was that they returned some of their earnings from racing the cars to Lotus Components.

The Lotus 16 looked good from the outside, but according to Jim Endruweit, it was far too complicated in respect of its drivetrain layout. Not only did the Climax engine have to be canted over one way or the other to fit in the chassis, the routing of the transmission through the cockpit caused the drivers some discomfort. By this time, Len Terry was on the Team Lotus staff as designer, and he drew up certain modifications for the works 16s. These included a fire-proof bulkhead ahead of the cockpit, which also stiffened the chassis amidships, and this was subsequently augmented by a perforated sheet hoop around the dash panel. The top chassis rail was bent in order to ease carburettor access when the engine was removed. On the works cars, the front anti-roll bar was located to the rear of the suspension set up, and bent appropriately to avoid the engine. Wider rear wishbones were substituted for the earlier radius rods.

Cliff Allison having departed to drive for Ferrari, Team Lotus's Formula 1 squad for 1959 started off as Graham Hill and the American Pete Lovely, but after failing to qualify for the season's opener, the International Trophy at Silverstone, the latter

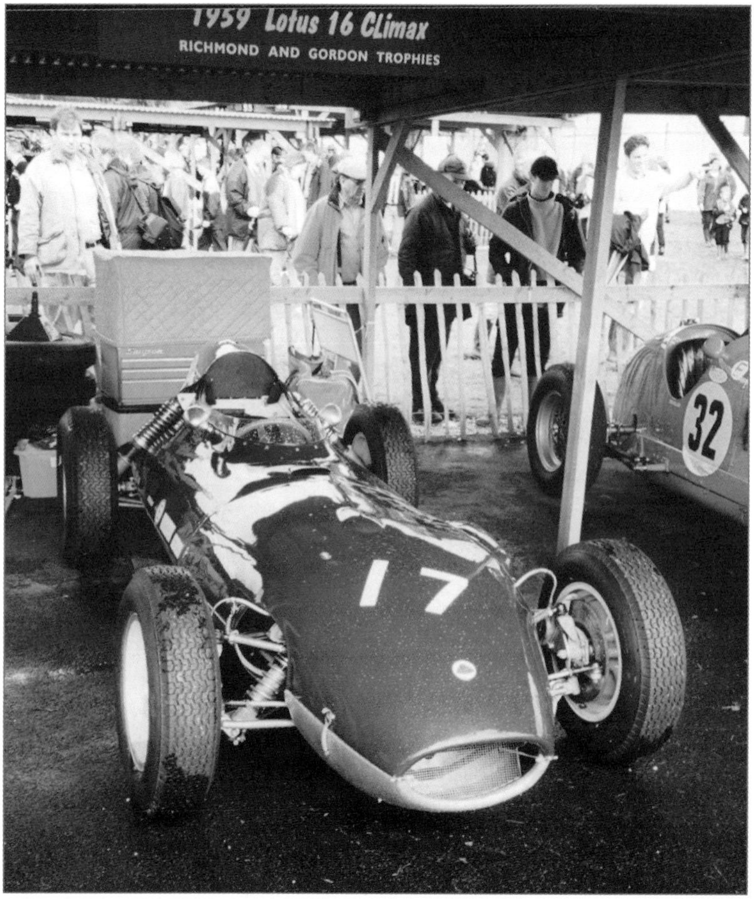

This 16 has the 2495cc Climax engine fitted and is entered by Jeremy Agace at the 1999 Goodwood Revival meeting.

returned to the States. This elevated Innes Ireland to the status of F1 driver. Season's highlights included fourth and seventh places for Ireland and Hill at Zandvoort, and a sixth for Ireland at Sebring in the US GP. Mk 16s scored a couple of second places in Formula 2 events – David Piper in the Preis von Zeltweg race in Austria, and Graham Hill in the Kentish '100' at Brands Hatch. During the course of the year the 16's bodywork was modified and external exhaust systems were fitted. However, in view of Brabham's World Championship success with the Cooper, it was clear to Chapman that he must follow suit with a rear-engined design, and the Mk 18 was in gestation. The 18 wasn't quite finished though: two works 16s were shipped laboriously to South America for the Argentine Grand Prix and the non-championship race at Cordoba for Stacey and Rodriguez Larreta, together with the prototype 18 for Ireland. He finished sixth at Buenos Aires but at Cordoba, having retired his new 18, Ireland ended up in Stacey's car when its driver got sunstroke, but he too pulled out. And finally, Alan Stacey drove a Formula 2 model at Siracusa in March 1960, retiring with a blown head gasket.

LOW FLYER

The replacement for the excellent Eleven in the 1100cc class was the Mk 17. It was created at Chapman's instigation to combat the new Eric Broadley-designed Lola 1100, which was lighter (at less than 750lb) than the Lotus Eleven and featured a stiffer and more elegant chassis. The Lotus 17 was sufficiently distinctive to be readily identified from the Eleven and Mk 15, being slightly smaller, lower and more aerodynamic. There was also a characteristic panel gap where the aluminium and fibreglass body sections met, running straight between the front and rear wheelarches. The four main body panels – front, rear and two doors – hinged outwards, making accessibility very good. And yet, despite the promise of its low-slung looks, the 17 goes down as one of Lotus's few failures.

The Mk 17 was basically a Len Terry design and was based on the familiar spaceframe, fashioned from ⅝ inch and ¾ inch square and round 20-gauge tubing. In the same way as the 15, the floor and transmission tunnel of the 17 acted as stressed chassis members. The 17 was the first Lotus to feature strut-type front suspension, with lower wishbones and upright coil springs and damper units, while the rear end featured the same strut set-up as the 15. The 17 ran on Lotus's wobbly-web magnesium-alloy wheels, with Girling disc brakes all round, inboard at the rear. According to ex-Lotus development engineer Hugh Haskell in his book *Colin Chapman: Lotus Engineering*, the 17's Achilles heel was the strut suspension that made handling so unpredictable, especially on a bumpy surface, that it could go sideways with no warning. The struts were too short, and the system couldn't work freely, causing the front wheels to tuck under in cornering. Len Terry recalls that he identified the problem from the outset, and the Chapman struts were soon changed for a wishbone set up.

The 1098cc Coventry-Climax FWA unit in stage 3 tune, or the 747cc FWM version were the designated engines for the 17, canted over at 10 degrees to the left. They were mated to the BMC B-series transmission housed in a Lotus casing that was in unit with the engine, and customers could specify a ZF limited slip differential as a cost option. Ancillaries, including the SU fuel pump and battery, were located at the rear of the car, along with the aluminium alloy 8-gallon petrol tank.

The 17's competition career was nothing like as notable as its celebrated predecessor the Eleven, largely due to its handling. It was adequate in the wet but

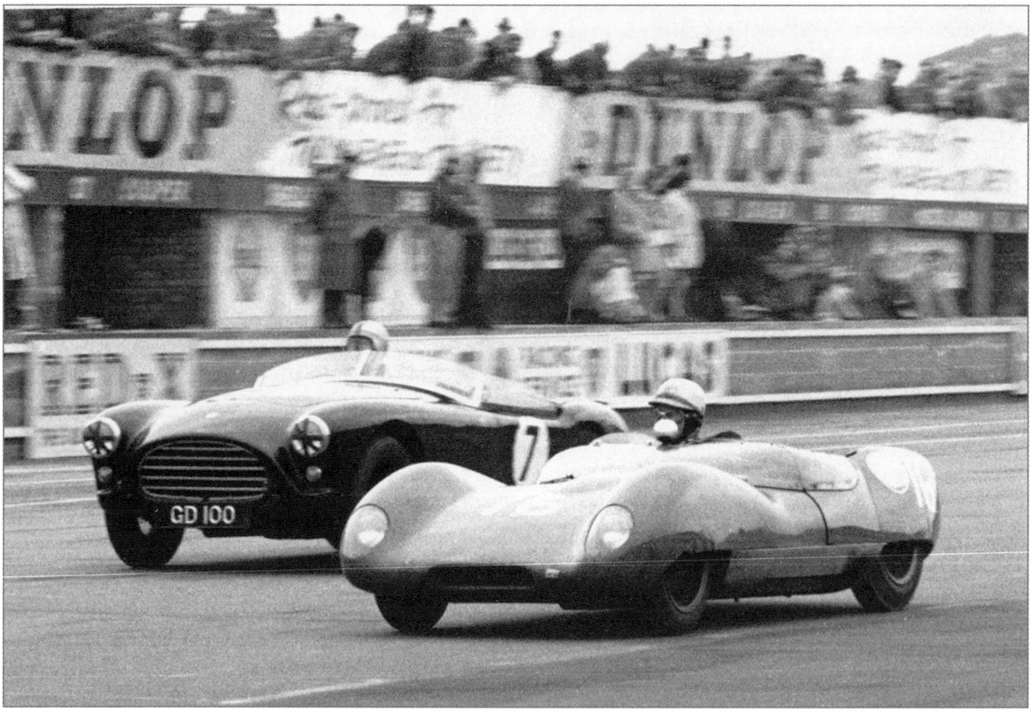

The 17 came out in 1959 and was intended to replace the Eleven. Although it was lower, lighter and had a stiffer chassis than its predecessor, it was regarded as a failure because its front strut suspension system made for unpredictable handling. This was only cured by fitting a wishbone set-up. This is Graham Eden's 17, dwarfed by Dempsey's AC Ace in the 1960 Martini International at Silverstone.

prone to treacherous oversteer in the dry. Its first race was at Aintree in April 1959, driven dramatically from the back of the grid into the lead by Alan Stacey, only to be demoted after loss of second gear. That rather sums up the 17's reputation, and it's remembered by Alan Stacey's brother Phillip as having been largely unreliable. Alan was responsible for taming the 17's handling to an extent, by having the suspension lowered and stiffened to create a more neutral set-up. He gave an account of the process in *Sports Car and Lotus Owner* magazine of July 1959. But even with wishbone suspension, it couldn't match the Lola 1100, which dominated the class on the domestic racing scene that year.

Team Lotus entered a pair of 17s for Le Mans, both fitted with the 745cc FWM engines, driven by Alan Stacey/Keith Greene and Mike Taylor/Jonathan Sieff. The former pair looked set for a win in the Index of Performance category once the leading OSCA had retired, but in the event, both 17s were sidelined by overheating problems.

Real Rivals

Although Lotus's traditional rival on the race-track was Cooper, there were a number of other small operations that produced sports racing cars in Britain in the two decades after the Second World War. Among them were Kieft, Dellow, Elva, Lola, TVR, Gilby Diva, Tojeiro GSM and Mallock, and while not necessarily rivals to Lotus, they characterised the similar innovative spirit of the period when innovation and adaptation reigned. Colin Chapman never completely lost the belief that anything was possible – and achievable – and his drive and ambition carried him and his company forward where others faltered. In the main, these minor players used spaceframe chassis clad in fibreglass or aluminium body panels, relying on stock engines and suspension components. They were really not so different from the more exalted BRM or Vanwall teams, apart from degrees of financial backing and ambition. In the main, the minor cars were created in workshops by small-time specialists and volunteers, rather than in the factories that spawned Jaguars or Aston Martins.

In international events, Lotus faced opposition in its classes from Porsche, the French D.B. Panhards, and the Italian makes Stanguellini and OSCA – the latter set up by the Maserati brothers in the early 1950s. But during the latter half of the decade, it was Cooper that presented the real challenge to Lotus.

Together with his father Charles, John Cooper had been building racing cars since 1946, and their almost ubiquitous T5 and T7 Formula 3 cars placed the 500cc or 998cc JAP engines behind the driver even then. Pioneers both, Cooper and Chapman were amiable acquaintances rather than close friends, and although they had colluded in overcoming mutual mechanical problems, they were undoubtedly rivals in sports car racing from the mid-'50s. While Cooper was dismissive of the fragility

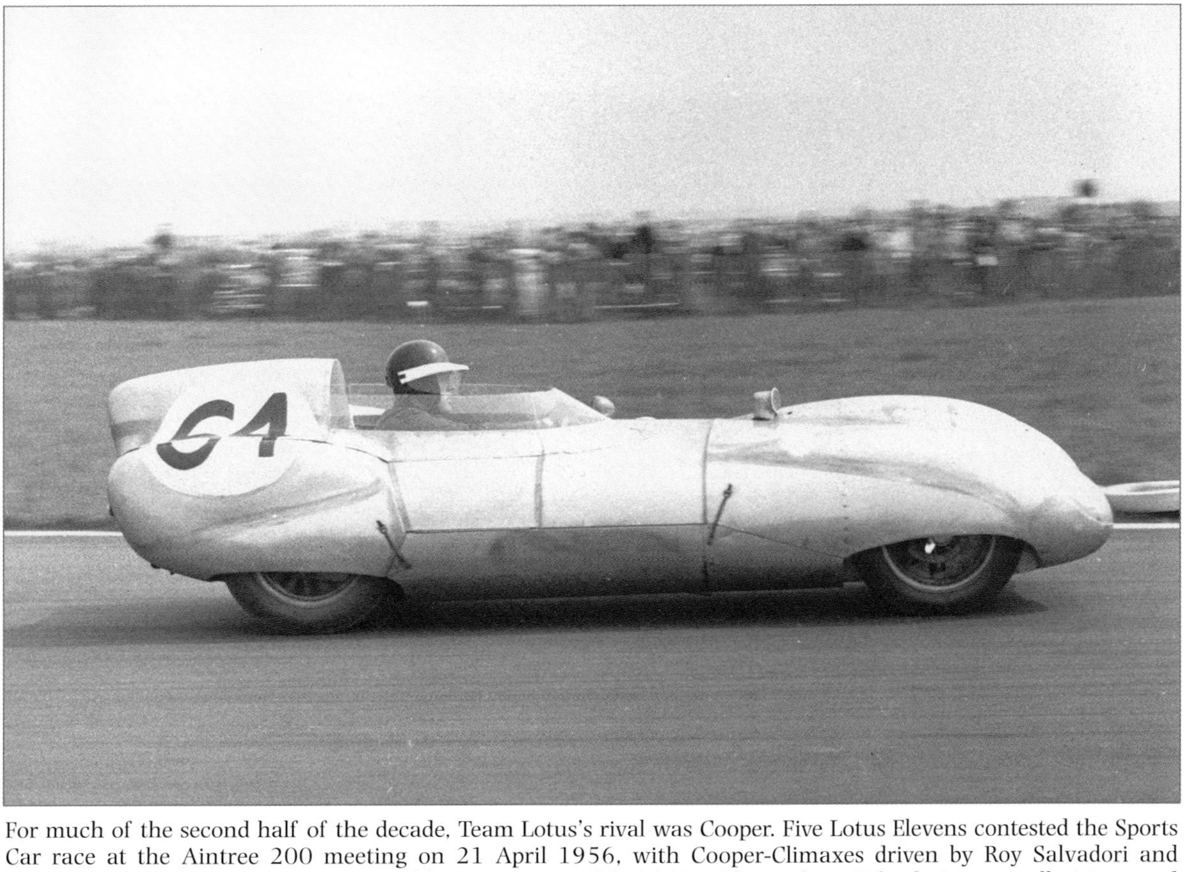

For much of the second half of the decade, Team Lotus's rival was Cooper. Five Lotus Elevens contested the Sports Car race at the Aintree 200 meeting on 21 April 1956, with Cooper-Climaxes driven by Roy Salvadori and Stirling Moss. Mike Hawthorn, pictured here in Ivor Bueb's 1460cc Eleven, beat Salvadori to overall victory and first in the 1100–2000cc class. Chapman beat Peter Gammon's Cooper to win the up to 1100cc category.

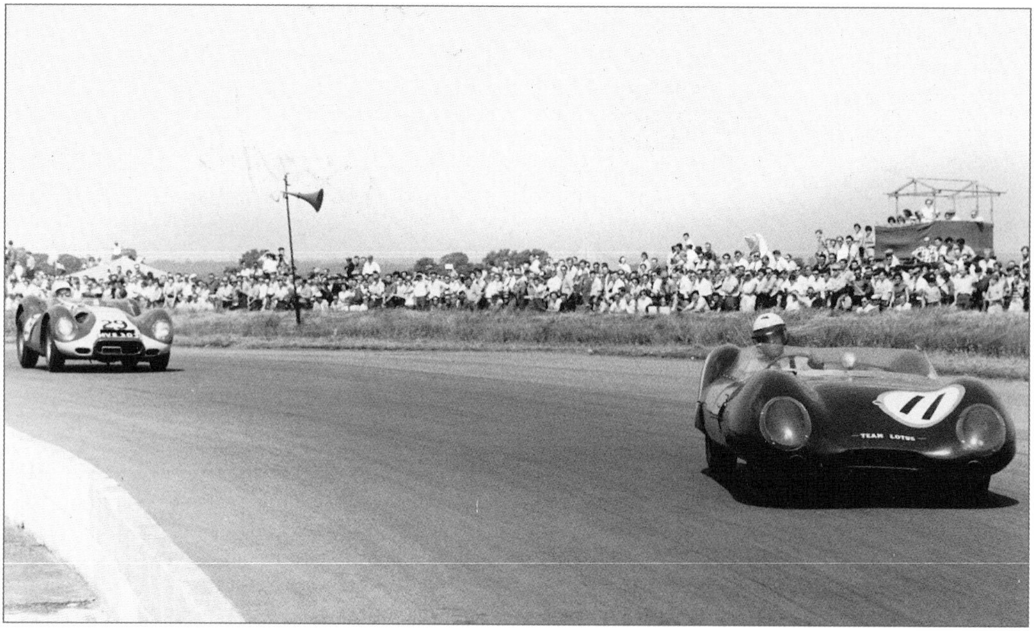

While the Lotus Eleven was challenged by Elva and Tojeiro as well as Cooper in the 1100cc category, the Team also had to contend with the big guys. Here at Silverstone during the 1958 British Grand Prix meeting, Stacey is about to be lapped by Moss's works' Lister Jaguar, and was involved in a scrap with Halford's 3.0-litre Lister as well.

of Chapman's lightweight chassis, Chapman considered Coopers to be the work of blacksmiths. However, based on his F3 experience – Harry Schell entered an F3 Cooper T5 for the 1950 Monaco GP – it was only natural that Cooper would go mid-engined with his T41 and T43 Formula 1 and 2 cars, and the Manx-Tail and Monaco sports racing cars too.

Having run the single-seater F2 cars in F1 events in 1957, Cooper's first GP win was in January 1958, when Stirling Moss won the Argentine race in a T43. The gap that separated Cooper and Lotus at this point can be exemplified by the grid positions at Monaco in May 1958, where Chapman's cars made their Grand Prix debut. The front-engined Mk 12s of Allison and Hill were at the back of the grid, while Coopers occupied four out of six places up at the front. It says much for Lotus that Allison's car held together for the entire race – he came sixth and last – while Maurice Trintignant in Rob Walker's Cooper T51 was the winner. However, it wasn't until 1959 that Chapman concluded that the way to go was with a rear-engined car. In spite of the success enjoyed by Cooper with the mid-engined configuration, Chapman had remained convinced for some time that the front-engined layout was still correct. The main reason for building the 18 was the need for a simpler solution to the mechanical layout. Placing the engine behind the driver did away with the propshaft and allowed a more compact car to be built.

The Mk 18 started life as a Formula Junior car, conceived only three weeks before the final Grand Prix of the 1958 season at Sebring. The view of Lotus personnel at the time, according to Jim Endruweit, was that Coopers were pretty agricultural compared with Lotuses. They were robust and screwed together well enough, but rather lacking in delicacy and sophistication.

CHAPTER FOUR

LOTUS GOES REAR-ENGINED

The Mk 18 was so adaptable that it could transcend the three significant single-seater formulae of the period – Formula 1, Formula 2 and Formula Junior. It was based on a triangulated spaceframe chassis comprised mainly of 18-gauge tubing, with a small amount of 16-gauge mild steel round tube. It was a much simpler car than the Mk 16. The 18's front suspension mounting points were fronted by a sub-frame that housed the radiator, oil tank and body mounting points. The engine bulkhead behind the cockpit was braced – with a rudimentary and unconvincing rollover hoop – while cockpit sides were strengthened. To the rear, two straight chassis tubes ran down to the lower suspension pick-up points, a

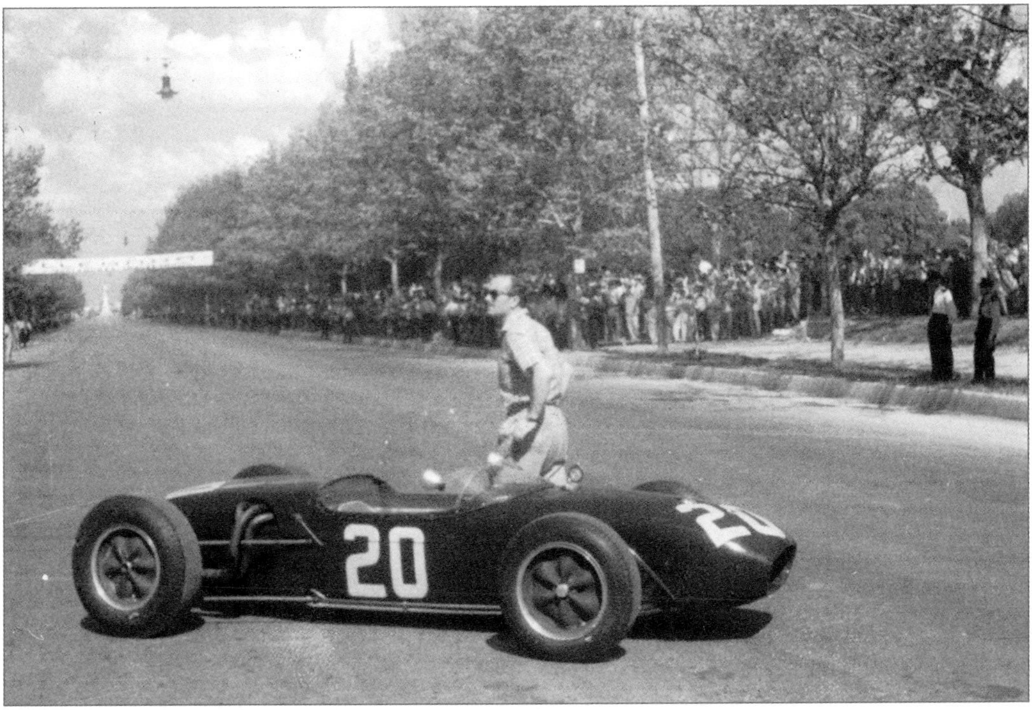

Lotus's first rear-engined car, the 18, made its debut at the Argentine Grand Prix in Argentina. While a pair of 16s were shipped out, the hastily completed 18 was flown direct to Buenos Aires. Innes Ireland led briefly, finishing a valiant sixth as one of the front wheels threatened to come off. Team Lotus moved on to Cordoba for the non-Championship Buenos Aires GP, but the 18, pictured here with Chapman, broke a driveshaft after three laps. Ireland then took over Stacey's 16 as its driver had sunstroke.

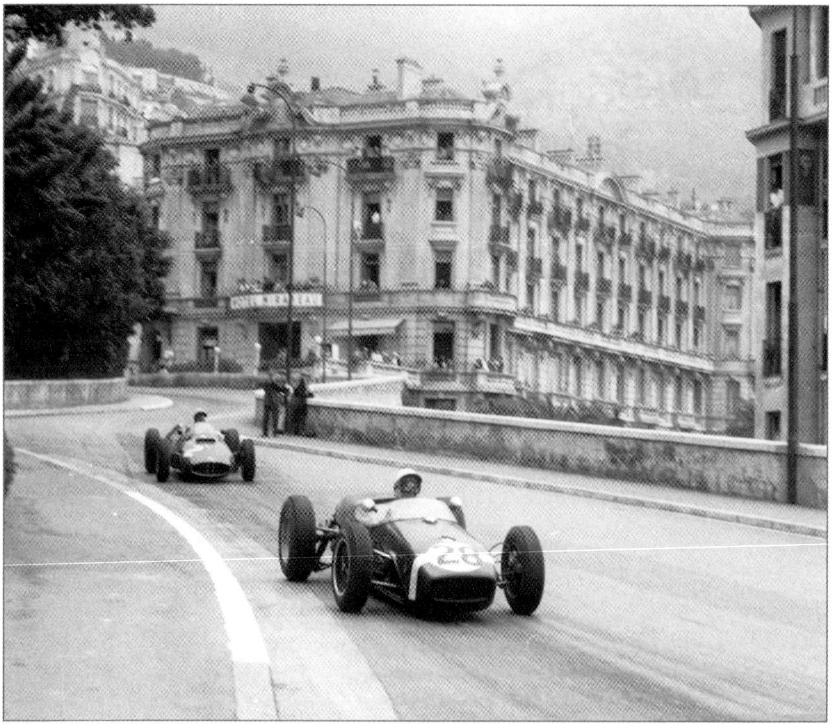

That first World Championship victory came at the slippery 1960 Monaco Grand Prix, and it was gained for Lotus by Stirling Moss driving Rob Walker's 18. Moss rushes down to the Station Hairpin followed by Bonnier's BRM – the two traded places as Moss pitted to have a plug lead replaced.

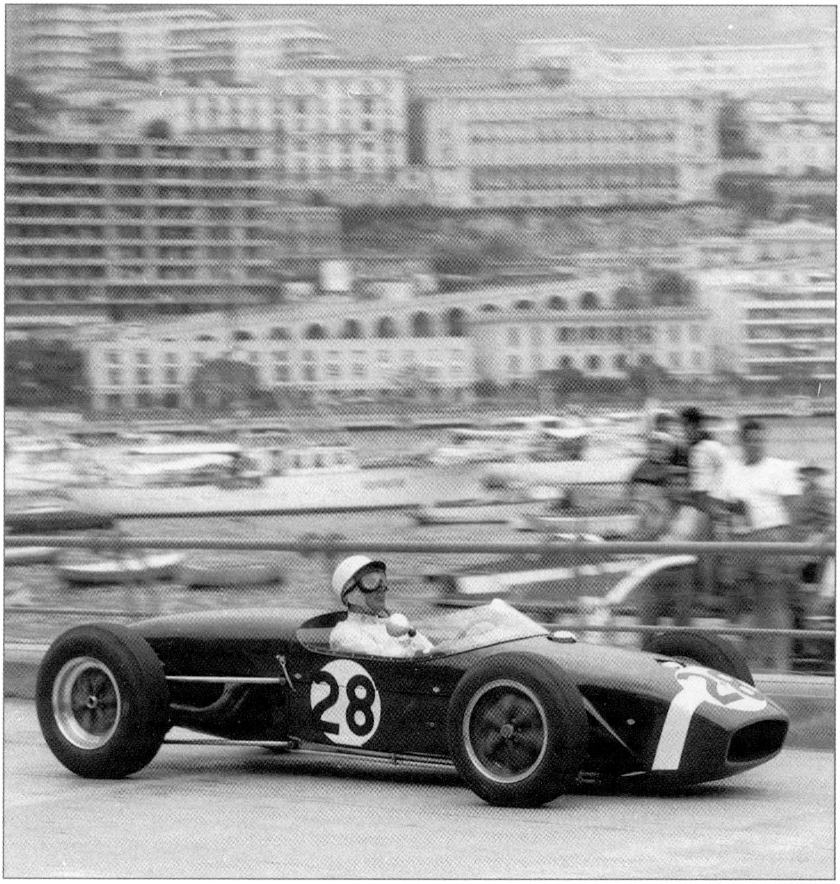

The Rob Walker 18 used a five-speed Colotti gearbox, which meant that its rear disc brakes had to be mounted outboard with thicker rear radius arms. Moss, here on his way to victory at Monaco in 1960, liked to have the 18's front wheels' camber angle set up to give slight oversteer.

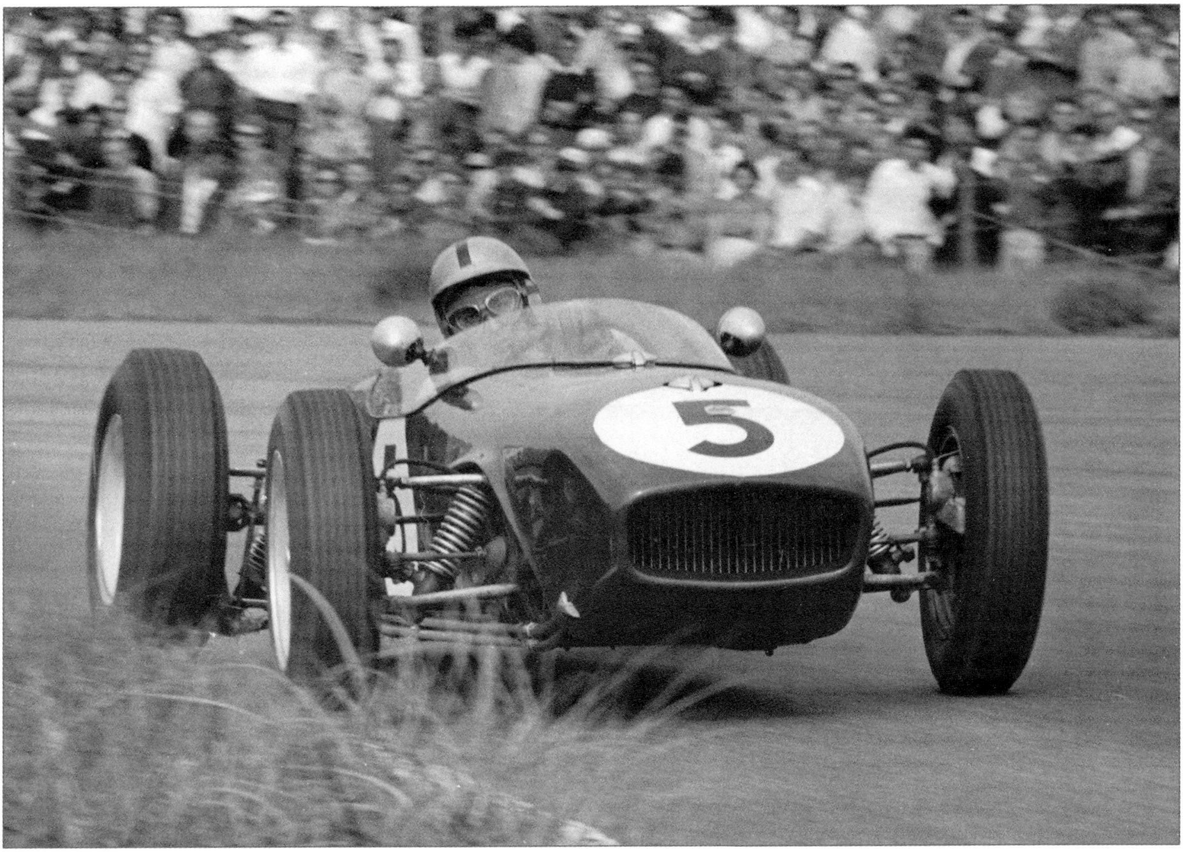

This dramatic shot shows Alan Stacey at Zandvoort for the 1960 Dutch Grand Prix, where he retired with transmission failure while running fourth. The 18s of Ireland, Moss and Stacey were quickest of all through the speed trap, at 237kph, and the race was also notable for being Jim Clark's first drive in a World Championship event.

Y-shaped diagonal piece went across the top of the engine and could be detached for extracting the engine and gearbox, while an ovoid steel sheet and tubular subframe made up the rear end. Above the driver's legs was the main 22-gallon aluminium fuel tank, with an auxiliary 9.5-gallon tank to the right of the seat backrest. Although the prototype 18 had an aluminium body made by Williams and Pritchard, the production versions were clad in fibreglass sections.

The 18's front suspension was virtually the same as the Series 2 Eleven's, by unequal-length wishbones, coil springs and damper units. It had rack-and-pinion steering, and at the back it had reversed lower wishbones, twin radius rods, combined coil springs and damper units, with unsplined half-shafts. There were anti-roll bars both front and rear, and former Chief Engineer Jim Endruweit remembers Chapman wanting to race the 18 in Buenos Aires without the rear anti-roll bar, just to confuse John Cooper – the idea being that the bar would be fitted once back in Europe for what were perceived as the more significant races. In the event, Ireland debuted the F1 car with the bar in place. The 18 was fitted with disc brakes front and rear, mounted inboard at the back. Its F1 specification featured the 237bhp Coventry-Climax FPF 2495cc engine, allied to the five-speed Lotus 'queerbox'.

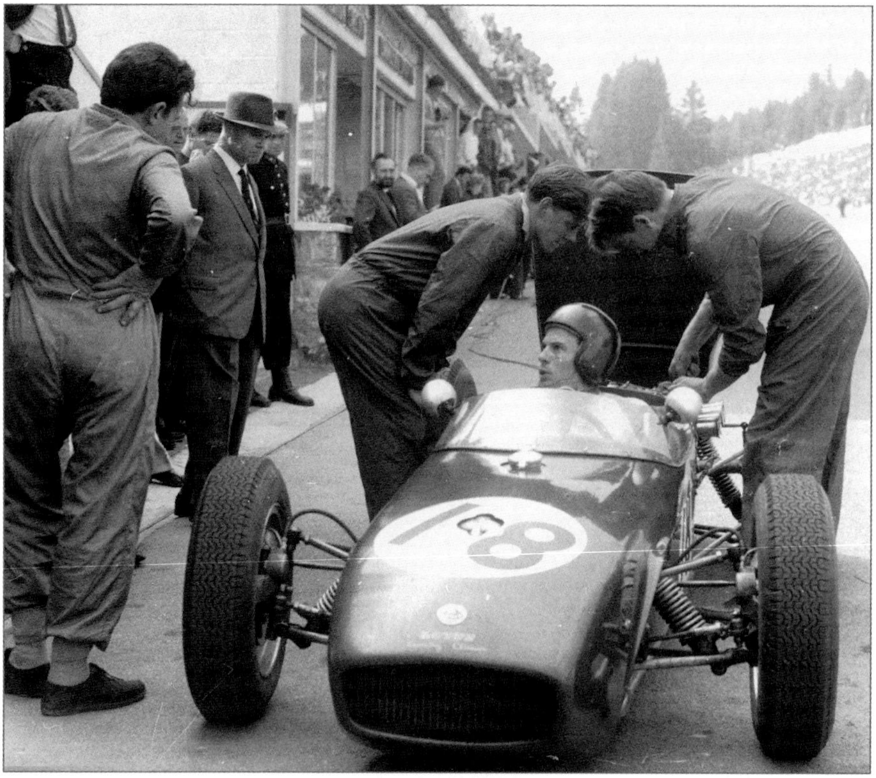

Jim Clark talks to Mal Simpson (left) in the pits at Spa-Francorchamps as Dick Scammell and Jim Endruweit work on the 18 – it suffered from blocked jets during the race. This event was marked by tragedy, with Alan Stacey dying after being hit in the face by a bird at high speed. The talented Chris Bristow also lost control and was killed in a separate accident, and both Stirling Moss and Mike Taylor were seriously injured in practice crashes due to mechanical failures on their 18s. Clark's was the only Lotus running at the finish and he came fifth.

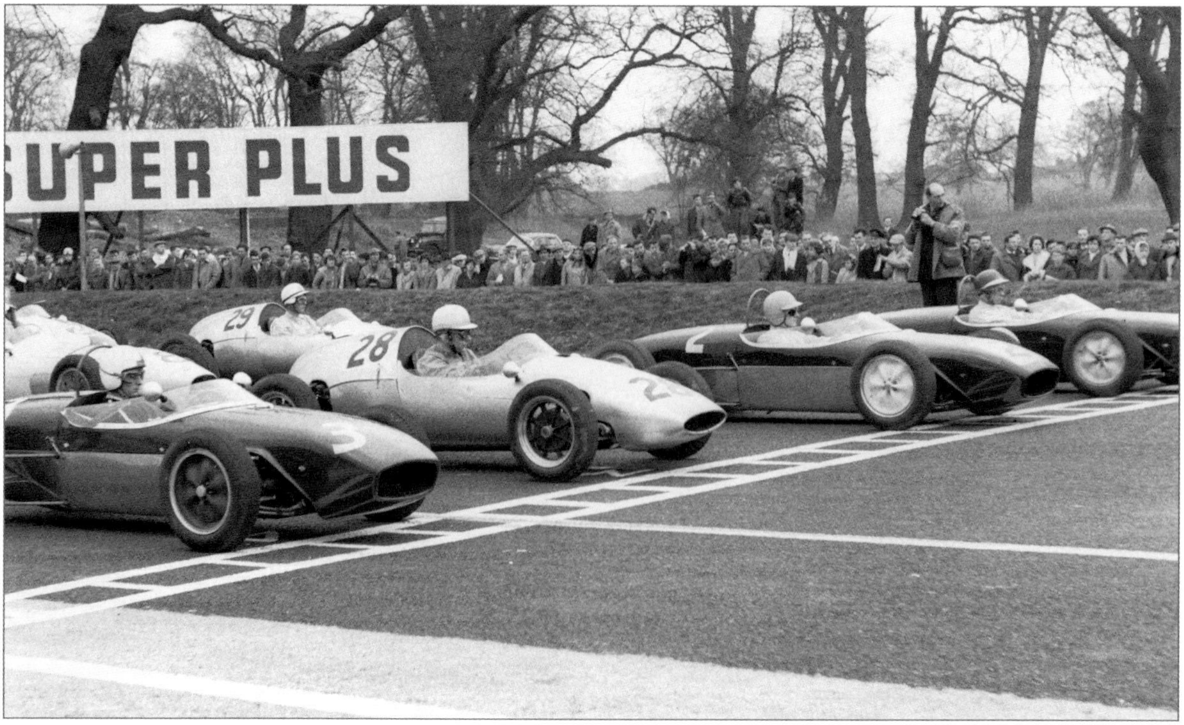

The Formula Junior event at the BARC meeting at Oulton Park on 2 April 1960 provided plenty of action, and came to be dominated by the works Lotus 18s of Jimmy Clark and Trevor Taylor, and Mike McKee in the Jim Russell 18. John Surtees in the Cooper-Austin (28) provided the major threat but bent his gear lever and retired, while Clark and Taylor exchanged places until Jimmy won by 1.2 seconds, setting fastest lap at 87.5mph.

The Mk 18 was launched at the 1959 Boxing Day Brands Hatch meeting as a Formula Junior car driven by Alan Stacey. He finished third after a spin, which was a promising result, although customers at the subsequent Racing Car Show queued to buy copies of the Elva that had won the race. The Mk 18 would make its Grand Prix debut shortly afterwards in South America. After a nightmare journey to Argentina and sundry problems with local customs officials and police, Innes Ireland was second fastest in practice for the Grand Prix and even led before deranging a front hub against a kerb. Sixth place was an excellent result under the circumstances. At Oulton Park on 2 April, Ireland beat Surtees' Cooper to the chequered flag in the F2 event by 16 seconds, while Jim Clark won the FJ race and set a new lap record. Over the Easter weekend on the domestic calendar, the 18 raced in its three different formats. At Goodwood on Easter Monday Clark's works F Junior car and Ireland's 18 in F1 and F2 guise won the Formula 1 Richmond Trophy and Formula 2 Lavant Cup from Moss's Porsche. At Silverstone, Ireland won the *Daily Express* International Trophy – often a form guide for the rest of the season – and Stirling Moss gave Lotus its first Grand Prix victory at Monaco in Rob Walker's 18. It wasn't widely known that the engine mountings had broken and by the end of the race the front of the engine was only secured by a water hose. The Rob Walker car was fitted with a Colotti gearbox and outboard rear brakes, with slightly different front camber angles at the front compared with the works cars.

Transmission failure sidelined the works 18s of Stacey and Clark at Zandvoort when both were going very well and in the top four, while Moss lost two laps replacing a damaged wheel and finished fourth. Innes Ireland was second to Brabham. Tragedy marred the Belgian race on the daunting Spa-Francorchamps circuit. In practice, a rear stub axle failed on Moss's 18 and the car shed a wheel. He was severely injured, and missed several races as a result. Then Mike Taylor's privately entered Taylor and Crawley 18 (the prototype) crashed when the steering column broke, with terrible injuries as a result. In the race, as the charging Ireland dropped out, Stacey was lying a comfortable sixth when he was struck in the face by a pheasant. Its driver stunned and still travelling flat out, the car somersaulted and the 26-year old Stacey was thrown out and killed. Another of my boyhood heroes, Chris Bristow, also crashed fatally in his T53 Cooper, so it was a deeply sad day for motor sport.

For the next race, at Reims, Team Lotus had three Scotsmen driving the 18s – Clark, Ireland and Ron Flockhart. To improve the cars' straight-line speed they were fitted with tapered nose-cones, and Flockhart's brand-new vehicle had enclosed carburettors, achieved by tilting the engine over to the right. The Lotuses still lacked the straight-line speed of the BRM and Ferraris, and finished fifth, sixth and seventh.

At the British Grand Prix, the works 18s of Surtees and Ireland came second and third, while at the Portuguese Grand Prix Surtees set pole position time but retired in the race. Clark drove to third place behind the works Coopers of Brabham and McLaren, and Ireland was sixth. Still on the mend from Spa, Moss spun off and was disqualified for trying to bump-start his 18 by going downhill on the pavement – but against the direction of the race. In view of his benevolent support of Hawthorn under similar circumstances two years earlier, that penalty seemed somewhat harsh.

Back in the UK, Ireland won the Lombank Trophy race from Clark at Snetterton, while Moss won the Oulton Park Gold Cup. In the States, Stirling took the autumnal

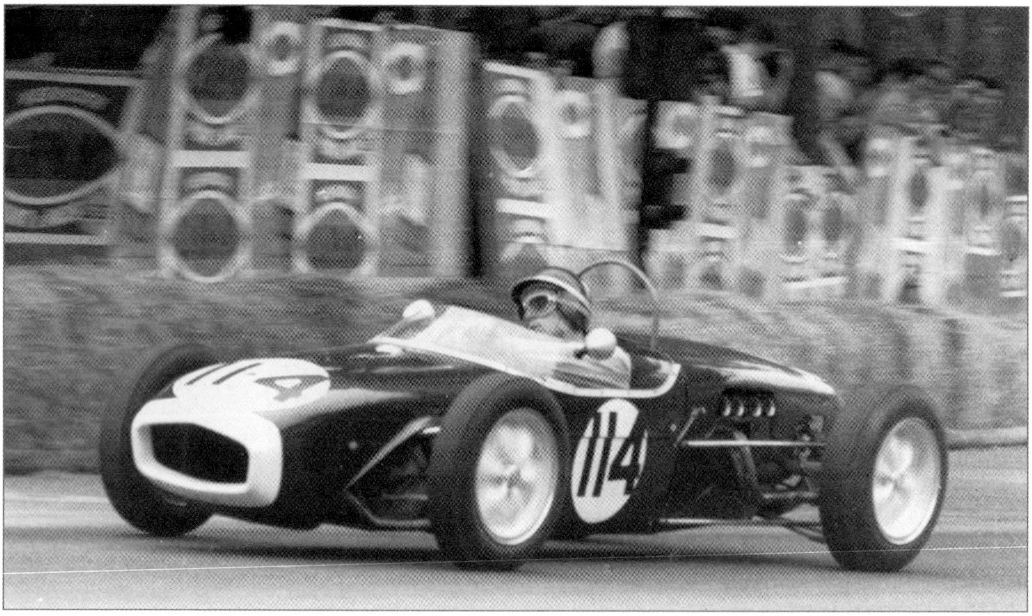

Jimmy Clark in the Team Lotus Formula Junior 18 at Monaco in 1960, where he was uncatchable until slowed by an electrical problem. In this category, the 18 was powered by the 997cc Ford 105E engine mated to a five-speed Renault Gordini gearbox. The first gear was usually blanked off to provide closer ratios than the four-speed option, although the latter was the more reliable transmission.

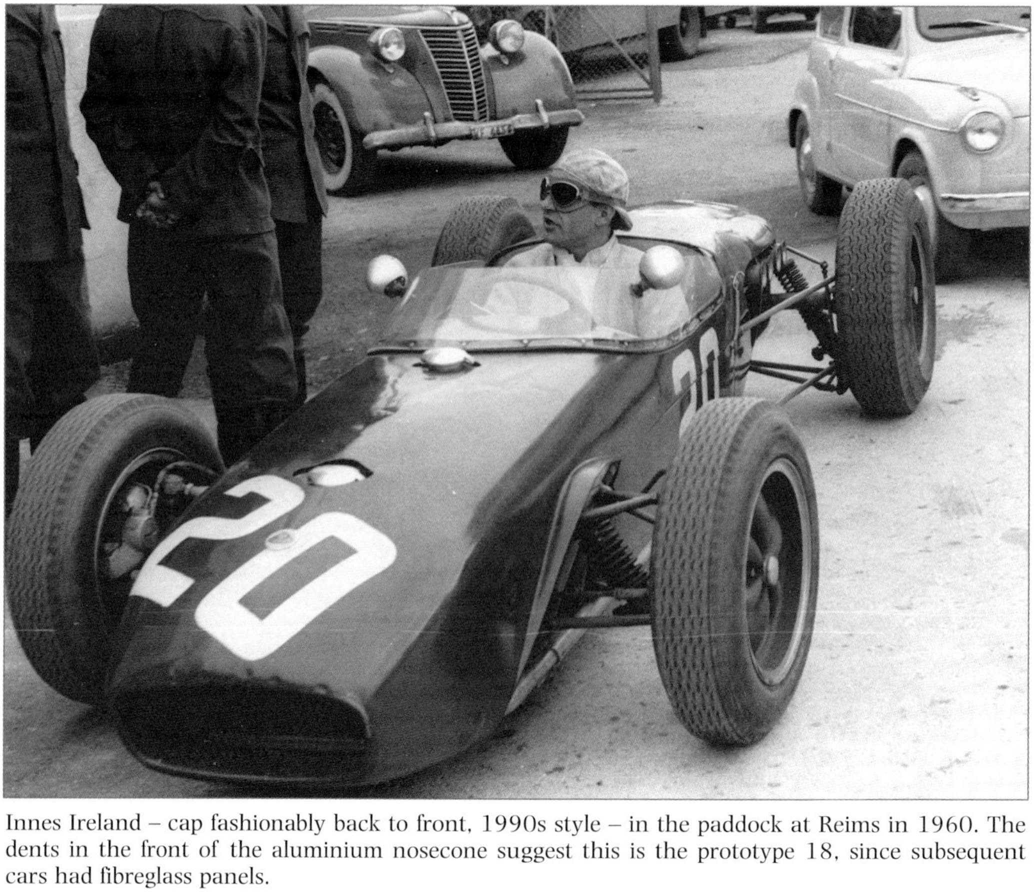

Innes Ireland – cap fashionably back to front, 1990s style – in the paddock at Reims in 1960. The dents in the front of the aluminium nosecone suggest this is the prototype 18, since subsequent cars had fibreglass panels.

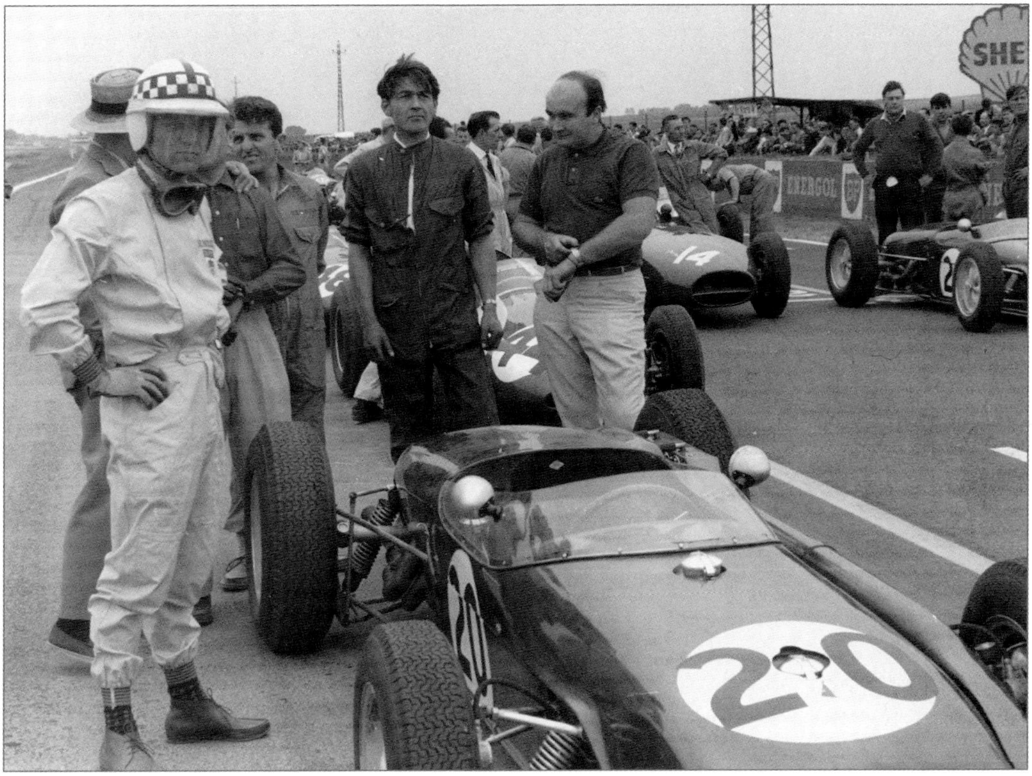

Innes Ireland contemplates his tactics before the start of the 1960 French Grand Prix at Reims-Gueux. Behind the 18 are mechanics Mal Simpson, Jim Endruweit and privateer David Piper (a non-starter in his 16). On the left-hand side of Ireland's car is the snorkel tube that extended from the nose to the Weber carburettor trumpets to aid induction of air. Another Team Lotus 18, probably Ron Flockhart's, sits on the third row.

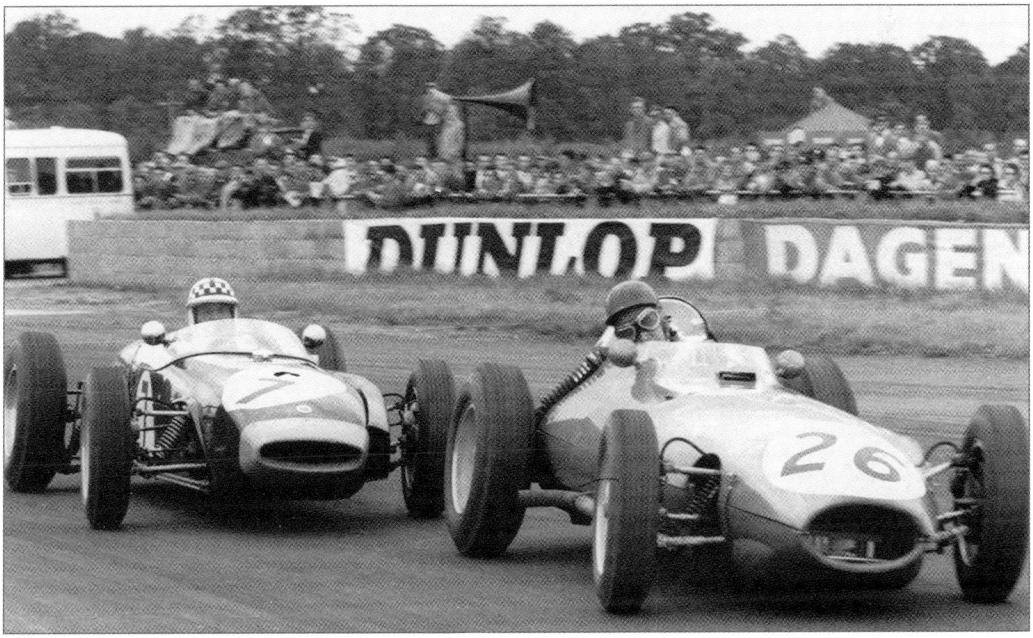

The old meets the new, as David Piper's 16 prepares to yield to Innes Ireland in the 18 during the 1960 British Grand Prix at Silverstone. Ireland was third to Brabham and Surtees, while Piper was lapped.

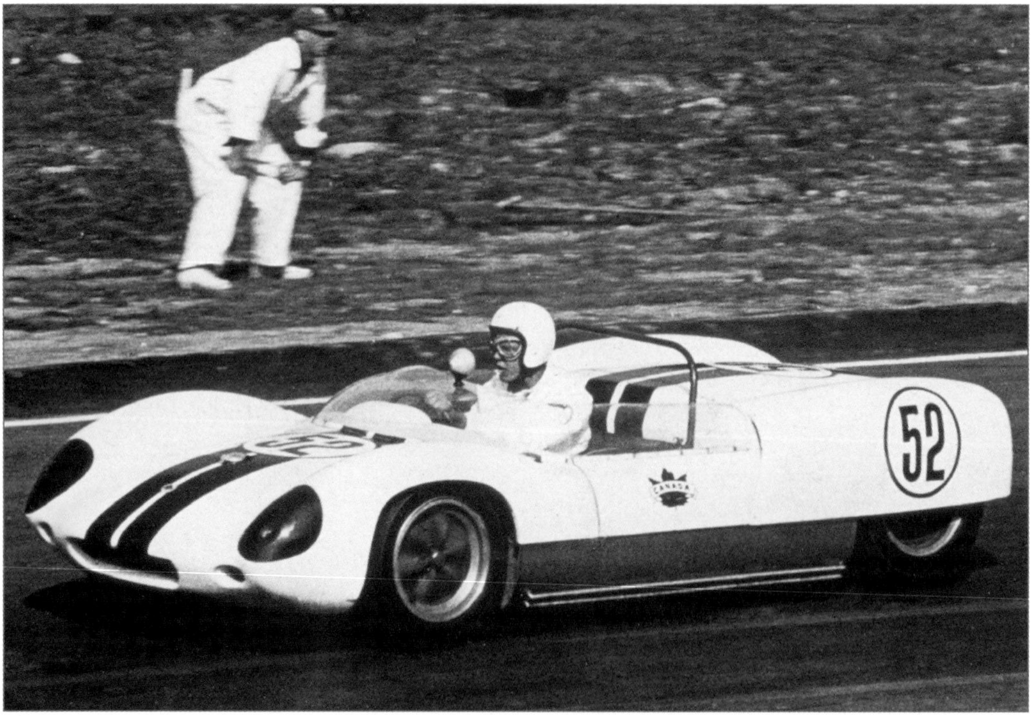

Peter Ryan on his way to victory at Mosport Park, Canada, in the Lotus 19 in 1961. The 19 (or Monte Carlo, as it was known, in celebration of Moss's victory in the 18 – and also as a dig at Cooper's Monaco) was announced in 1960 and was the sports racing equivalent of the 18. It was also far more aesthetically pleasing, but made use of the same suspension componentry and similar tubular spaceframe construction. It was Lotus's first rear-engined sports car and could be powered by any of the Climax FPF engines from 1475cc to 2751cc. One of the most successful 19s was the UDT/Laystall car, driven by Stirling Moss, and he and Dan Gurney trounced the Cooper Monacos in the States. Some examples ran with Maserati and American V8 power-units fitted, and Gurney's Ford V8 car of 1962/3 was designated the 19B.

honours at the Watkins Glen Formule Libre race and was first ahead of Innes Ireland in the US Grand Prix at Riverside. It was after this particular victory that Stirling was presented with a cake topped by a confectionery version of his Lotus 18. Flush with success, but smarting from crashes caused by lost wheels and broken stub axles, Stirling cut a wheel off the car-cake and presented it to the Team Lotus boss. Chapman was not amused.

In Formula 2 events, Trevor Taylor won with the 1.5-litre Climax FPF-engined 18 at Crystal Palace and Jimmy Clark won the 'Kentish 100' at Brands Hatch. In Formula Junior, Team Lotus entered cars for Clark, Taylor and Pete Arundell, and between them Clark and Taylor cleaned up completely, with Henry Taylor in the Ken Tyrrell 18 taking the British Empire Trophy. In fact every FJ event of note in 1960 was won by a Lotus.

The new Lotus 21 was in the frame for the 1961 F1 season, but delays with the designated engine meant that, until it was ready, Team Lotus campaigned the 18s, and Clark won the Pau Grand Prix in April. Fitted with 2.5-litre Climax engines, 18s were raced in Intercontinental events and two F1 races in the States, when 21s were sidelined through crashes. Significantly, Stirling Moss drove Rob Walker's 18 from pole position to victory in the Monaco Grand Prix, setting fastest lap and vanquishing the sharknose Ferrari 156s in the process. Moss was also fourth at Zandvoort, first at Brands in the Silver City Trophy race, and first again at the Nürburgring for the German Grand Prix. Unable to get a works-built 21, the Walker équipe fitted Stirling's 18 with body panels for a 21 that transformed its appearance into the 18/21 hybrid. Moss went on to win the Swedish Kanonloppet race at Karlskoga, the Danish Grand Prix at Roskilde Ring and the Modena Grand Prix, and came second to Clark at Kyalami in the Rand Grand Prix as well as the Natal Grand Prix and the South African Grand Prix at East London. It was in a much-modified 18 that Stirling's F1 career came to an abrupt end when he crashed – inexplicably – out of the Richmond Trophy at Goodwood in April 1962.

During the 1961 season, the UDT/Laystall team fielded 18/21s for several drivers including Tim Parnell, Henry Taylor, Masten Gregory, Olivier Gendebien and Wolfgang Seidel, while other 18s were raced by Ernesto Prinoth, Gerry Ashmore and Gaetano Starraba.

The Mk 19 represented the sports racing version of the 18, being mid-engined, and with a tubular chassis designed as a two seater. Most componentry was identical to the 18's and its normal engine was the FPF 2495cc Climax unit mated to the Lotus five-speed transmission. Body panels were in aluminium and glass fibre, and it was very much state of the art in appearance and handling. Between 1960 and 1962, just twelve units were built, with nine cars including the prototype going to the States and three raced in the UK under the banner of the UDT/Laystall finance team. The 19 enjoyed a measure of longevity, partly because some of the US cars were retro-fitted with American engines. An example was the one-off 19B, chassis 966/19B, commissioned in 1963 by Frank Arcerio for whom Dan Gyrney had driven a 2.5-litre Climax-powered 19 (chassis 951) between 1960 and 1962. Gurney drove the V-8 engined 19B to victory in the 1963 Daytona Continental 3-Hours race.powered by a Ford V8 unit, which Dan Gurney drove to victory in the 1962 Daytona Continental 3-Hours Race.

Contemporary with the Mks 19 and 21 was a single-seater Formula Junior car, the Mk 20, unveiled at the Racing Car Show at Olympia in January 1961. The 20 was produced and marketed under the Lotus Components subsidiary, and its basic underpinnings were similar to the outgoing 18. The chassis and suspension systems had not altered, except that the top right and lower left longitudinal chassis members doubled as water pipes running between the engine and radiator. The height factor was addressed by storing fuel in fabricated side tanks instead of the tank that straddled the scuttle on the 18. In appearance, though, the 20 was low and lithe, clad in much more rounded bodywork that made the boxy 18 look positively outdated. If the 18 was the vehicle that got the engine in the 'right' place, the 20 was the means of refining the aerodynamics. Its body sections were produced from three fibreglass moulds by Williams and Pritchard.

At first, the 20 was fitted with the 85bhp 997cc Cosworth-Ford engine, derived from the Anglia, but was offered later in the year with a 1340cc 109E Classic unit linered down to 1100cc. This yielded 100bhp allied to a Renault four-speed transmission, and first powered Trevor Taylor's car to victory at Reims. The 20 ran with drum brakes all round, and used Lotus's cast magnesium wobbly-web wheels that were 13 inches diameter at the front and 15 inches at the back. The 20B was a derivative fitted with the 1498cc Ford Cortina engine and disc brakes, available for Formule Libre and Formula B in the States. A total of 118 units of the Mk 20 and

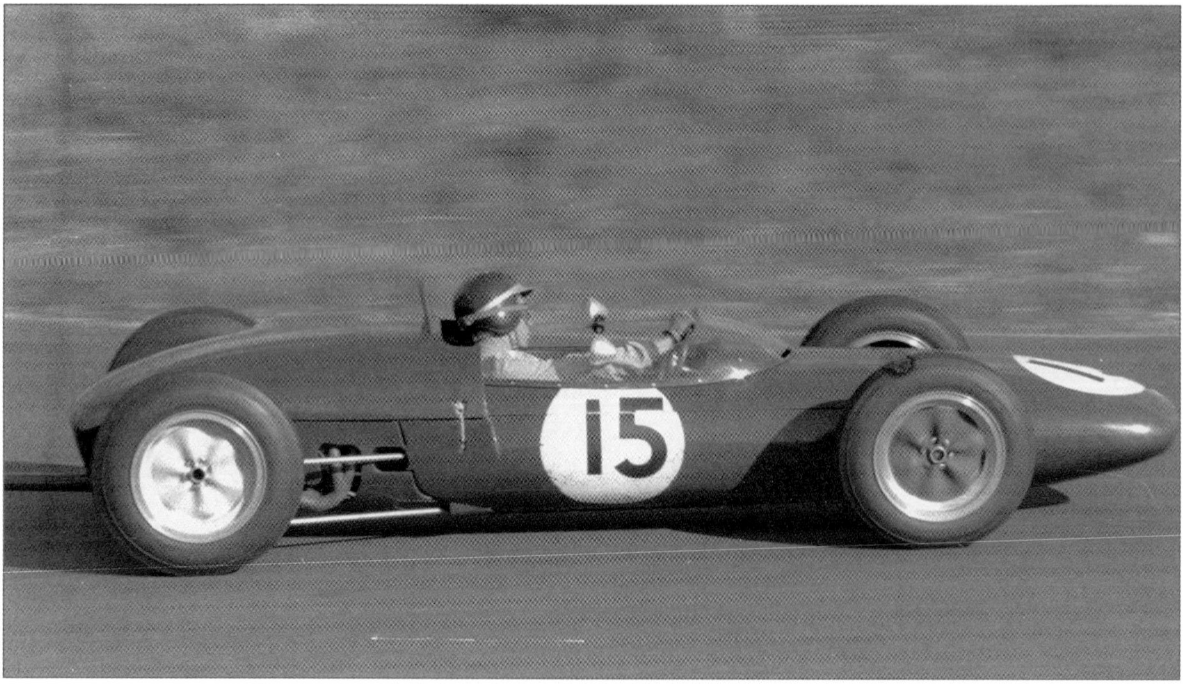

Jimmy Clark in the Formula 1 Lotus 21 at Zandvoort in 1961. This was a spaceframe car powered by the 1475cc Climax engine in F1, and also by the 2495cc version for the Tasman series. The 21's best result was Innes Ireland's win at Watkins Glen in 1961, and Jim Clark's third places at Zandvoort and Reims. Stirling Moss also drove the Walker 18/21 to victory at the Nürburgring.

20B were produced by Lotus Components, who would sell you one as a kit for £1,450. It was with the 20 that Team Lotus drivers Trevor Taylor and Peter Arundell consolidated their reputations as the young lions. During 1961, Taylor won eight races to Arundell's seven, although the latter's tally included the prestigious Prix Monaco Junior. Another up-and-coming driver to star in a Lotus 20 was the Swiss Jo Siffert, who took seven victories as a privateer.

Key to the Door

The 21 was designed with the same aerodynamic concepts and configuration as the 20, but with the intention of carrying the new Coventry-Climax V8 engine to contest the incoming 1500cc Formula 1. There were serious delays in producing this engine (and the BRM unit), however, centring on the ambivalence of British constructors towards the new formula. To get the 21 mobile, Lotus redesigned it to accept the updated four-cylinder 1497cc FPF unit, incorporating the stronger bottom end of the 2.5-litre version allied to a ZF gearbox. The engine was leaned over at 18 degrees to the right to enable the lower engine cover to be used. The car's 30-gallon aluminium fuel cells were attached to the chassis rails by bungee straps.

Like the 20, the 21's chassis was based on the geometry of the 18, but in order to comply with the minimum weight limit of the new F1 regulations, the 21 weighed in at a heavier 450kg. As with the 20, certain frame tubes were used to carry oil and water. The driving position was more recumbent than ever, so the driver's head was a full 2 inches lower in the car than was the case with the 18.

The Lotus 22 was an elegant Formula Junior car for the 1962 season, developed from the 20 with upgraded suspension and using a 997cc Ford engine. Altogether, Lotus Components produced seventy-seven units of this model. The Team Lotus car was driven by Peter Arundell, winning eighteen out of twenty-five races and coming second three times. Team Lotus successes were at the root of the big-engine slur by Richard von Frankenberg, and Arundell's reprise of his quick times at Monza elicited a retraction from the German journalist.

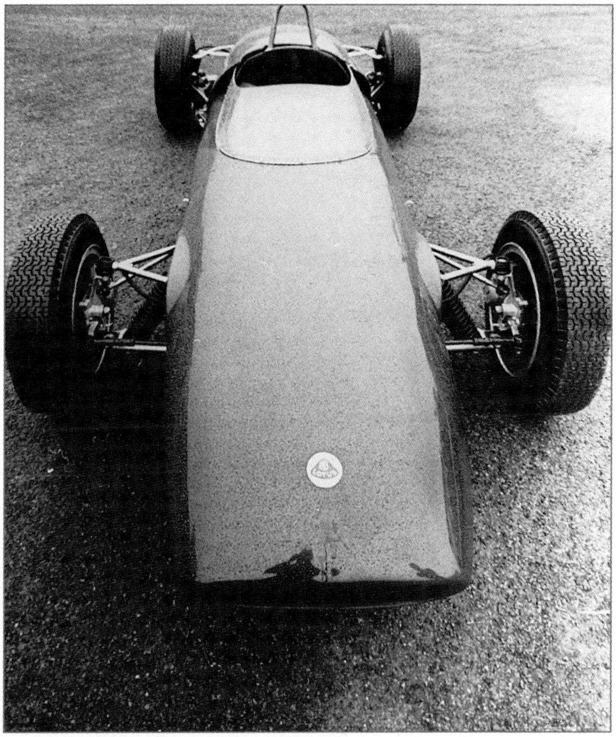

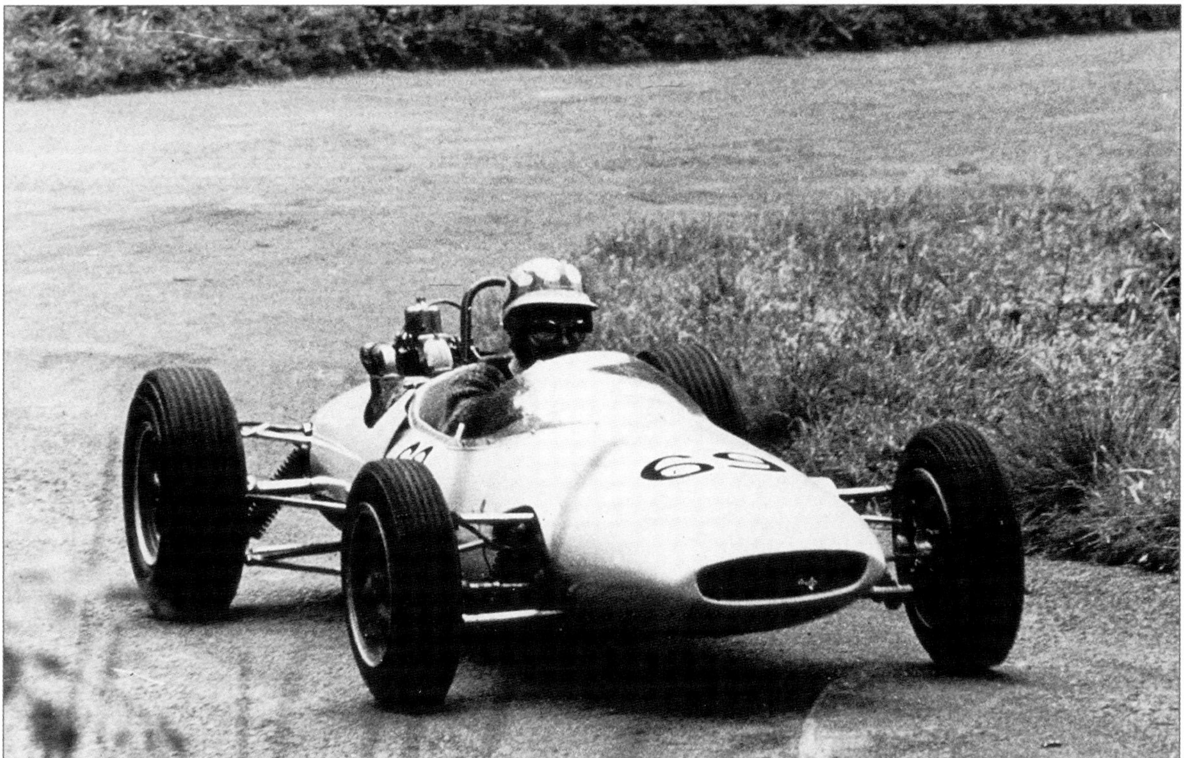

Hillclimb specialist Peter Boshier Jones takes his Lotus 22 Formula Junior up Shelsey Walsh in 1964. The regular 22 was powered by a modified 1097cc Ford 105E engine mated with a four-speed Renault gearbox, but Boshier Jones's car had a supercharged Coventry-Climax engine.

There were bigger changes to the suspension than had taken place for some time. Up front, fabricated upper rocking arms operated inboard-mounted combined coil spring and damper units, while the rear set up now incorporated new Metalstik rubber couplings, and featured a top link that defined the roll-centre height. Disc brakes were mounted outboard back and front, and the 21 was shod with 15 inch wobbly-web wheels rather than the 13 inch ones fitted to the 18.

The 21 made its debut at the 1961 Monaco GP – won by Moss in his updated 18 that had the side panels removed but Ireland non-started because of a practice crash, and Clark came tenth. The cars to beat that year were the Ferrari 156s, and there were sometimes as many as five of them running. So Ireland's win in the non-championship Solitude Grand Prix was the 21's first success. He went on to win the Flugplatz Rennen race at Zeltweg, as well as the United States Grand Prix at Watkins Glen in October, which the World Championship winners Ferrari didn't contest in the wake of von Trips' death in the collision with Clark at Monza. The US GP win was Team Lotus's first World title win in its own right, and a morale booster after the Monza tragedy. Meanwhile, Clark went on to win the South African non-Championship races at Kyalami, Westmead and East London, but he wasn't officially cleared of culpability in the von Trips accident for two years. By this time, Ireland had been sacked as Chapman saw that he didn't exactly get on with his promising understudy, nor with Chapman himself for that matter – Ireland was a bit of a playboy, contrasting with the more serious and reserved Clark. There was something of a puritan streak in Colin Chapman; for example, he neither smoked nor drank much at all. Although he would let his hair down at an event such as the annual Team Lotus Ball, there was a very serious intent about his racing, which was passionately shared with Clark. Thus, Jimmy was elevated to team leader, with Trevor Taylor as number two.

Junior Winner

For the 1962 Formula Junior season, Lotus Components came out with the 22. It was essentially an updated version of the 20, with the high-revving 100bhp 1098cc Cosworth-Ford power unit tilted at 30 degrees to the right to reduce the height of the rear bodywork and improve induction. Transmission could either be four-speed Renault or Volkswagen. As well as the radiator feed, the oil system also passed through the chassis tubes. The 22 used the Mk 20 front suspension and the 21's rear suspension set up, and was equipped with Girling disc brakes all round, production components homologated on the Triumph Herald, and mounted outboard at the back because of lack of space for inboard rear callipers. It reverted to 13 inch wobbly-web wheels, the track was also slightly wider, and a half-inch rear anti-roll bar was fitted. Once again, Chapman sought to lower the driving position to improve aerodynamics and centre of gravity, and accordingly introduced his 'theory of the compressibility of bums'. This allowed him to take out the curvature of the base of the seat, so the driver's backside was literally on the car's floorpan. There was thus no need to recline the seat or extend the cockpit, and ground clearance remained the same.

The three Team Lotus cars were campaigned during the 1962 season by Pete Arundell, Alan Rees and Bob Anderson, and Arundell won no fewer than eighteen out of twenty-five races, with three second places, making him the Junior king that year. The quasi-works Ian Walker Racing Team ran 22s for Mike Spence and the promising

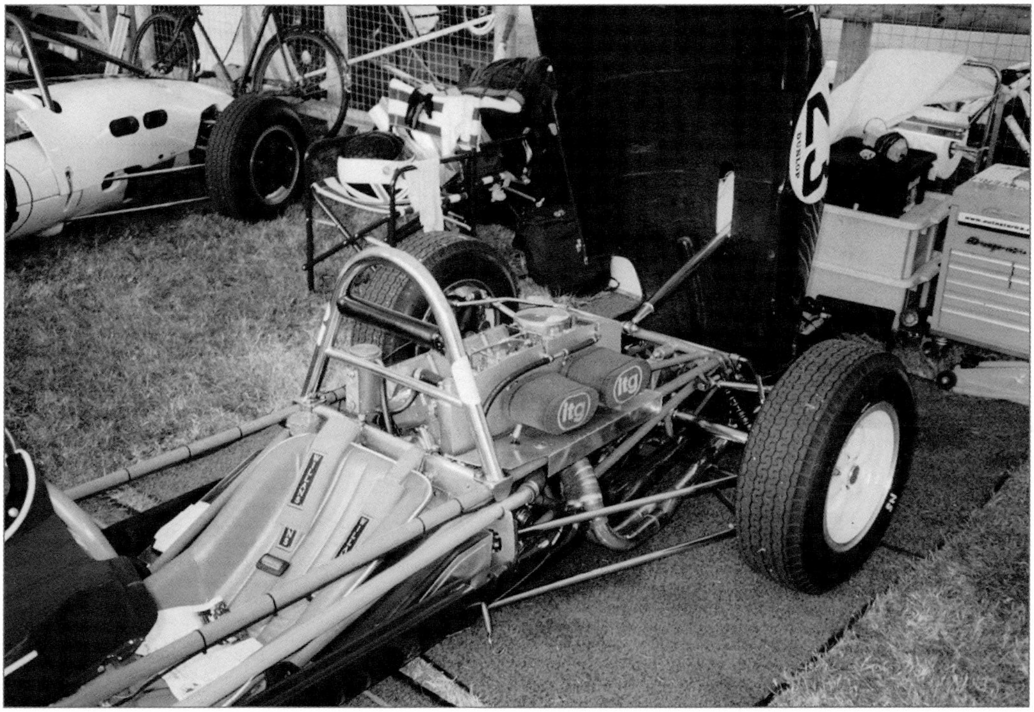

The chassis and rear suspension detail and the 1100cc Ford engine revealed in Martin Walford's 22 Formula Junior car at Goodwood 1999. Malcolm Ricketts' 22 is to the left of the shot.

Canadian Peter Ryan, who unfortunately got killed at Reims. European privateers in 22s included Alfa Romeo star Giacomo 'Geki' Russo, who won at Enna Pergusa, and Moises Solana. In the States, Pete Lovely and Pat Pigott starred in the Team Rosebud 22s.

Such was Lotus's dominance that the editor of *Auto Motor und Sport*, Richard von Frankenberg, was moved to suggest in the magazine that Lotus was using 1450cc engines in the FJ 22 – his comments came on the basis of an interview with a concussed Alan Rees in his hospital bed following a crash at the Nürburgring. In order to set the record straight, Chapman offered to take a 22 to any circuit the German – an ex-Porsche racer – chose; repeat the race-winning speeds and then give full access to scrutineers. In addition to expenses, there was also a £1,000 stake to be settled by the loser. Von Frankenberg chose Monza, always a power circuit, and Pete Arundell was required to match his lap times from the recent Monza Lottery Grand Prix in which he'd averaged 113.47mph over 30 laps and set a fastest lap of 1 minute 50.8 seconds – 115.99mph. It was 2 December, frosty and cold, with patches of ice beneath the trees. Those present threw sand about to thaw it. Chapman knew he was on safe ground, however, as Arundell went out and duly averaged 115.16mph for 30 laps, with a fastest lap of 1 minute 50.4 seconds. The engine was stripped down, and its capacity registered at 1092.348cc (85 × 48.15mm), and 3kg over the minimum weight limit. It was a nice piece of justice, as von Frankenberg's allegation had been wildly over-the-top – 'the biggest disgrace in international motor sport', he called it. He retracted the slur and paid up.

It was becoming a familiar exercise to create a sports racer and a single seater using the same componentry and powertrains, and like the 19 which was based on

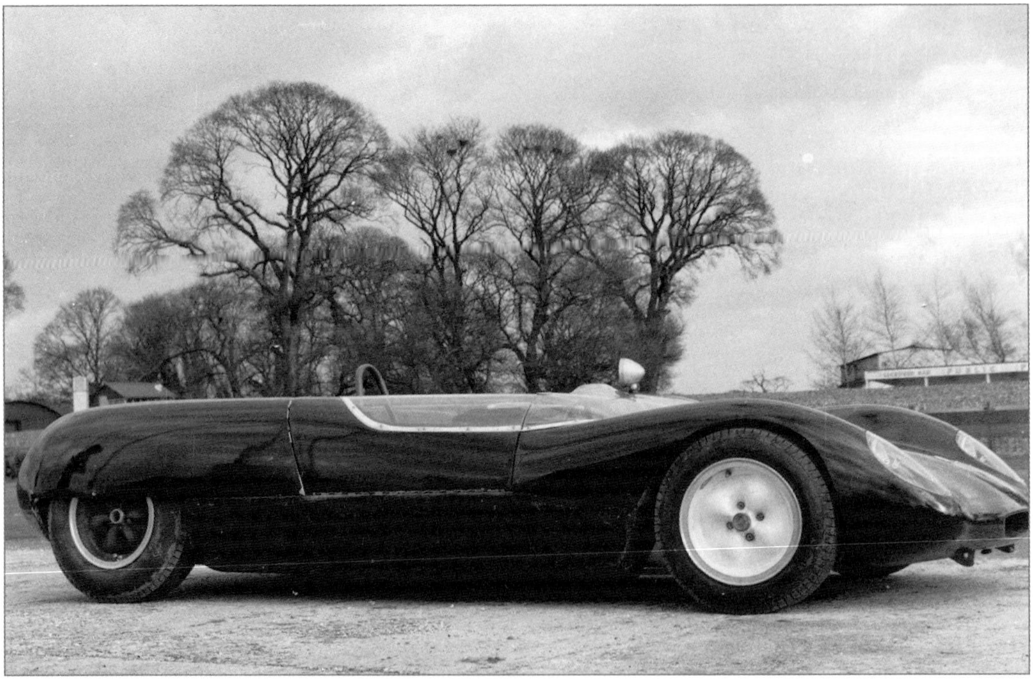

A Lotus 23 sports racing car in the Goodwood paddock, showing its low, almost flat profile. It was based on a widened version of the single-seater 20/22 FJ spaceframe chassis. There were three series: the S1 that used the 997cc sohc, 1098cc FJ unit, and 1470cc sohc engines; the S2 that used the 997cc and 1498cc twin cams; and the S3s, which were the 997cc twin-cam Le Mans car and 750cc UDT twin-cam Le Mans car. Subsequent 23Bs used 1598cc Lotus-Ford twin-cams. Between 1963 and 1966, production totalled 131 units.

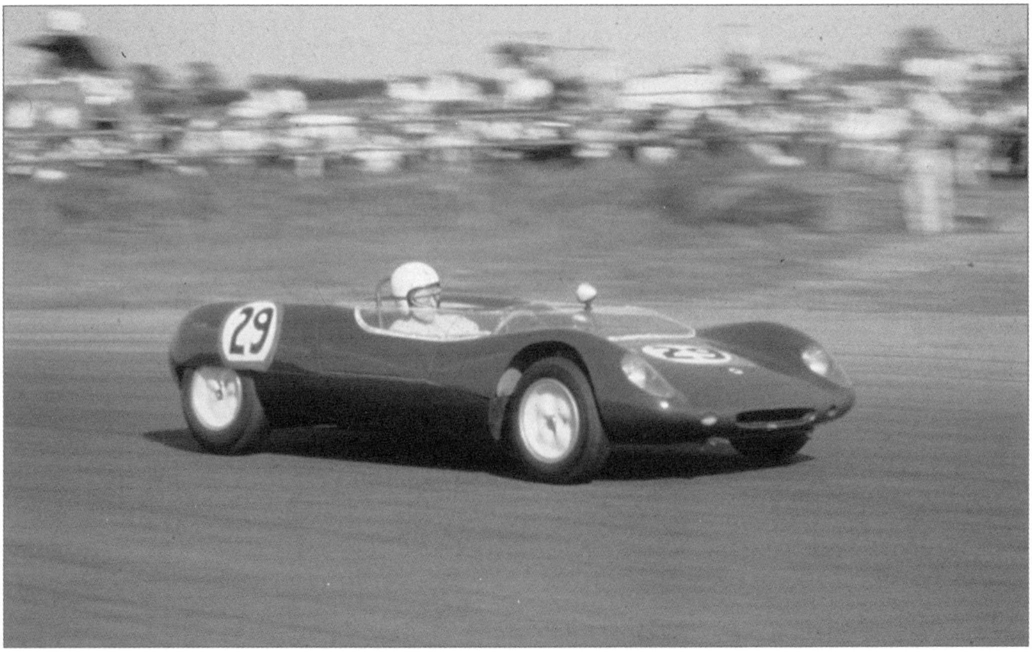

Peter Arundell in one of the works' 23s at Silverstone in 1963. Probably the 23's finest hour was Jimmy Clark's effort in the wet at the Nürburgring 1000km in 1962 when the prototype Lotus-Ford twin-cam was fitted, and he led all the big Ferraris, Astons and Porsches going away. Fumes from a leaking fuel pipe caused him to make a rare error, but the point was proved.

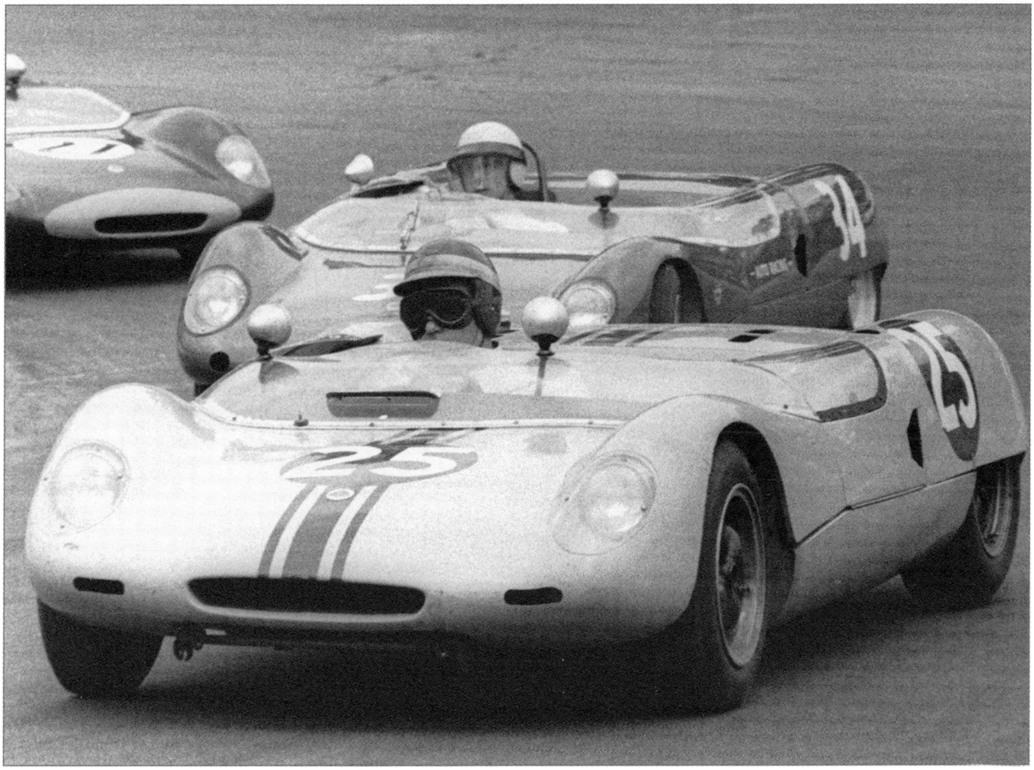

The Ian Walker Racing 23 driven by Paul Hawkins heads another 23 out of the Mallory Park Hairpin in July 1963. Another successful outfit running 23s that year was Team Normand, whose 1594cc twin-cam engined 23Bs were piloted by Mike Beckwith, Tony Hegbourne and sometimes Jimmy Clark. Other drivers to shine in 23s were Alan Rees, Keith Greene, Jack Pearce and Rodney Bloor.

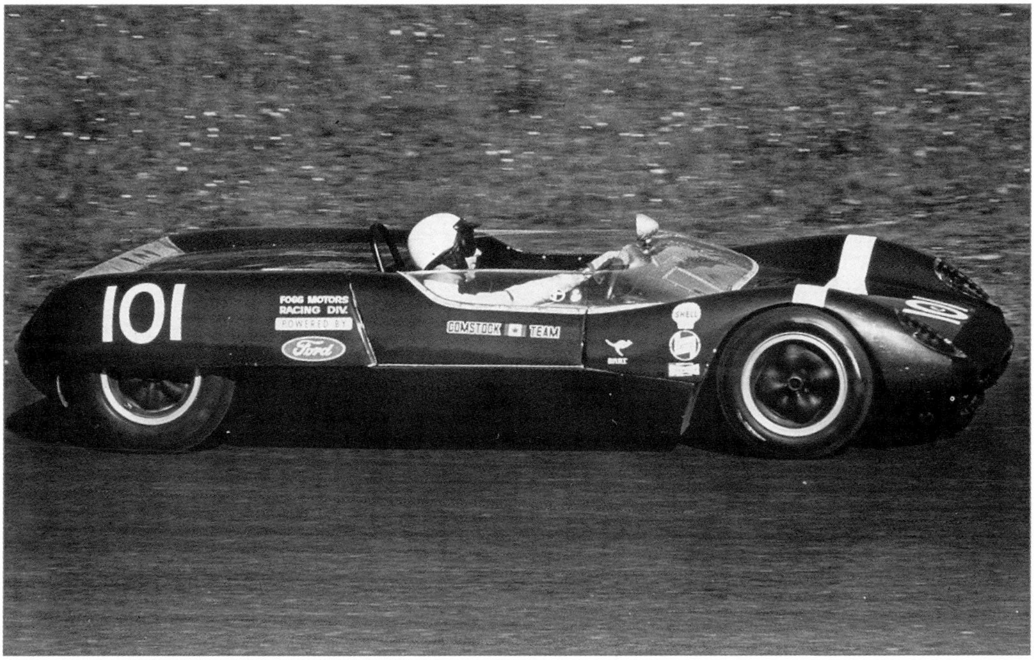

Bob McLean drove the Comstock Team's Lotus 23B with much success in the Canadian sports racing series in 1965.

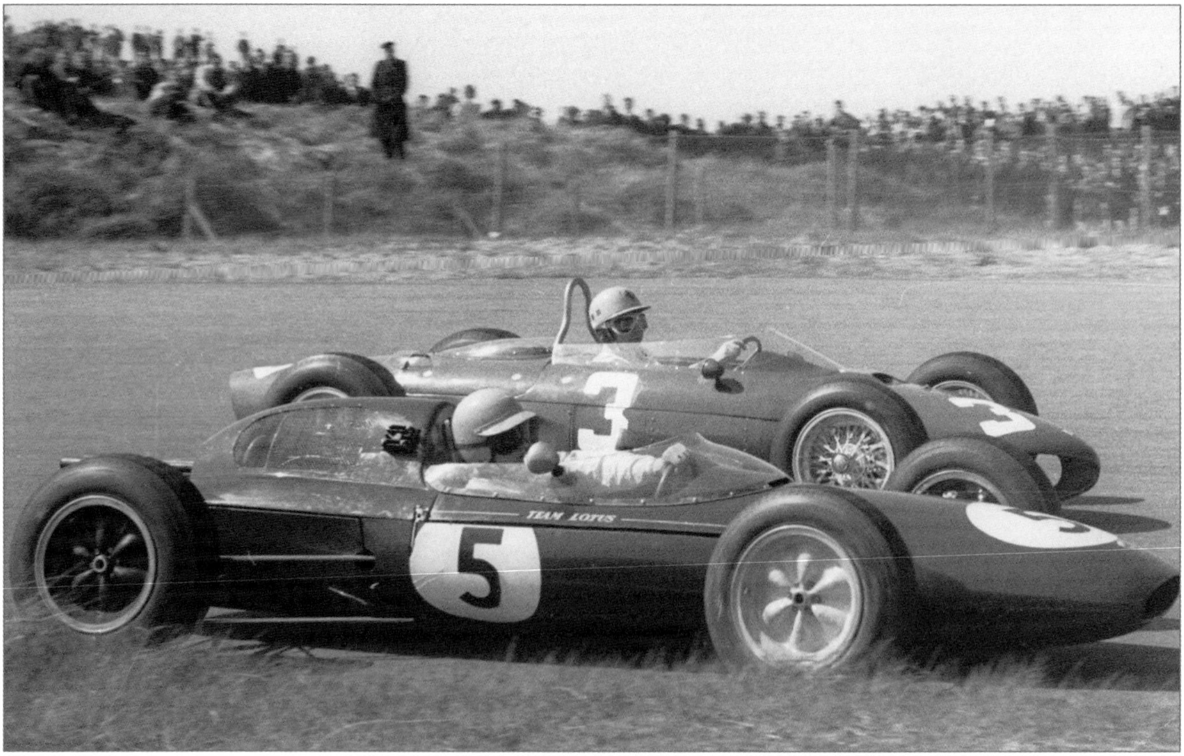

Trevor Taylor's 24 dives inside to lap Giancarlo Baghetti's Ferrari 156 at Zandvoort in 1962. Taylor is on his way to second place and his first six World Championship points. The 24 was the first Lotus to accommodate the 1498cc V8 Coventry-Climax and BRM engines, and it was also the last spaceframe Lotus F1 car.

the 18, the 23 of 1962 was founded on the 22 Formula Junior car. It was generally powered by the same 103bhp 1097cc Cosworth-Ford engine, although 997 and 1599cc motors could also be specified, allied to Renault or Volkswagen four-speed gearboxes and final drive units. As with the 19, the 23's multi-tubular space frame was broader, befitting the two-seater configuration, and both vehicles had similar suspension set-ups. The 23's sleek low bodywork was entirely fibreglass, and it rapidly gained the similar status accorded the Eleven in the late '50s. Race of the year for this model was the Nürburgring 1000km race, where the works car was entered under the Essex Racing team banner and supervised by Mike Costin. It ran with a 140bhp twin-cam unit based on the 1498cc Ford 116E block, which was destined to become the Lotus production engine. Drivers were none other than Jimmy Clark and Trevor Taylor. The old 14-mile circuit was wet and slippery, yet at the end of the first lap the Scot had pulled out almost half a minute from the hoards of Porsches, 4.0-litre Ferraris and an Aston Martin. Clearly Clark and the 23 were in their element in such conditions. But with the circuit drying out and plagued by a leaking exhaust, Clark made an uncharacteristic error and dropped the Lotus into a ditch. Nevertheless, it was a hugely promising start, bolstered by Clark's class lap record, and one of the Ian Walker Racing 1.0-litre 23s winning its class and the other setting fastest lap.

Chapman was thus tempted to enter a 997cc twin-cam engined 23 for Le Mans, supported by a 750cc version run by the UDT/Laystall team. Clark and Taylor were to drive the work's car and Leston/Johnstone the UDT car. They never raced. In an outrageous display of nationalistic bigotry the scrutineers threw out the Lotuses for having unmatched wheel studs – four at the front and six at the back. This meant that the spare wheel could only be used on one axle. When Chapman standardised all four wheels with four studs, they claimed the cars were no longer safe, and despite the intercession of Lotus aficionado and noted writer Jabby Crombac, that was the end of the matter as far as the Automobile Club de l'Ouest was concerned. Quite clearly the Lotuses posed enough of a threat to the Panhards and René Bonnets that normally contested the Index of Performance to warrant their exclusion. Chapman swore never to return, and in his lifetime, the works Lotuses never did. Strangely enough, there was a postscript to the 1962 Le Mans affair. The Team Elite cars of Hobbs/Gardner and Wyllie/Hunt came eighth and eleventh overall in the 24-Hours, and first and second in the Index of Thermal Efficiency.

An evolved version of the 23 came out later in the year. It was known as the 23B, had a strengthened chassis and ran with the Lotus-Ford 1594cc twin-cam engine. A total of 131 Mk 23s was made between 1962 and 1964, and they remained active on the club scene for much of the decade. Notably successful entrants were Ian

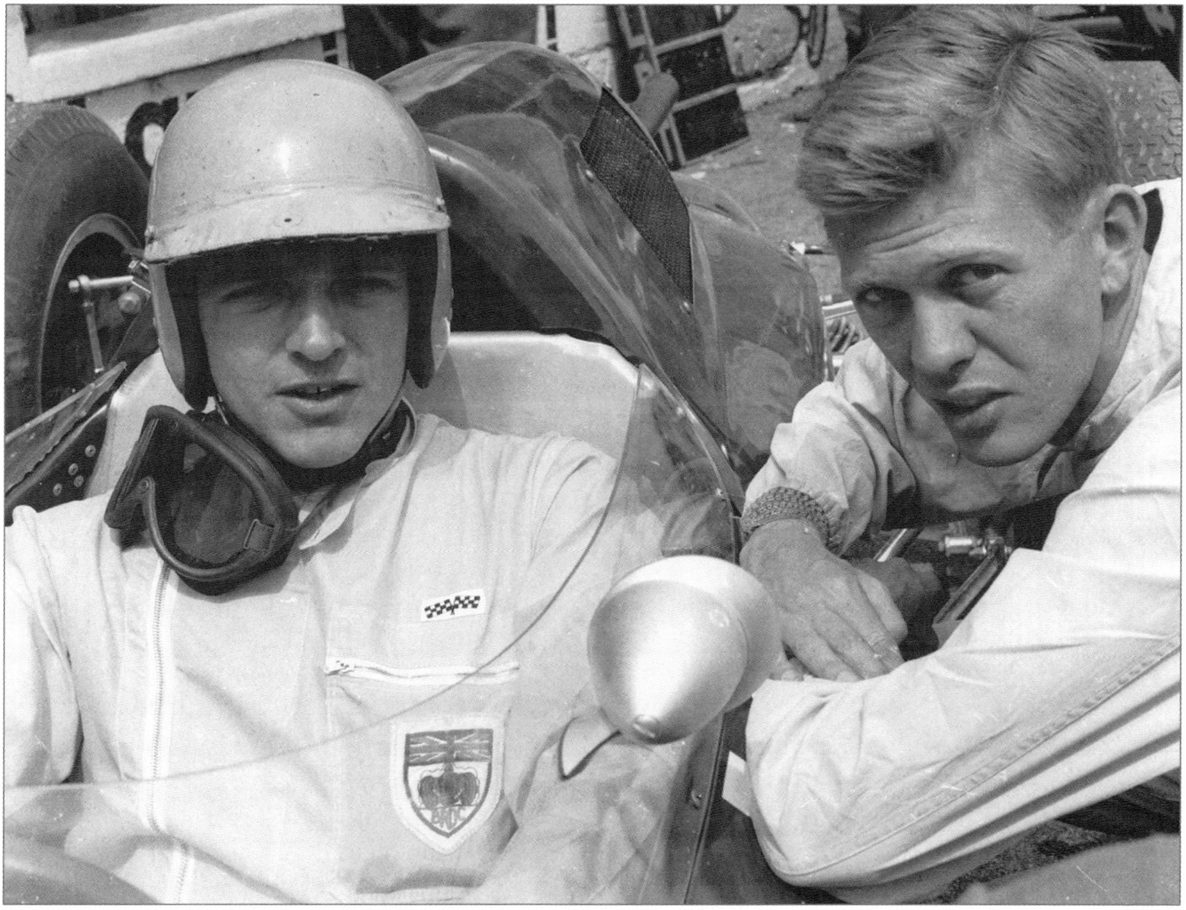

Trevor Taylor in the Team Lotus 24 has a pitlane conversation with works Cooper-Climax driver Tony Maggs in 1962. Although Cooper was by now past its heyday, the T60s of McLaren and Maggs were often among the front runners. Cooper's only victory in the 1.5-litre F1 era was Bruce's win at the 1963 Monaco Grand Prix.

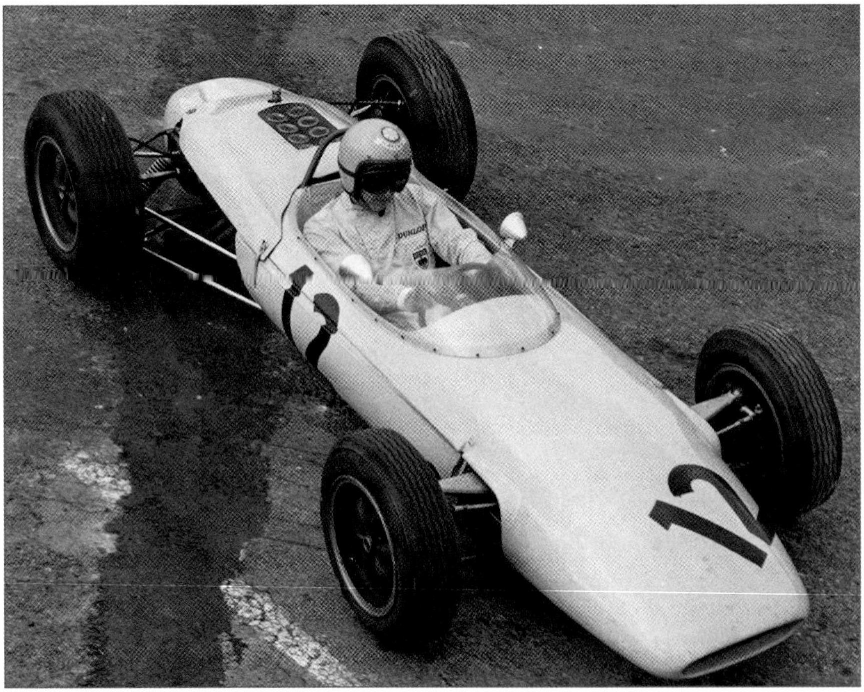

By 1964, Trevor Taylor was driving for the British Racing Partnership BRP team alongside Innes Ireland. The pale green Lotus 24 is fitted with a BRM V8 and Colotti gearbox. The same engines were used in Tim Parnell's privately entered Lotus 25s for Chris Amon, Mike Hailwood and Peter Revson. On the whole they retired more often than not.

Walker Racing and Normand Racing, who ran Mike Beckwith, Tony Hegbourne and Jim Clark, and UDT/Laystall, whose 23 was powered by the 997cc Lotus twin-cam. This car was driven by José Rosinski/Bernard Consten to a class win at the 1962 Montlhéry 1000km in 1962.

The 24 Formula 1 car was notable for having the first chassis to be built by Arch Motors, who still make the chassis for the Caterham Seven to this day. Until now, Lotus spaceframe chassis had been made by Progress, but it built just one 24. The model was introduced for the 1962 season, and was designed to accept the new Coventry-Climax FWMV V8 four-cam engine. The front suspension uprights were now cast items, and the 24 also had new rack-and-pinion steering. The driveshafts had rubber doughnut couplings on their inboard ends. The 24 reverted to a scuttle fuel tank over the driver's legs, which was necessary because of the V8's higher fuel consumption.

Team Lotus raced the 24 at the opening of the 1962 season, with wins for Jim Clark at Snetterton in the Lombank Trophy and the Aintree 200 race in April, and a second place in the Silverstone International Trophy. At Zandvoort, Trevor Taylor finished second in the Dutch Grand Prix, and was destined to race the 24 for the rest of the season, which included an accident at Spa. Clark, on the other hand, was given a pearl of a car – the new 25.

Probably 15 units of the 24 were made and run privately by such luminaries as Jack Brabham, then on the threshold of launching his own car, and Maurice Trintignant in Rob Walker's car. Scuderia Filipinetti ran a 24-BRM version for Jo Siffert, and the UDT/Laystall team ran a Climax-engined car for Team Lotus exile Innes Ireland, as well as BRM V8-powered cars for Masten Gregory and Pete Arundell. Ireland had lost little of his old flair, and had a win at Crystal Palace and third places at Reims, the Roskilde Ring and the Mexican Grand Prix.

CHAPTER FIVE

THE FIRST MONOCOQUE

There is no question that the revolutionary 25 Formula 1 car with its monocoque tub was built under conditions of extreme secrecy. As one of the handful of Team Lotus personnel involved with its construction, Jim Endruweit tells of the close-knit coterie that surrounded Chapman at the time. There was designer Len Terry, team manager Andrew Ferguson, Jim as chief mechanic, and that was about it. It was constructed in the design and development area to the rear of the Lotus-Cortina floor, and access to factory personnel was strictly forbidden. The monocoque chassis comprised the whole car, in the same way as a spaceframe, and it consisted of two stressed pontoons united by the undertray, with bulkheads front and rear, and a stressed panel locating the instruments. Solid rivets were used instead of pop rivets, which could work loose. It had additional tubular cross-bracing at the front, and the whole package imparted a degree of torsional strength unknown in contemporary Formula 1. After the prototype 25, subsequent cars were fashioned in lighter gauge aluminium sheet. As well as chassis design, other aspects of Lotus's racing car construction were influenced by the aviation industry. Fuel was carried in aircraft-type rubber bag tanks inside the monocoque's two pontoons, saving a lot of weight, and it was down to Jim Endruweit's Fleet Air Arm connections that such items as filler necks and caps were sourced easily.

The business of building Lotus racing cars was a progression of new ideas and innovations. If you take the Type 25 as an example, Chapman's latest ideas would go into a car just being built, and these modifications would not necessarily be put into an existing car. The 25 was raced for three seasons, and continuously subjected to tweaks and updates. More major evolutions included the adoption of fuel injection and air-deflector windscreens. There was a minor difference in the 25's suspension in that the rear coil spring and damper units were located ahead of the driveshafts, rather than behind them. The power unit was the Coventry-Climax FWMV V8 twin-cam 1497cc engine, transmission was by a ZF five-speed gearbox and final drive with synchromesh on second to fifth gears via a twin-plate Borg and Beck clutch. Metalastic bonded-rubber couplings were used rather than sliding-spline drive-shafts. Pedals were pendant, on account of the tight cockpit, and fitting a small 12 inch diameter steering wheel also made for a bit more space. By the time the last of the 25Bs was being made, that car was probably closer in specification to its successor, the 33, than to the original 25.

In reality, the cars that made up the three Lotus type numbers 24, 25, and 26 – the Elan – were all conceived at about the same time. Chapman was considering the idea of a backbone chassis like that of the Elan that would accommodate the driver in the middle, and it was a short step from this idea to the notion of a monocoque. Because Lotus could hardly sell an unproven concept to its customers, the 24 was marketed with its tried and tested spaceframe chassis, but with identical suspension and powertrain to the 25. Teams to race the 24 with either Climax or BRM engines

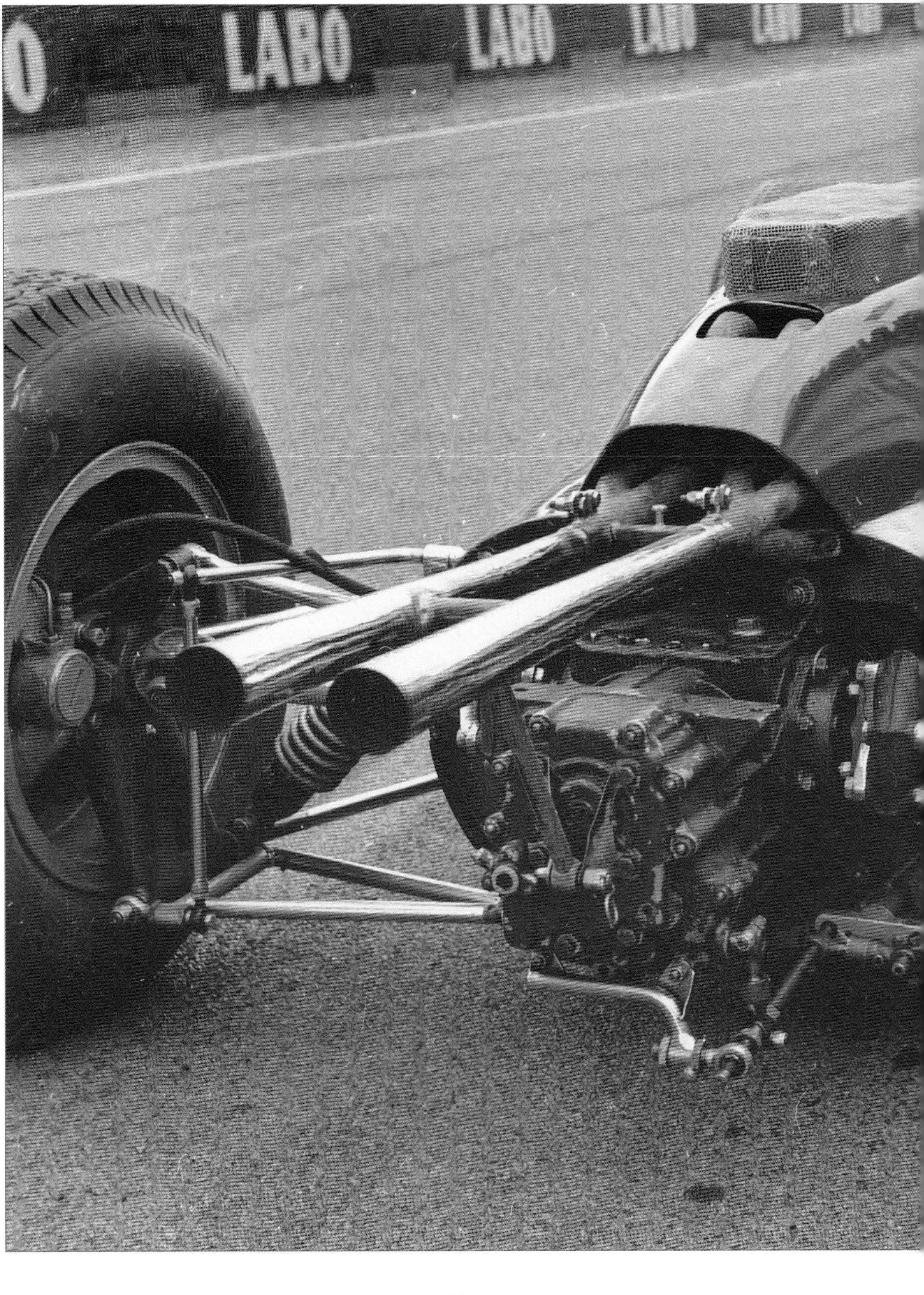
LABO
LABO
LABO
LABO
LABO

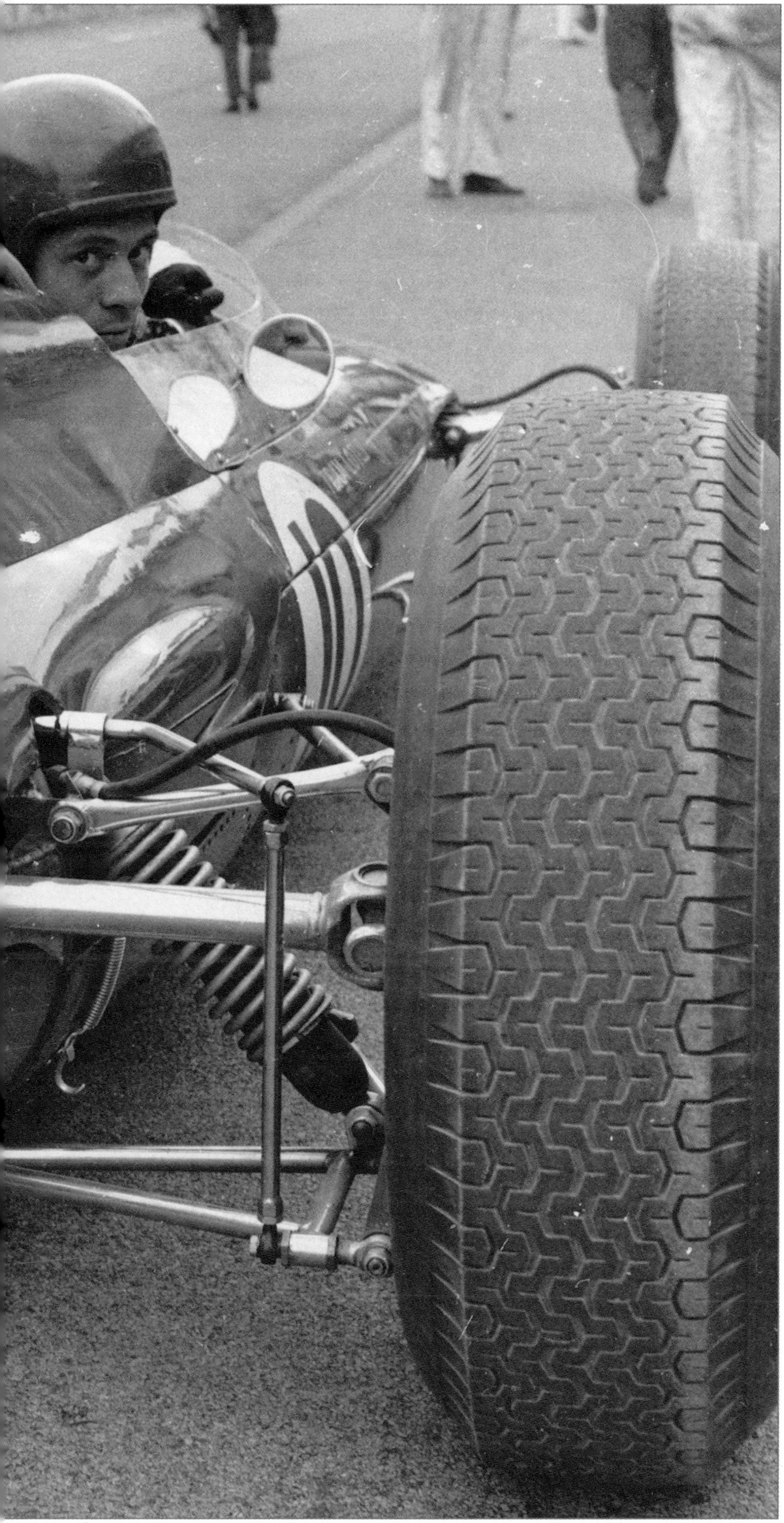

Fastest in practice for the 1963 French Grand Prix at the high-speed Reims-Gueux circuit, Jim Clark's Lotus 25 sports the short-stroke fuel-injected Climax V8 and ZF gearbox. There were discrepancies in the spec of the other Team Lotus cars – Taylor's injected 25 used the Colotti gearbox, while Arundell's 25 was still on carburettors. In the race it was clear that Clark would win, barring mechanical mishap, and despite the V8 going off song and rain beginning to fall, he nursed it home to win by over a minute from Tony Maggs' Cooper-Climax. Apparently there was little on hand to drink except champagne. . . .

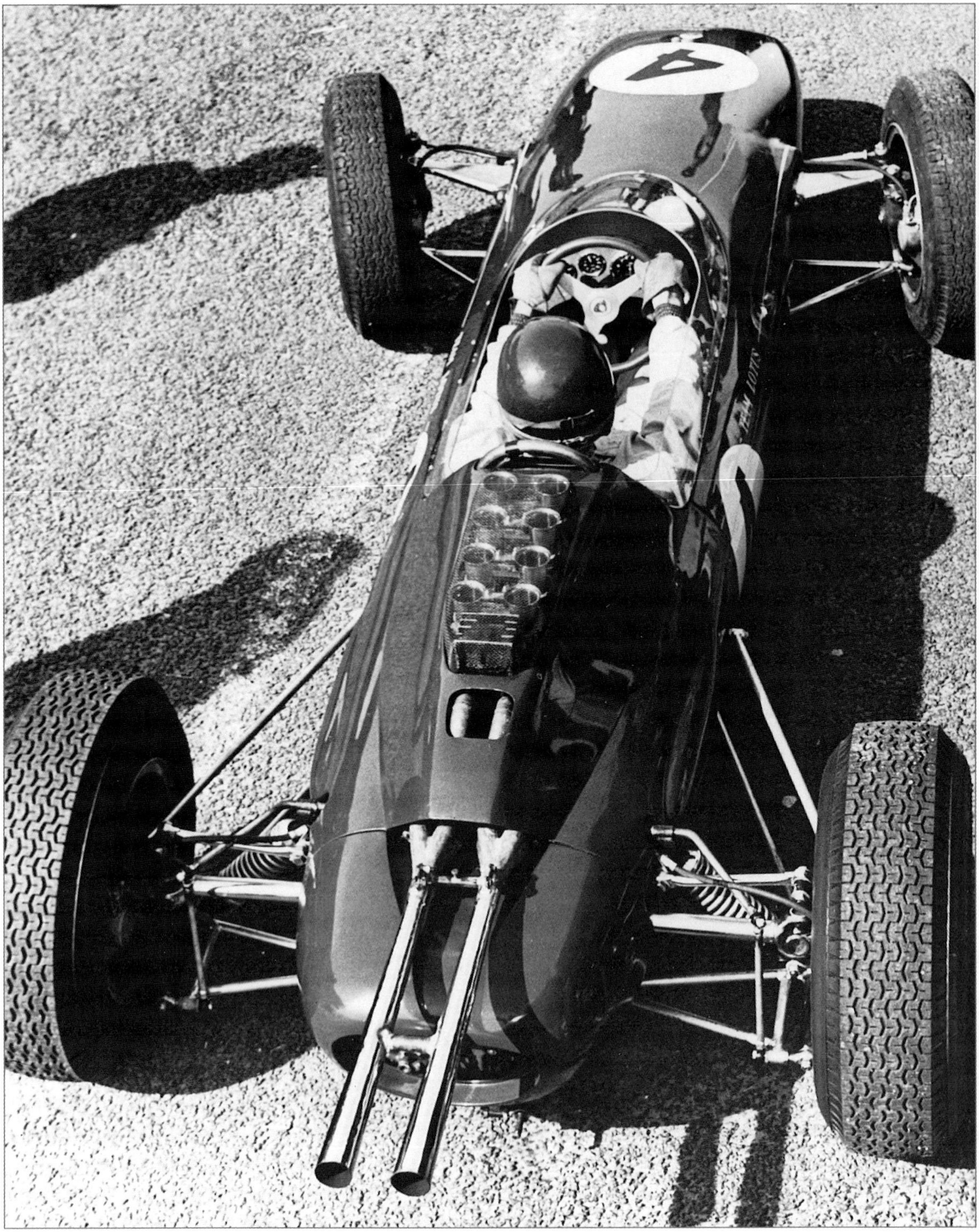

Jim Clark gave the 25 its Grand Prix debut at Zandvoort in 1962. The 25 was the first racing car in the world to have a full monocoque chassis, and its suspension was developed from the 21. The power unit was the 200bhp 1.5-litre Coventry-Climax V8, and the intake trumpets and exhaust manifold were always a distinctive hallmark. For its first year the 25 ran with a regular windshield, and the air-deflector version didn't appear until a year later at Zandvoort in 1963.

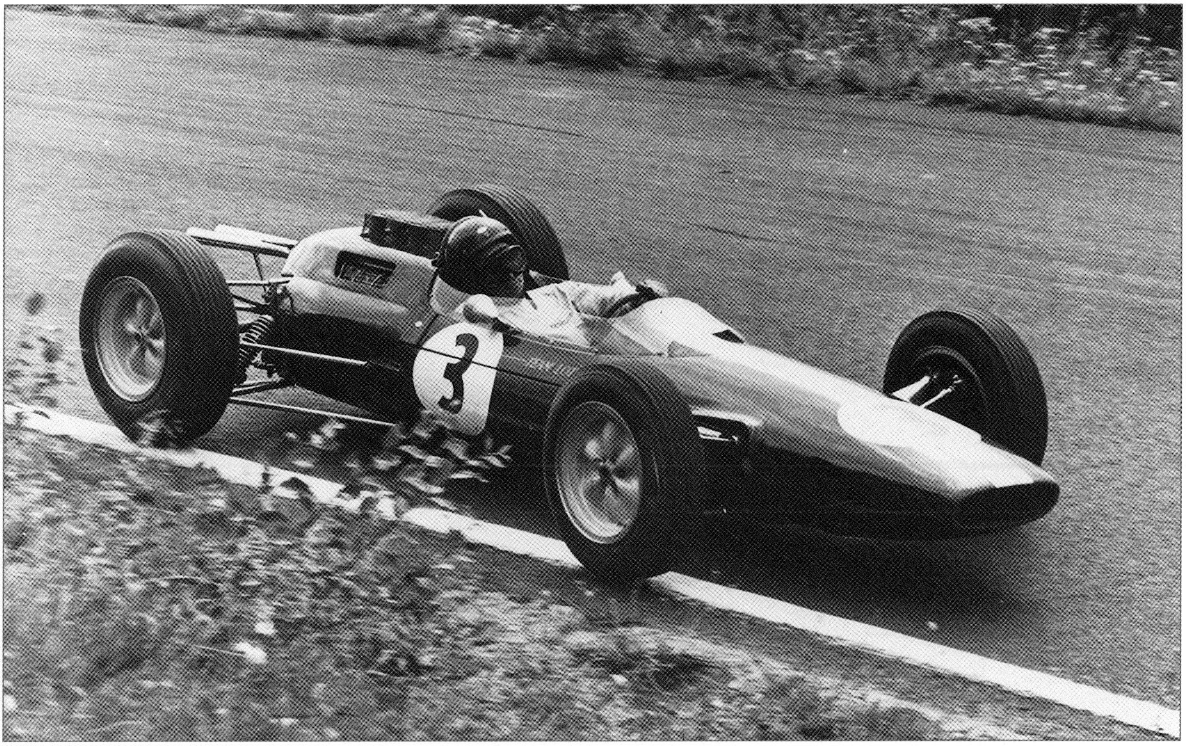

The maestro at work: Jim Clark's car control was so good that he could do four races on the same set of tyres. During 1963 he won seven out of ten World Championship rounds and was second and third in two more, which was more than enough to take the title.

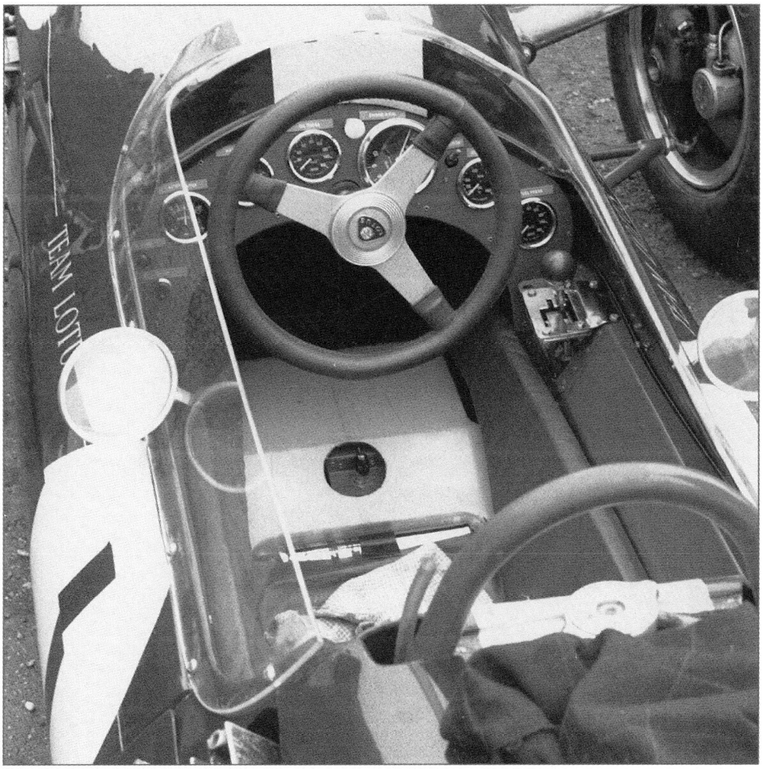

Clark's workplace, the cockpit of Lotus 25 R6, photographed at the *Daily Mirror* Trophy meeting at Snetterton in 1964. For this meeting the car had new suspension uprights and hub carriers, and smaller brake discs to accommodate new 13 inch wheels. Steering and suspension geometry had to be altered to suit, and it was known as the 25B. This was something of an interim measure, as the 33 was being readied at the time. In this very wet race, Jim's new Dunlops aquaplaned, the rain found its way down the Climax injection trumpets, and he had to retire.

included Rob Walker, the UDT/Laystall équipe, which ran four cars during 1963/4, the Bowmaker team, Scuderia Filipinetti, and Tim Parnell, who built up three cars from components sourced elsewhere.

Jimmy Clark first appeared in the 25 at Zandvoort for the Dutch Grand Prix, and led for the first few laps until clutch problems relegated him to ninth place. The first win for the 25 was at Spa-Francorchamps, a circuit Clark disliked, and he beat Graham Hill's new BRM by 44 seconds. Taylor had his first drive in the 25 at Rouen-les-Essarts, but an altercation with Trintignant wrote the car off. Further Grand Prix wins for Clark came at Aintree, Watkins Glen, Mexico – in Trevor Taylor's car, but he was disqualified for having received a push-start. He won the non-Championship Rand Grand Prix, a round of the Springbok series, and also the Oulton Park Gold Cup. But Trevor's best results with the 25 were third at Solitude, second in the Rand Grand Prix and first in the Springbok series' Natal Grand Prix. Jim Clark was second to Graham Hill in the F1 Drivers' Championship, and Lotus was second to BRM in the Constructors' table.

For 1963, Team Lotus made only slight alterations to the 25s, involving ducts on the sides of the fibreglass nosecones. The carburettors of the original Climax V8 engine were superseded by fuel-injection. Clark was quick to notch up some race wins in the non-Championship events at Pau and Imola. He was second to Hill in the Lombank Trophy race at Snetterton, third at Aintree in Taylor's 25 – they were allowed to switch cars in mid-race in those days – and first at Silverstone in the *Daily Express* International Trophy. Along with the works 25s, privateer Lotuses were abundant in these events, with 24s, 22s, and even an 18 still running. A new 25 raced at the non-Championship Austrian Grand Prix at Zeltweg, fitted with the latest flat-plane crank Climax V8 engine with low-level exhaust system and Hewland five-speed Volkswagen-based gearbox. A new innovation was seen for the first time at Zandvoort, and this was the deflector screen. The driver's best view ahead is above rather than through the windscreen, and in order to get that as low as possible Chapman introduced the deflector screen, which had a channel at the front that forced air rapidly through a vortex between screen and bodywork and thence over the driver's head. The screen could thus be lower than a conventional one.

Team Lotus built seven examples of the 25, and the first two were written off in accidents involving Trevor Taylor in 1962, and the fifth chassis 25 R5 was also wrecked by Taylor. When 25 R3, R4 and R7 were sold off to Tim Parnell in 1964 he replaced the Climax engines with BRM V8 units. Jo Bonnier bought 25 R6 in 1963, and sold it to MGM to be an extra in the John Frankenheimer movie *Grand Prix*. You can see a 25 in Team Lotus trim on display at Tom Wheatcroft's Donington Collection. While the works cars were racing at Aintree, Siffert won at Syracuse in his Lotus 24-BRM.

In the F1 World Championship, Clark retired at Monaco, but was unstoppable thereafter. Wins came at Spa, Zandvoort, Reims, Silverstone, second at Nürburgring, first at Monza, third at Watkins Glen, first at Mexico and East London, South Africa. The World Title was Clark's by twenty-five points from Hill and Ginther. Taylor, on the other hand, was down in fifteenth place with just one point. Such was Clark's form that at Zandvoort, for example, he lapped the entire field. Now that Clark had won the World crown so convincingly, Lotus's reputation and status was suitably consolidated. Quite rightly, the Team and the company were on a roll – the Elan was in production, and things had never been so good.

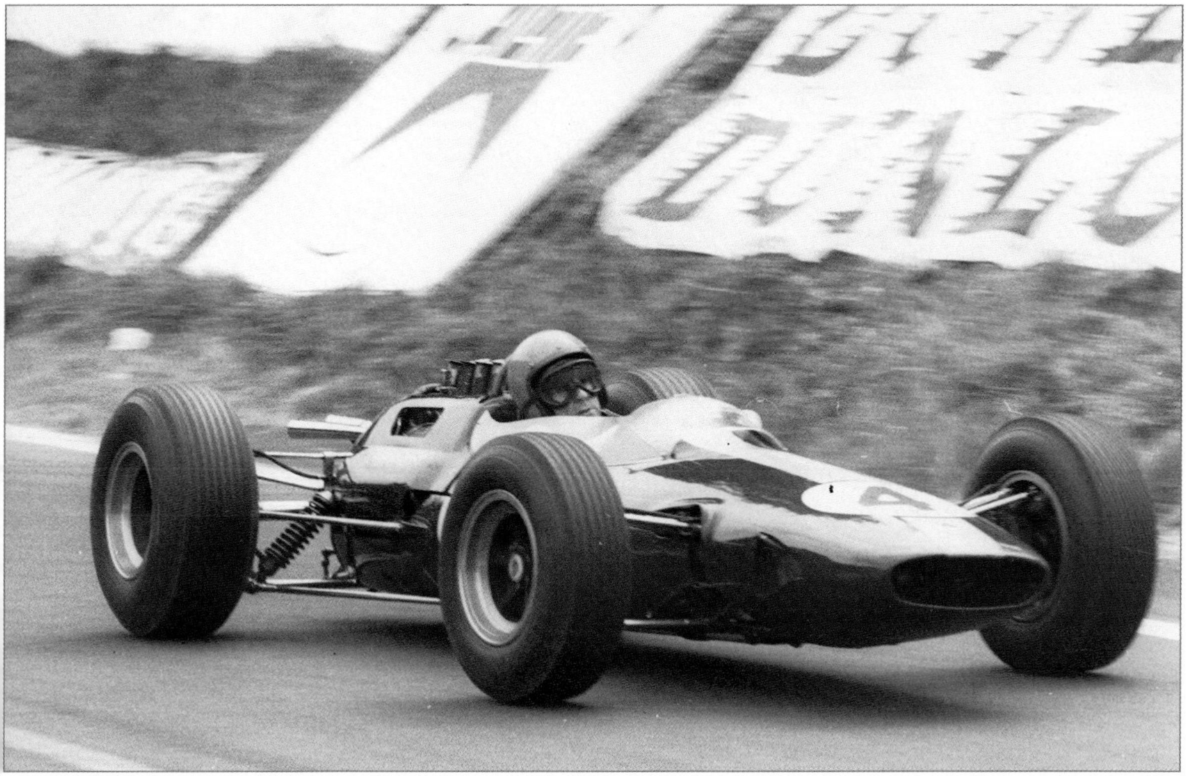

Driving Lotus Climax 25 R4, Peter Arundell came fourth in the 1964 French Grand Prix at the scenic Rouen-les-Essarts circuit on 28 June. Dan Gurney won after Jim Clark retired with a broken piston, giving the Brabham team its first World Championship points.

THE BRITISH INVASION

Meanwhile, Lotus had made its first appearance at Indianapolis back in May '63. The USAC establishment's attitude was somewhat blinkered to the design of the Formula 1 'funny car', sticking doggedly to the traditional front-engined 'roadster' layout. The typical Indianapolis racing car was a Kurtis-Offenhauser – big, brash and brutish, powered by a four-cylinder 2.8-litre supercharged or 4.2-litres unsupercharged Meyer-Drake Offenhauser engine putting out masses of horsepower but lacking the lightness and efficiency of the contemporary F1 chassis. These Indy cars also needed three or four fuel stops to the F1 cars' one. In 1960, John Cooper pioneered the European attack on Indy with an adapted T53 Lowline model for Jack Brabham, powered by a 2.7-litre Climax FPF motor with Indy-style two-speed gearbox – sponsored by Kleenex. Despite Brabham's promising ninth place and masses of prize money, Cooper never returned to the 'Brickyard'.

Early in May 1963 veteran Texan driver Lloyd Ruby paved the way for Lotus to an extent by showing what an ex-works Lotus 18 with a 2.7-litre Climax engine was capable of in USAC racing. He led for forty laps at Trenton's 1 mile oval until he was sidelined with transmission failure, having set fastest lap in practice. Although one or two far-sighted US constructors had made rear-engined chassis in 1962, the establishment was far more reliant on the relatively ponderous Offenhauser-engined roadsters that were, admittedly quick in a straight line. A few wondered what the funny

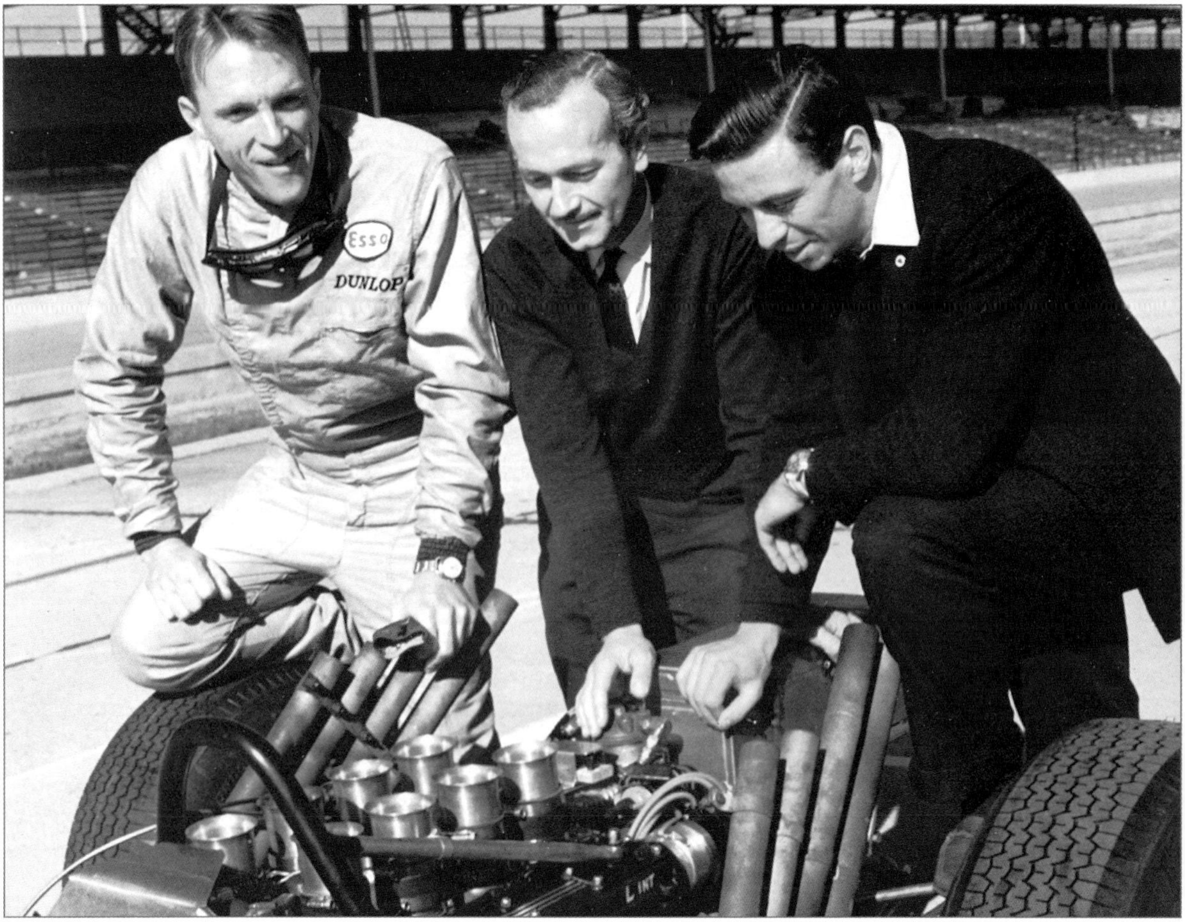

Dan Gurney was largely responsible for influencing Colin Chapman to consider an attempt at the Indianapolis 500. Here Dan (in racing kit), Colin and Jimmy contemplate the 29's 4.2-litre stock-block Fairlane-derived pushrod V8, with its bank of 58mm Weber carbs and stack-pipe exhausts. The latter would be replaced with an arrangement known as the 'snake-pit' system, so called because of the intertwined pipes. The 29 was based on the same monocoque chassis layout as the 25, and the Ford V8 engine developed 370bhp.

European cars might be capable of. A racer with feet in both camps (F1 and USAC) was Dan Gurney, and he was largely responsible for encouraging Chapman to go racing at Indianapolis, and a deal was done with Ford to run two Lotuses at the circuit in 1963.

The Indianapolis 500 occupies the entire month of May, and teams are closeted in designated garages within the confines of the oval circuit to prepare their cars. Times set in the first weekend's qualifying counted for the race, and there was much feverish activity at Lotus, as Clark, Dan Gurney and Chapman commuted to and fro across the Atlantic to qualify the cars. At the same time, fresh cars were constructed at Cheshunt and Ford tested race engines at its Dearborn headquarters. The mechanics, meanwhile, beavered away in their garage, constantly adjusting suspension settings and rebuilding the cars, all the while striving for the optimum set up. Jim Endruweit headed the mechanical effort and he recalls that in most of the other garages, tuning operations were limited to polishing the bodywork, despite the comprehensive array of tool and equipment on hand. Green was considered an unlucky colour at the 'Brickyard', and the establishment was horrified that Lotuses were not only painted green, but that the sleep-deprived, grease-stained mechanics were also clad in green overalls.

Lotus's Indianapolis challenger was the 29, designed by Len Terry, and it was based heavily on the 25 monocoque, with an extended wheelbase to comply with the USAC minimum specification of 8 feet. The suspension was strengthened, and built with the asymmetrical offset to the left that was *de rigueur* for racing on banked oval circuits with no right-hand corners. The 29's Ford V8 engine was a push-rod 4260cc unit, based on the production Fairlane model. It was rather more sophisticated than that, however, having a reduced bore size, an alloy crankcase, cylinder block and heads, and mechanically operated valves. It was mated to a Colotti Type 37 four-speed gearbox. Final drive and power output was in the region of 370bhp.

Following a shakedown at Snetterton, where it ran with stack-pipe exhausts and symmetrical suspension, the 29 was dispatched to Ford's Kingman, Arizona, test track. But in qualifying at Indianapolis, Gurney crashed heavily and was forced to race the prototype. In the actual race, Rufus Parnelli Jones ran out the winner in his Watson-

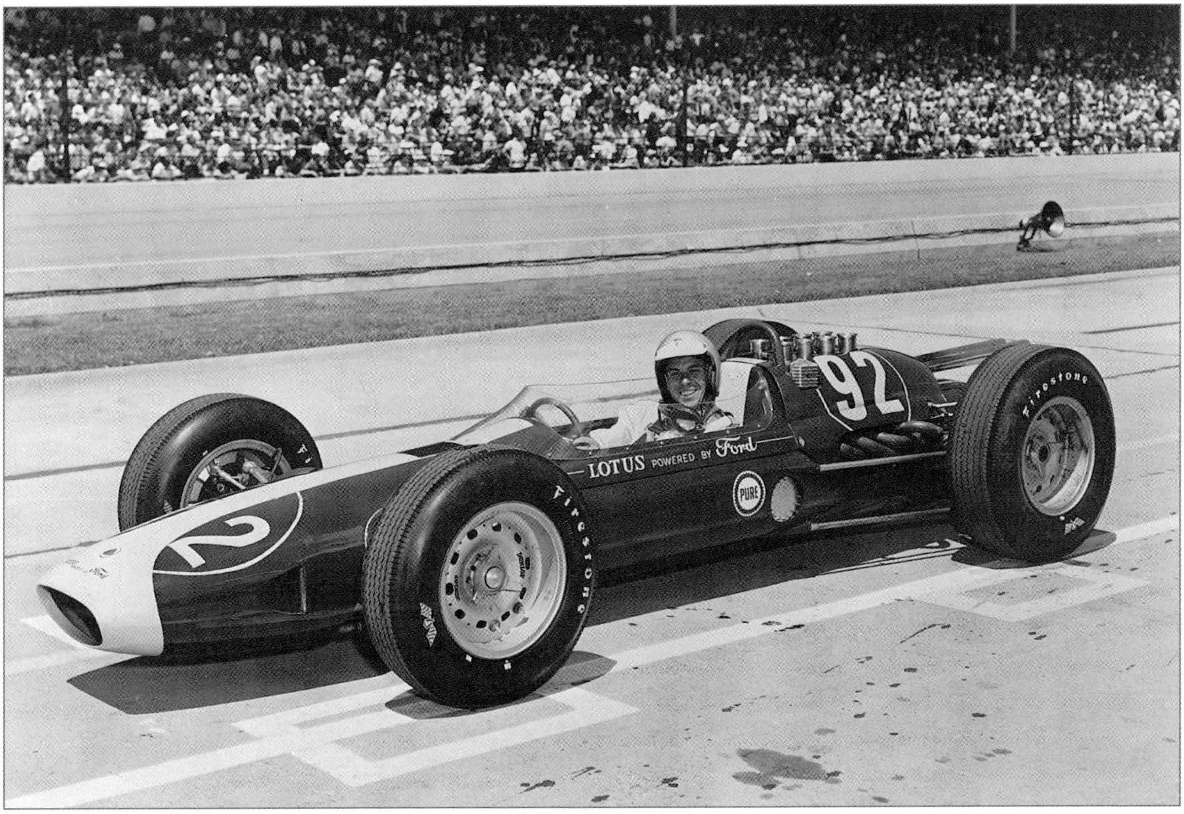

In a joint venture financed largely by Ford of America, Team Lotus brought the Lotus 29 to Indianapolis in May 1963 for the Team's first attempt on the Indy 500. Had it not been for Rufus Parnelli Jones's winning Agajanian Special leaking oil more profusely than should have been allowed, Jim Clark would have won. As it was, he came second and Dan Gurney finished seventh.

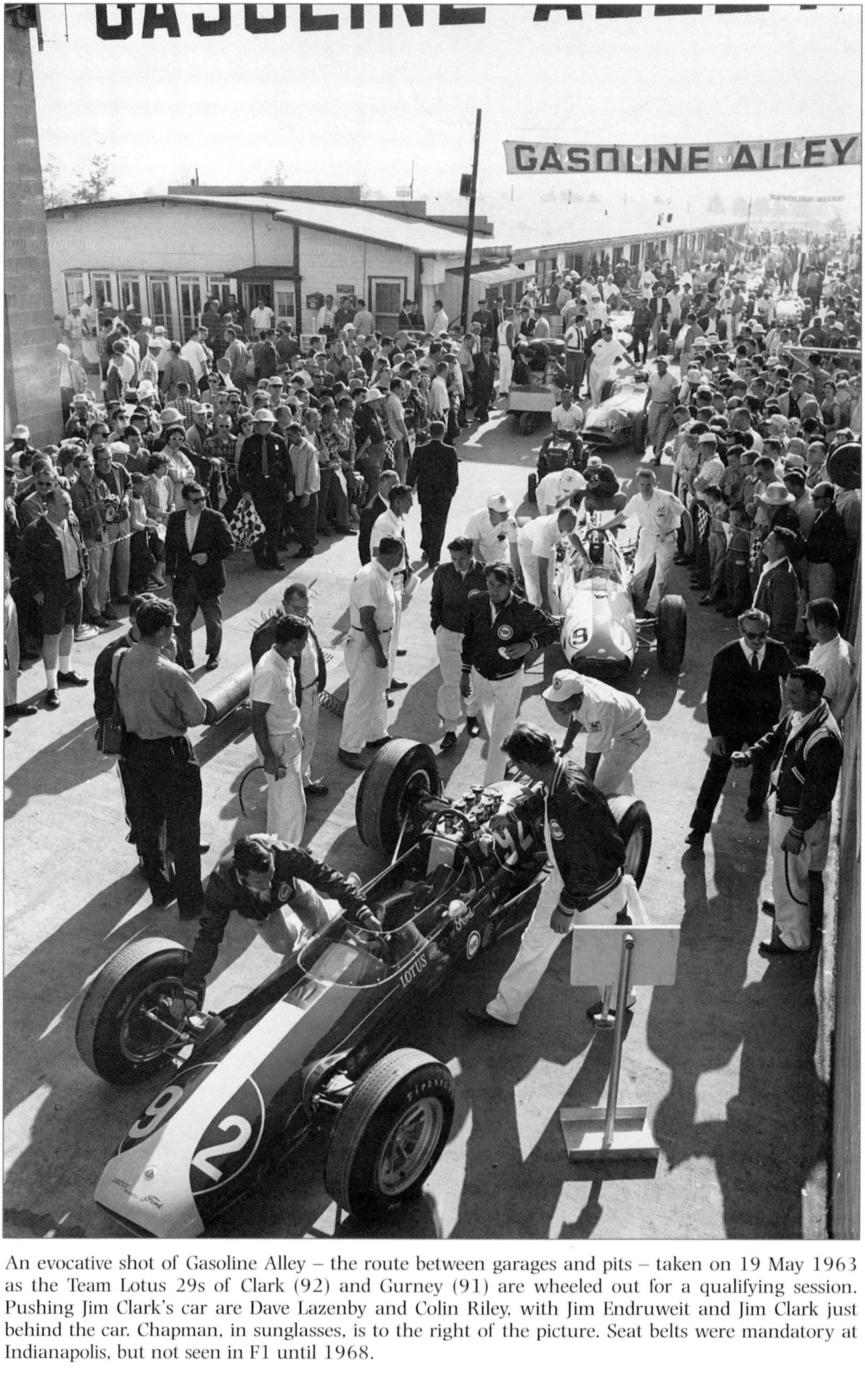

An evocative shot of Gasoline Alley – the route between garages and pits – taken on 19 May 1963 as the Team Lotus 29s of Clark (92) and Gurney (91) are wheeled out for a qualifying session. Pushing Jim Clark's car are Dave Lazenby and Colin Riley, with Jim Endruweit and Jim Clark just behind the car. Chapman, in sunglasses, is to the right of the picture. Seat belts were mandatory at Indianapolis, but not seen in F1 until 1968.

Offenhauser Willard Battery Special, with Clark second and Gurney sixth. Clark had led during Parnelli's fuel stops, and there was predictable mayhem as Parnelli's serious oil leak was overlooked by the stewards, while Clark was delayed by yellow flags in his pursuit of the ailing leader. Despite the second place finish, Jimmy had bettered all previous Indy winners' times, establishing race records on the seventieth, eightieth and nintieth laps, and, predictably, set the fastest average speed ever for a 'Rookie' entry as well as for a rear-engined car. The Lotuses were running on petrol rather than alcohol fuel, and needed just one pit stop against the three or four of the Offenhauser specials. Later in the year Clark won from pole position at Milwaukee, while Gurney finished third. Although the 29s dominated the Trenton 200, both cars retired.

Dan Gurney's crashed Indianapolis qualifying car enjoyed a chequered life thereafter. It was sold to Lindsey Hopkins and was fitted with a Ford four-cam engine, and in the following year's Indianapolis 500, Bobby Marshman drove it as the Pure Firebird Special, setting fastest race lap and leading until sidelined by a transmission oil leak. The 29 was raced by Al Miller in the 1965 Indy 500 to fourth place, but in 1966, Miller was involved in the multiple pile-up at the start of the race.

Lotus's arrival on the Indy scene wasn't quite like the Beatles' invasion of America, but the team had thrown down the gauntlet and proved that the Lotus Climax was considerably quicker than the regular Indianapolis cars. Two years later, only a single front-engined car was to be seen on the grid at Indianapolis.

Corporate Reorganisation

In 1964, Team Lotus Ltd regrouped as four separate operations. Team Lotus was the F1 wing, Team Lotus Racing with Ford of Britain ran the Lotus-Cortina saloons in the UK and Europe, while Team Lotus Racing with English Ford Line (Dearborn) ran Cortinas in North America. Confusingly, Team Lotus Racing with Ford of America ran the Indianapolis programme and the Lotus 30 Group 7 sports racers in the USA. To further complicate matters, Team Lotus's works F2 cars were run by Ron Harris, and its sports racing cars by Ian Walker Racing. The racing was largely financed by sales of road and race cars, with accountant Fred Bushell traversing an intricate fiscal web, augmented by funds from the Esso petroleum company and bolstered by tyre contracts, until the bonanza of the sponsor's bankroll arrived.

Formula Junior was on its last legs, and customers having a final fling were offered the Lotus 27. This was an aluminium monocoque chassis with fibreglass panels riveted on, powered by a 1097cc Cosworth-Ford motor. Following the demise of FJ, the 27 was run by a few entrants in F3 with sleeved-down engines – or not, in Harry Stiller's case, as he was found to have an oversize engine at Monza. Von Frankenberg would have had a field day. It also featured as an F2 car before the Lotus 32 came out.

Len Terry had been back in the design office since September 1962, and he was mainly responsible for the refinements that led to the 25 evolving into the type 33, while all the other F1 teams struggled to get their own versions of the 25 up and running. Although the new 33 was seen at Aintree in April, Team Lotus ran the 25 for most of the season. It was a revised version of its groundbreaking model, known as the 25B, and debuted at Snetterton in March 1964. Suspension updates included lower rear radius rods and lower rear links, new uprights, and the car was now fitted with 13 inch diameter spoked magnesium wheels, retarded by smaller brake discs. The ZF gearbox and final drive unit had an improved linkage. Jim Clark preferred to use chassis R6, which

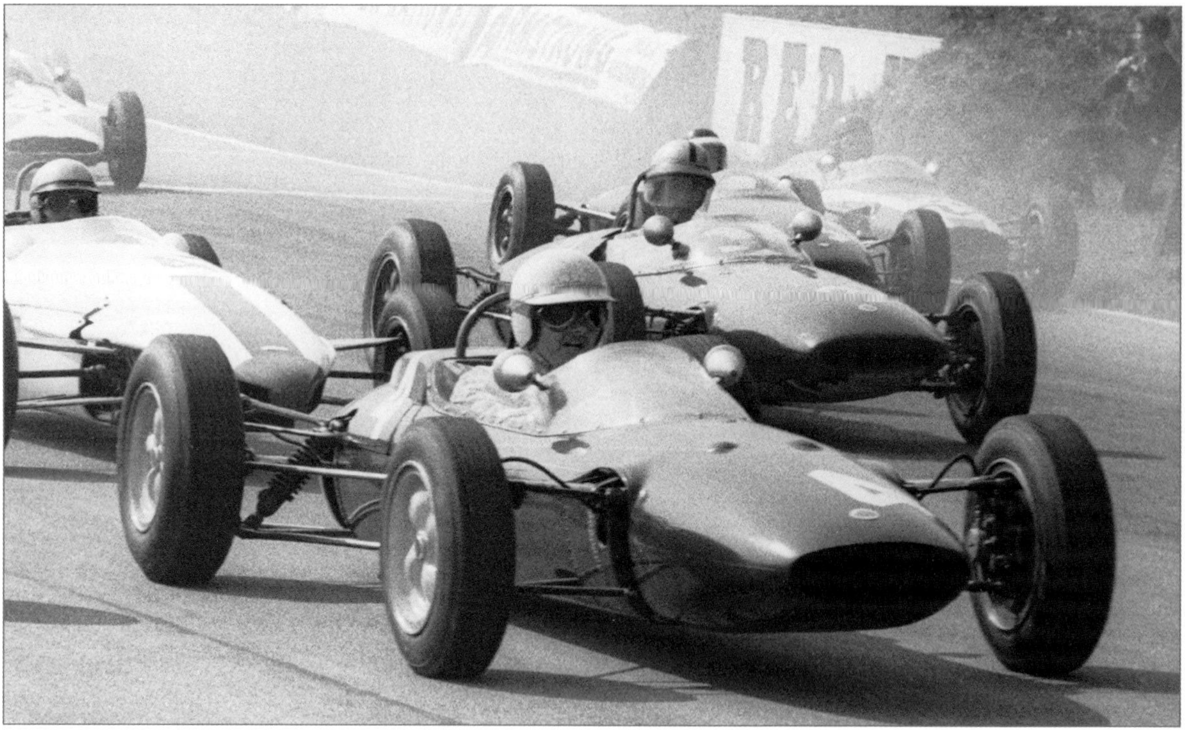

Lotus's offering for the final season of Formula Junior in 1963 was the 27, a monocoque tub with fibreglass outer panels pop-riveted on and fabricated steel bulkheads at either end. It was powered by the Ford 105E engine raised to 1098cc. Here, amidst engine fumes and tyre smoke, John Hine's 27 heads the FJ field away from the Brands Hatch grid. The most successful 27 driver was Peter Arundell in one of the Ron Harris-Team Lotus cars, pipping Denny Hulme's Brabham for the FJ title by just one point.

incorporated these modifications, as well as having the top radius rod located on the hub carrier and this car was known as a 25D. It was subsequently fitted with Mercedes-Benz sliding-spline driveshafts, although these too were replaced by the original rubber couplings. Clark's successes in 1964 were wins at Zandvoort, Spa, and Brands Hatch, placing him third in the World title stakes. He would have taken the title but for a last-lap engine failure while leading at Mexico. Jim also won the non-Championship *News of the World* Trophy at Goodwood and the Solitude Grand Prix at Stuttgart. His Zandvoort performance prompted Firestone to enter Formula 1, starting the first of many tyre wars.

With Trevor Taylor relegated to the BRP Lotus 24 BRM squad, Jim's new team-mate for 1964 was Pete Arundell. His 25 was chassis R4, known as 25C because it was something of a compromise in having the original suspension but running the smaller wheels and modified gearbox. Arundell's season was marred by an accident in a high-speed F2 race at Reims, but he had previously been third at Monaco and Zandvoort. His place was taken by Mike Spence, who was sixth at Monza and fourth at Mexico. Clark and Spence gave the 33 its World Championship debut at the Nürburgring, while American Walt Hansgen was fifth at Watkins Glen in a 33. Towards the end of the season, Reg Parnell Racing bought the two Team Lotus 25s. They were fitted with BRM V8 engines and Hewland transmissions, and there was a fair amount of back-up support from the works. The 25 was a masterpiece, in so far

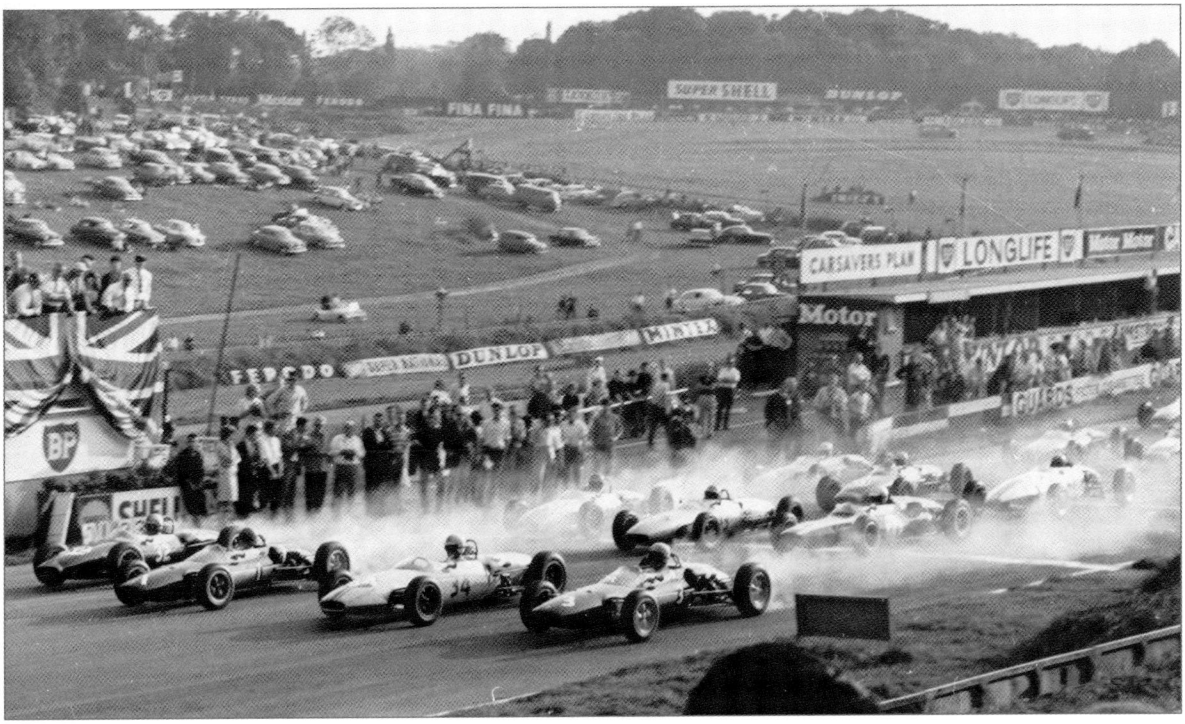

At the start of the 1963 Formula Junior final at Brands Hatch, the Team Lotus 27s of Peter Arundell (1) and Mike Spence (3) share the front row with the Brabham-Fords of Denny Hulme (32) and Frank Gardner. Other opposition consisted of Lolas and Coopers. Lotus was slow to gain successes with the 27 because of its complex structure, but Arundell eventually clinched the title.

as it was simple, aesthetically attractive, innovative and yet extremely effective. The 33 that superseded it was in many ways an evolved 25, and the 33's chassis numbers followed on consecutively from the 25 line.

Racing with Elan

The Elan took the sports car world by storm in 1962, and was a much more viable commercial proposition than the Elite had been. It was fitted with the Harry Mundy-designed twin-cam version of the Ford 1500 GT engine, rationalised to 1558cc from May 1963. These wonderful little cars were agile, powerful and fun to drive, and were instantly taken to the race track. However, Lotus hadn't done any race development on them at that stage, and they needed firming up a good deal before they worked as racers. Accordingly, Lotus produced fifty-two units of its 26R competition version in S1 form, and a further forty-five cars in S2 form. The 26R Elan had a lighter fibreglass body, modified steering and suspension, and delivered some 140bhp. A lot of the development work was carried out by Graham Warner's Chequered Flag garage, and Warner's car, which bore the same LOV 1 registration as his Elite, was also driven by Mike Spence and Jackie Stewart. Others were run by Ian Walker Racing, John Willment's Team for John Miles – using a 158bhp BRM-built version of the Lotus engine – and SMART – the Stirling Moss Automobile Racing Team. Subsequently a host of privateers campaigned race-prepared Elans in Club events well into the 1970s.

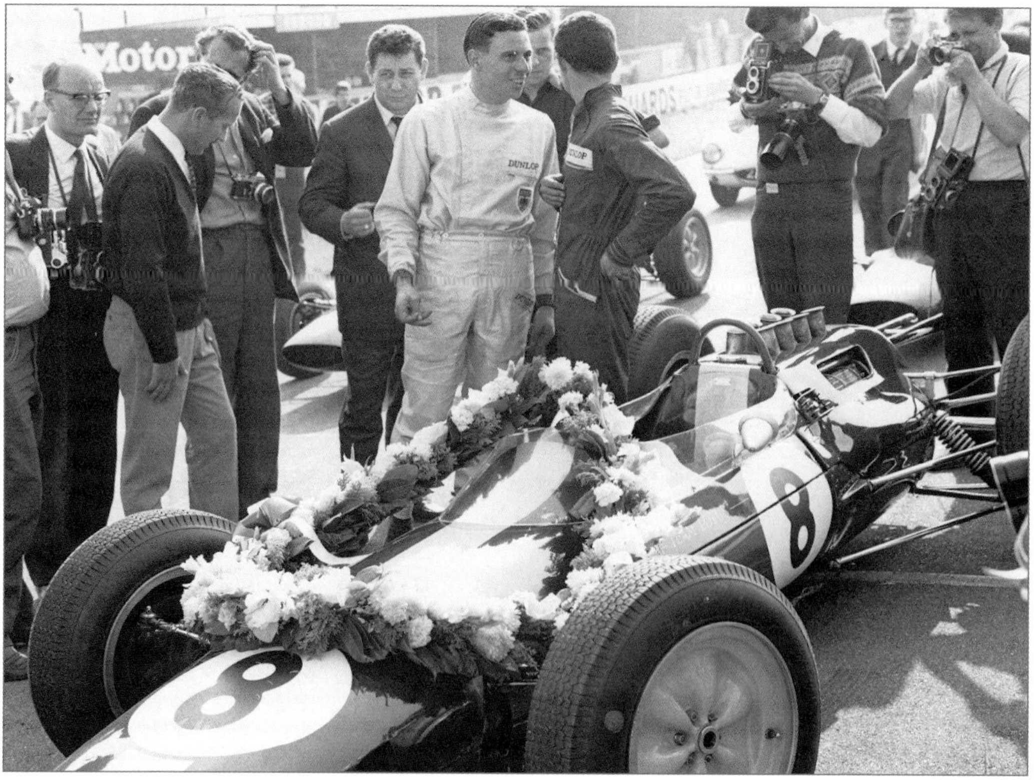

Lotus had such a good year in 1963 that a celebration was staged at Brands Hatch at the end of the season, featuring a display of the most successful racing cars. Here is the garlanded World Championship winning 25, with Jimmy talking to his mechanic Cedric Seltzer, and Colin in discussion with photographers to the left of the picture.

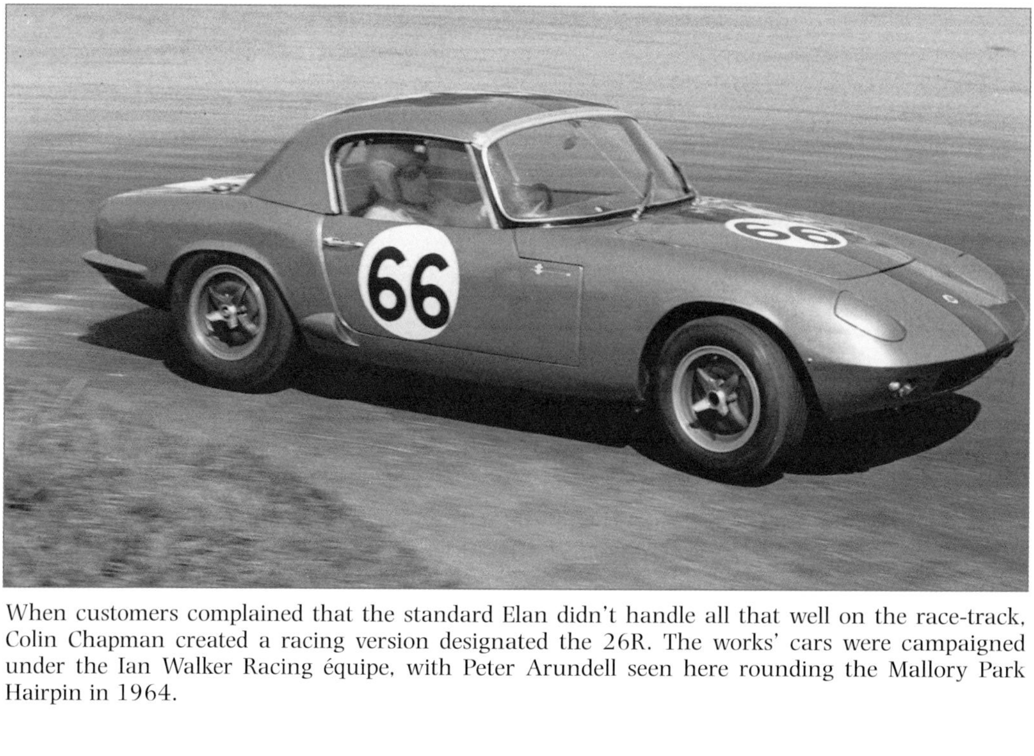

When customers complained that the standard Elan didn't handle all that well on the race-track, Colin Chapman created a racing version designated the 26R. The works' cars were campaigned under the Ian Walker Racing équipe, with Peter Arundell seen here rounding the Mallory Park Hairpin in 1964.

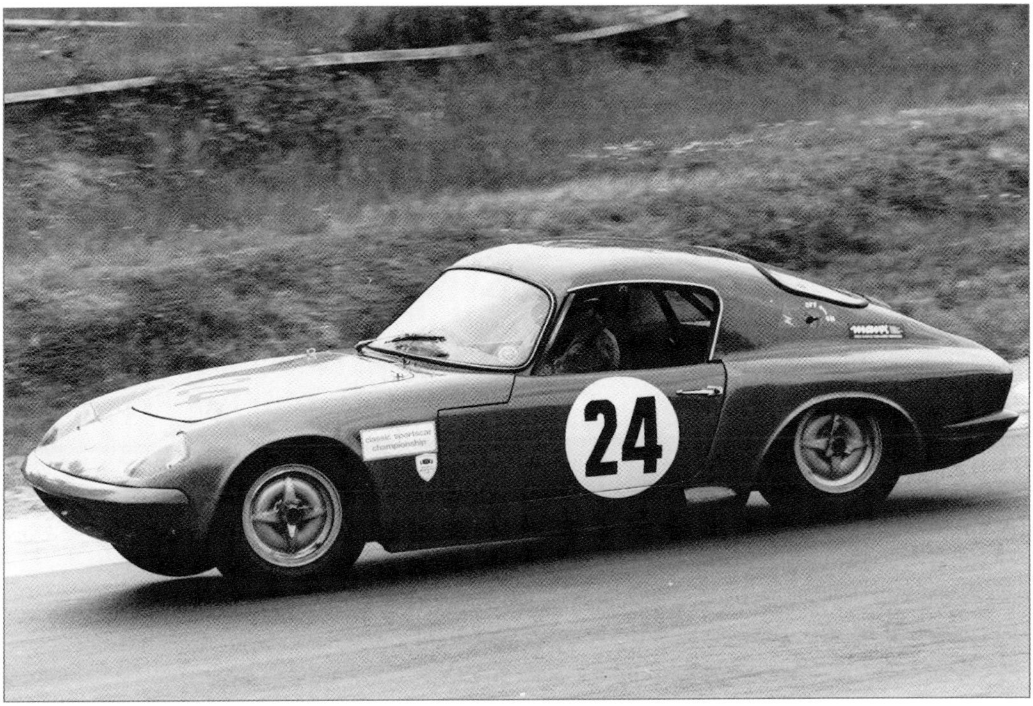

An elegant GT version of the Elan was produced by Ian Walker Racing, bonding on a Shapecraft fastback section, and mounting the headlights under perspex fairings. It is seen here cocking a front wheel and with the rear suspension under load at Brands Hatch in the hands of hillclimber David Good in 1965.

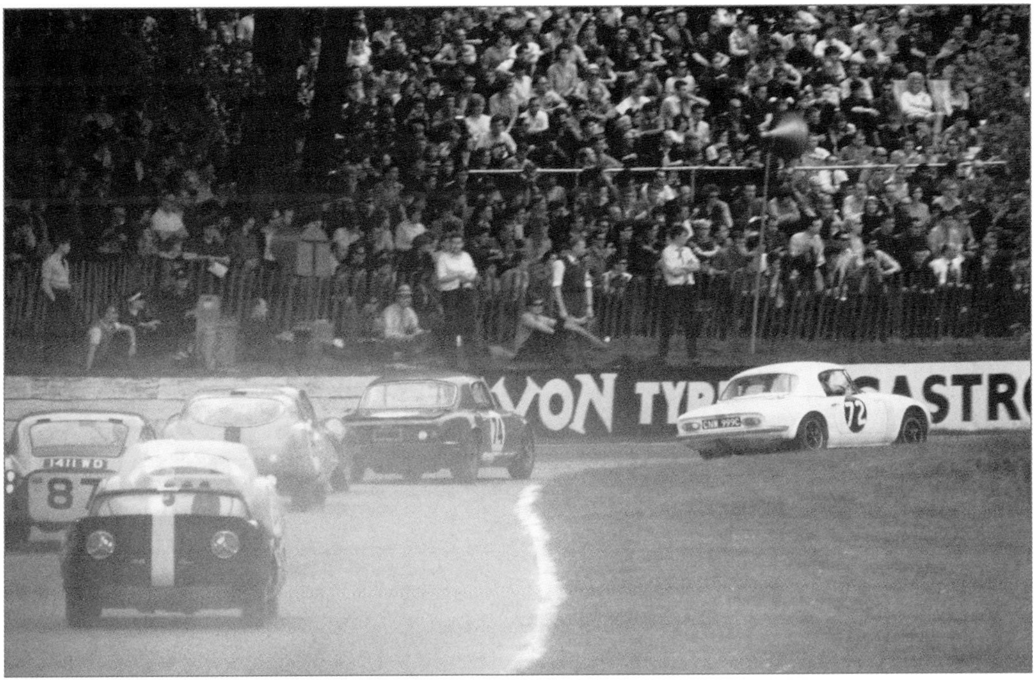

One of the supporting events at the annual International Formula 2 races at London's Crystal Palace on Whit Monday 1965 was the BRSCC Grand Touring Car thrash, with the Elans of Jackie Oliver and Eric Liddell leading an assortment of small GTs including a pair of Lotus Eleven GTs. The Elans were eventually split by the Alfa Romeo GTZ of Boley Pittard.

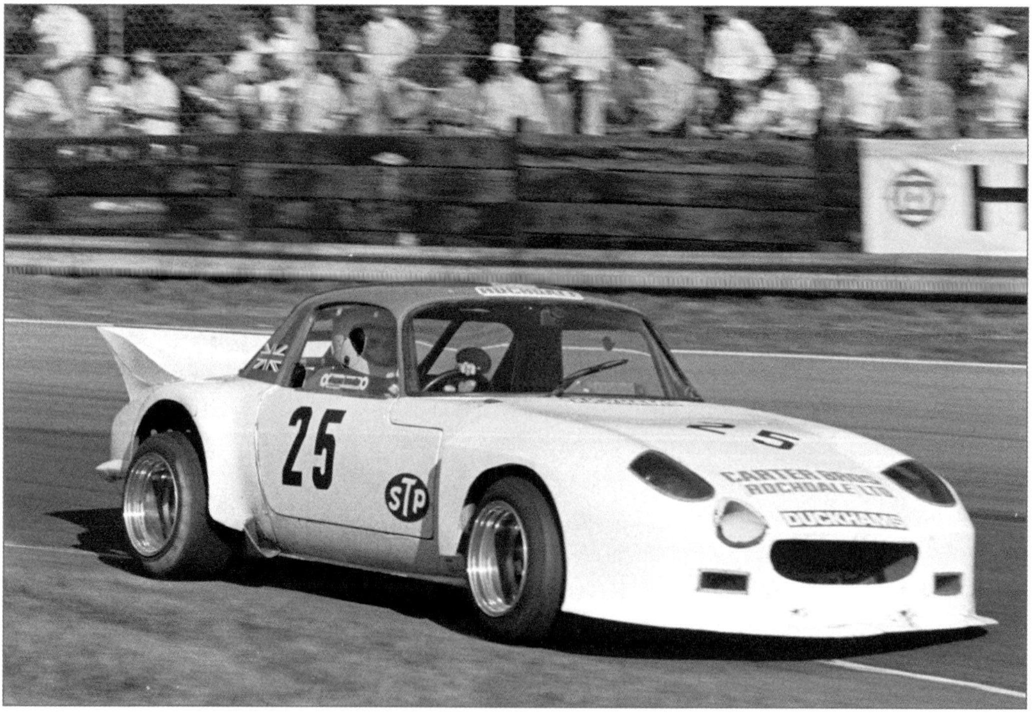

By 1970 the Elan had developed into a caricature of its former self, and Jon Fletcher's car has removable front and rear body sections and carries air dam and splitter, cooling ducts and whale-tail spoiler.

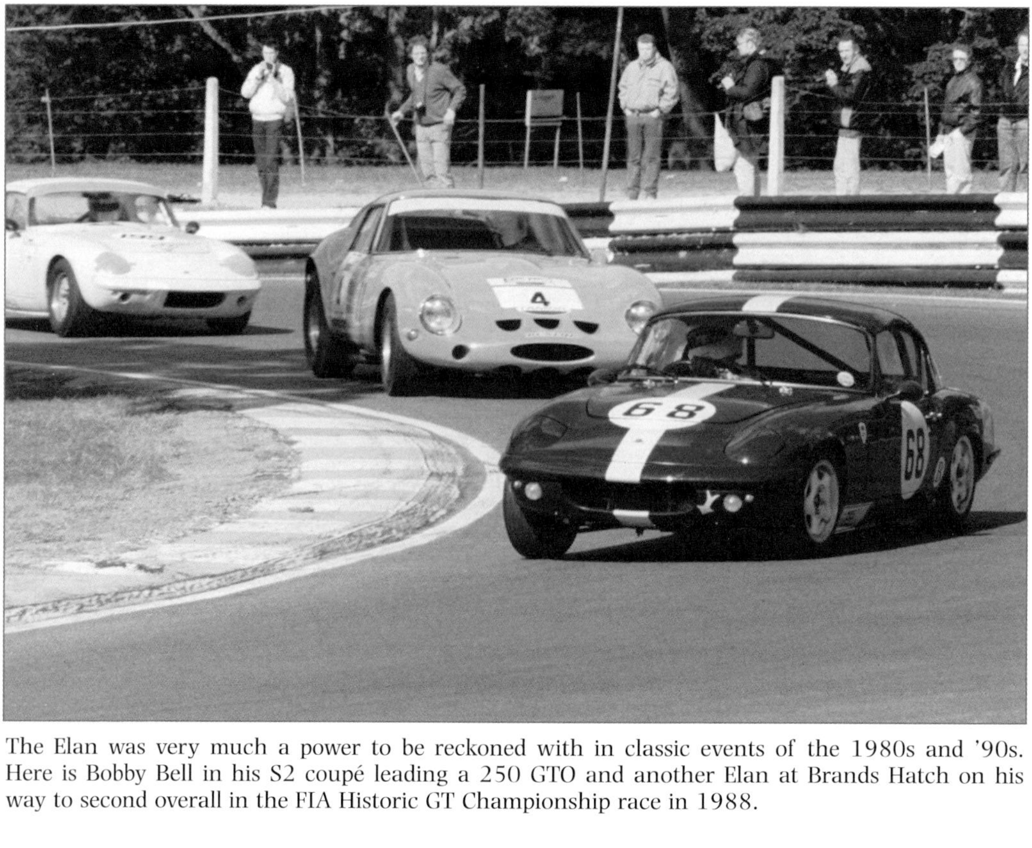

The Elan was very much a power to be reckoned with in classic events of the 1980s and '90s. Here is Bobby Bell in his S2 coupé leading a 250 GTO and another Elan at Brands Hatch on his way to second overall in the FIA Historic GT Championship race in 1988.

CHAPTER SIX

THE LOTUS-CORTINA

Ford's marketing strategy in the early 1960s was highly geared towards the lucrative family saloon market, and in order to publicise the new Cortina, a souped-up racing model was required. The venture was entrusted to Colin Chapman, and the Lotus-Cortina was fitted with Ford's 1499cc 116E engine with its five-bearing crank. The Harry Mundy design retained the bottom end of the Ford unit, while its twin overhead camshafts were driven by chain from the nose of the crankshaft, and it breathed through a pair of side-draught twin-choke 40 DCOE Webers. Engine size was soon rationalised to 1558cc, and power output of the racing versions was 140bhp, taking it to a top speed of 128mph. The first engines were built by Cosworth, and later by BRM. The twin-cam was mated to the close-ratio Ford four-speed all-synchromesh gearbox as found in the Elan. Initially the rear-end geometry featured a tubular A-bracket and radius arms, which resisted the twisting movement and created an oversteering car, and to make it work, Len Terry substituted channel-section components with the A-bracket located in ball-joints in

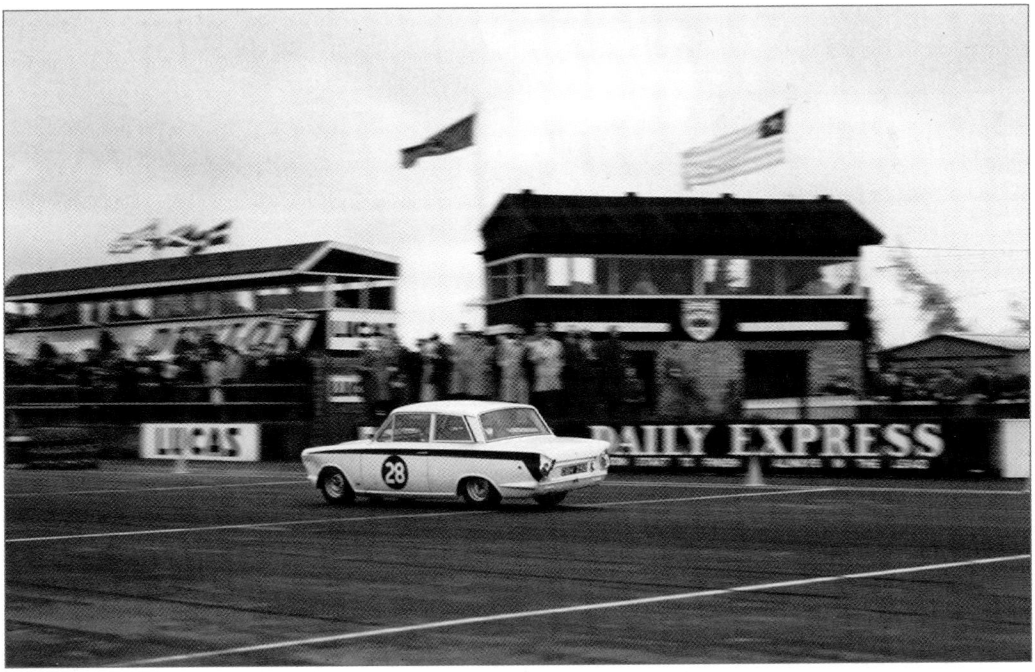

The Lotus-Cortina was homologated as a Touring car by the end of 1963 and a GT car in 1964. It was light, fast and agile, and could trounce the previously dominant Jaguars, especially on a tight circuit. Private entrants were quick to start racing the car, and this is D.B. Haynes at the International Trophy meeting at Silverstone on 2 May 1964. The typical Lotus colour scheme hasn't been extended across the rear panel, and the nearside front wing shows signs of damage.

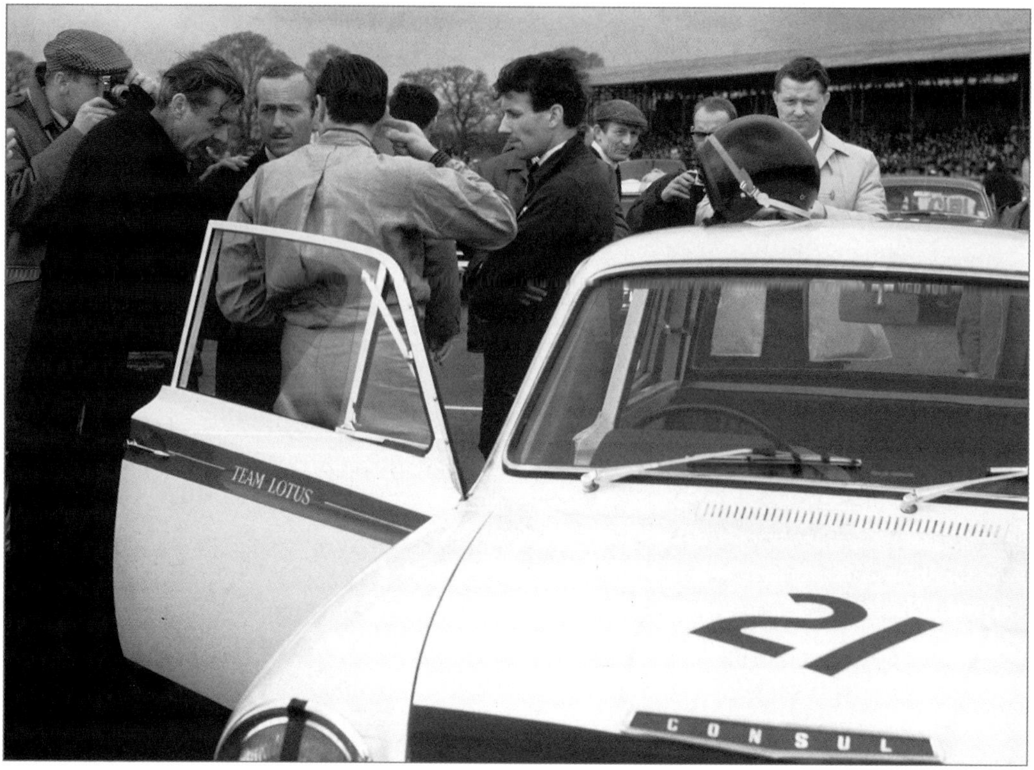

Startline conversation between Jimmy Clark, Colin Chapman and Chief Mechanic Bob Dance (closest to car) before the International Production Touring Car twelve-lapper at Silverstone on 2 May 1964. Note the absence of a roll cage. Team Lotus encountered certain scrutineering problems in the early days of the Cortina's homologation as a racing car. On one occasion, it was disqualified for having a corrugated cardboard backing to the rear seat – slightly stiffer than the standard flat cardboard!

the same way as his Terrier Mk 2 sports racer. They also had a rear anti-roll bar. There was a limited slip diff and a more direct steering-gear ratio. The power-to-weight ratio was improved by fitting aluminium door skins, bonnet and boot lid, as well as light alloy castings for the clutch housing, remote control gear shift extension and differential housing. Road cars also received different interiors. Upgrades to the spec included a two-piece propshaft in July 1964 instead of the original one-piece item, and the light alloy transmission casings were replaced by standard Ford cast iron, while the Cortina GT gearbox with an uprated second gear cluster was used instead of the Elan gearbox. In October 1965 the Lotus-Cortina got self-adjusting rear brakes and Corsair 2000E gear ratios. Safety considerations were minimal in the first instance. There were no roll-over bars initially, and no firewall between the cabin and side-to-side long-range fuel tank. Front hubs used to break in the early days, and wheels would fly off.

Lotus Components made ninety-seven racing Lotus-Cortinas, and although fewer than the necessary number of cars had been built, the Lotus-Cortina was homologated as a Group 2 car in September 1963. Its first international appearance in Team Lotus hands was at Oulton Park in the Gold Cup support race, when Jack Sears and Trevor Taylor came third and fourth, beating the once invincible Mk II

Jaguars in the process. The following year – 1964 – Team Lotus went racing in the UK and the red and gold Alan Mann Racing versions competed in Europe, while they were run by Ford in the States. Early successes included Jim Clark's victory in the British Saloon Car Championship. Jackie Stewart/Mike Beckwith won the rain drenched Marlboro 12 Hours race in the USA. The Alan Mann cars of Sir John Whitmore/Peter Procter and Henry Taylor/Peter Harper were first and second in the Brands Hatch 6-Hours. There were myriad successes in Europe, Australia, New Zealand and South Africa. For example, the Boreham-built Lotus-Cortina of Vic Elford/David Seigle-Morris came fourth and won its class in the 1964 Tour de France. The Willment team run by Jeff Uren raced Lotus-Cortinas to promote its tuning equipment business and car sales, with Frank Gardner driving.

The Team Lotus works' project was originally run by Aussie Ray Parsons, who was also responsible for preparing and driving the Team Lotus 26R Elan. When he was dispatched to the States to run the so-called English Ford Line Lotus-Cortinas there, Bob Dance became Chief Mechanic on the Team Lotus operation, concentrating on the British Saloon Car Championship. Bob had joined Lotus Components in 1960, having built his own 750 special and he was originally employed on assembly of the 'queerbox'. Working with Steve Sanville under Mike Costin on gearboxes and twin-cam engine development, Bob was brought into the Cortina programme via the twin-cam engined Lotus 23. When the 23 was banned

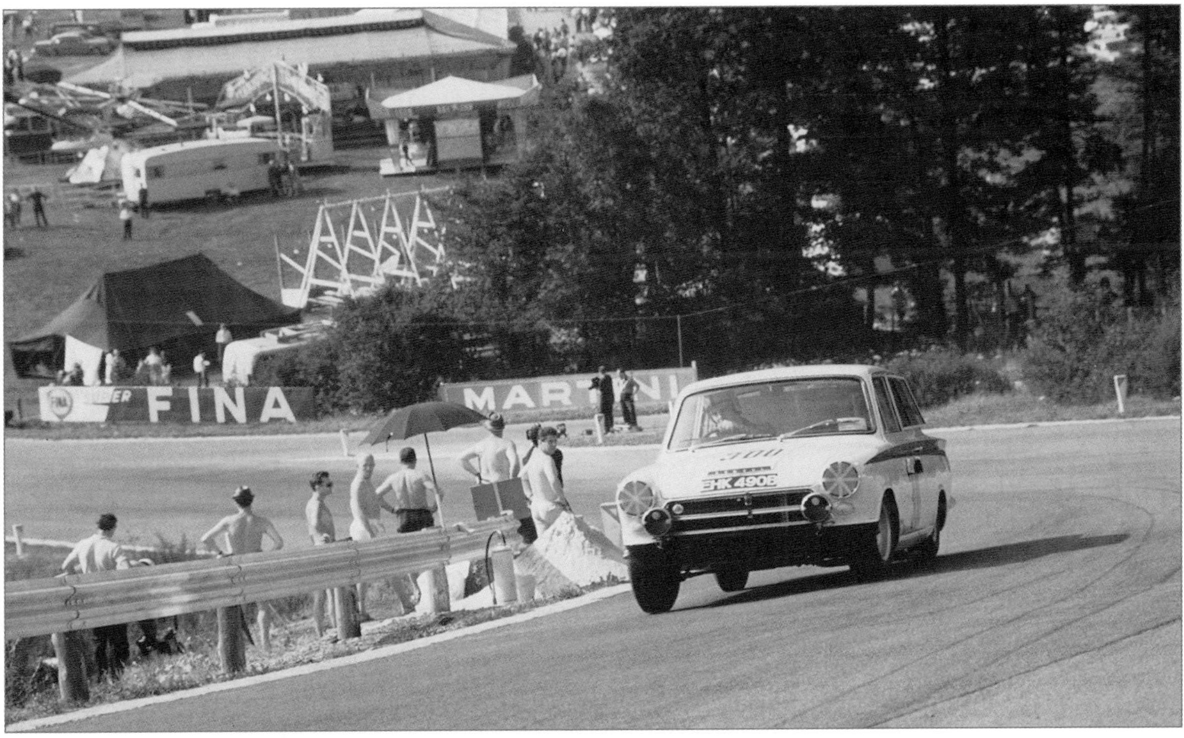

Its suspension working overtime – the Henry Taylor/Peter Harper Lotus-Cortina powers through Eau Rouge at Spa-Francorchamps in 1964. Alan Mann dispatched John Whitmore to act as the hare to break the Mercedes-Benz of Eugen Böhringer and the pair had a magnificent race for the first hour and a half, swapping the lead two or three times a lap. When Whitmore pitted, he got a standing ovation from the packed grandstands.

from Le Mans, Jabby Crombac borrowed a twin-cam engined 23 to 'show them what it could do' in a few French races, and Bob Dance went along as the mechanic. From there he was posted to the Lotus-Cortina programme.

Bob revealed his records of the modifications made to the Cortina to turn it into a racing car. The list is extensive, even if some of the items are small and mundane, like the Dzus fasteners to keep the bonnet shut. He explained that the original scheme was to homologate the car in 1963 in order to race against the Jaguar Mk IIs. To qualify as a Touring car, they had to make 1,000 units, which they struggled to do at first. By 1964 they were homologated as GT cars as well as Touring Cars. Although their main rivals were the Alfa Romeos, they could often match the big Falcons, Mustangs and Galaxies on circuits where handling counted for more than straight-line speed, and the BMW 1800s in the class above stood little chance.

'We were a small group of people,' said Bob. 'There was only one mechanic per car. Chapman would almost always be there, and maybe Jim Clark's girlfriend or Jack Sears' wife would do the timing. It was also a dynamic package, morale was good, and Chapman's charisma rubbed off on people. He could get the best out of people. Team Lotus were innovators in racing, with Chapman, Mike Costin, Len Terry and later Maurice Phillippe, and there was a lot of determination to do well, and there was always a buzz about it. If you had Jim Clark driving and Chapman as engineer, you could go out and expect to win. We worked hard, mind you, but we also had a lot of fun.'

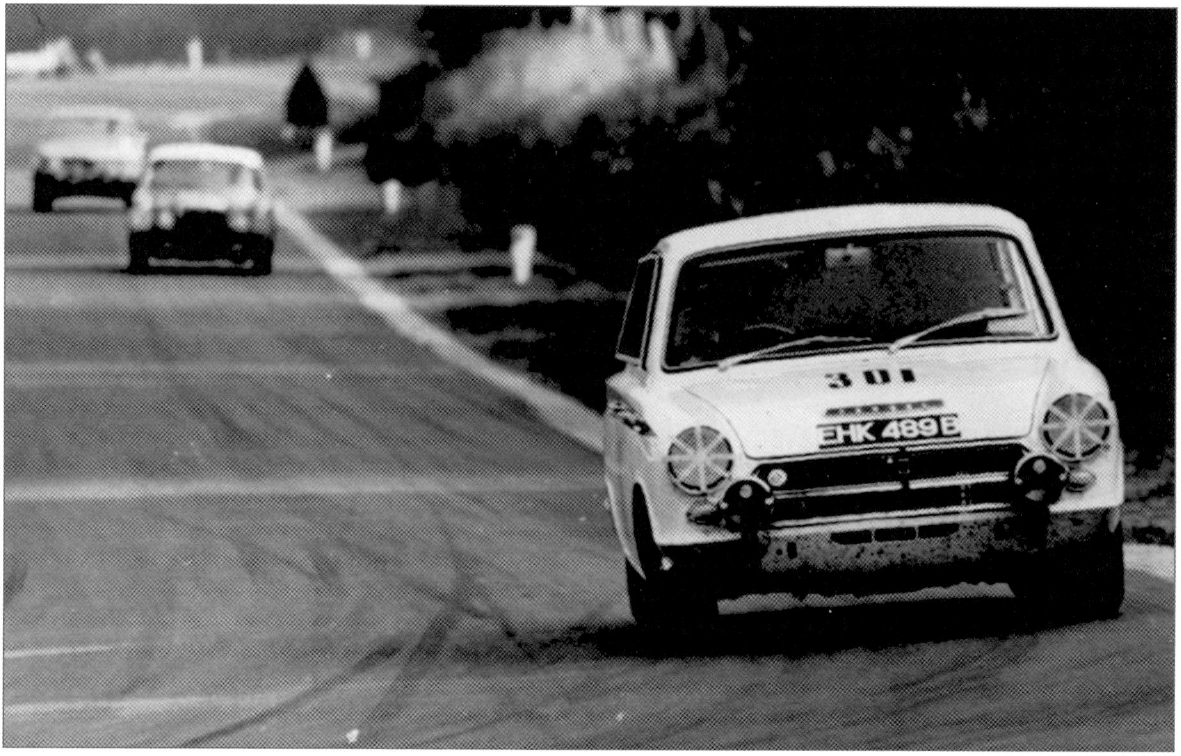

Spa 1964, John Whitmore's Lotus-Cortina approaches La Source hairpin ahead of Böhringer's Mercedes and a BMW.

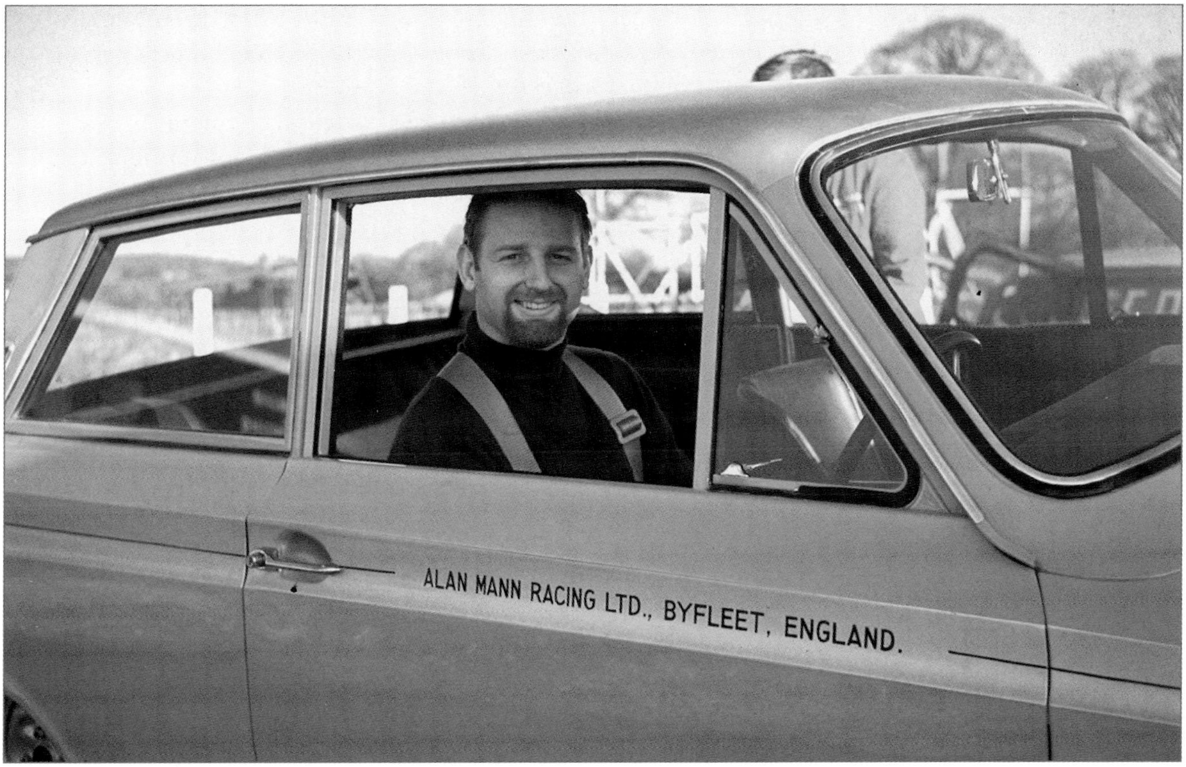

Sir John Whitmore was European Touring Car Champion in an Alan Mann Racing Lotus-Cortina in 1965, and was always a spirited contender. He had victories at the Nürburgring 6-Hours (with Sears), Mont Ventoux Hillclimb, Zolder and Snetterton, and also drove for the US English Ford Line Operation in the Indianola and Marlboro 12-Hours in 1964. By 1966 the Alan Mann cars of Stewart, Ickx and Whitmore were locked in combat with the Alfa Romeo GTA coupés of Rindt, Galli and De Adamich. The Autodelta squad protested against the Alan Mann cars of Whitmore and Ickx after Whitmore's Zolder win, but the red and gold cars were found to be perfectly compliant. However when Rindt's third-placed Alfa was stripped down, the scrutineers disqualified it.

'We were more fortunate than most teams, as we had an arrangement with Ford to take the cars to the local Ford dealership, like Quicks of Chester if we were racing at Oulton Park. There were no facilities at the circuits in those days of course. But we were always put up in hotels and we had two nice transporters as well – Ford chassis with six-cylinder diesels and two-speed transmission, built by Marshalls of Cambridge with no expense spared. One had a crew cab with seven bunks and carried just one car with a little workshop space, and the other carried two cars. We drove them day and night if we were racing abroad. In Formula 2 a typical journey would have been from Sweden to Sicily, and there were no motorways then. Morale was good, and luckily we all got on well.'

Team Lotus drivers in the British series in 1964 were Jimmy Clark, Mike Spence, David Hobbs, John Whitmore, Peter Arundell, Jackie Stewart and Ray Parsons, and Colin Chapman did some of the test driving. In the states, Trevor Taylor, Dan Gurney, Pete Lovely, Mike Beckwith and David Hobbs were the drivers – a very talented squad. Each driver had the same car – Bob was Clark's mechanic in 1964 – and Jim always drove the same car unless he was racing elsewhere, when Hobbs or Stewart drove the car. The same arrangement lasted through '65 and '66, and for

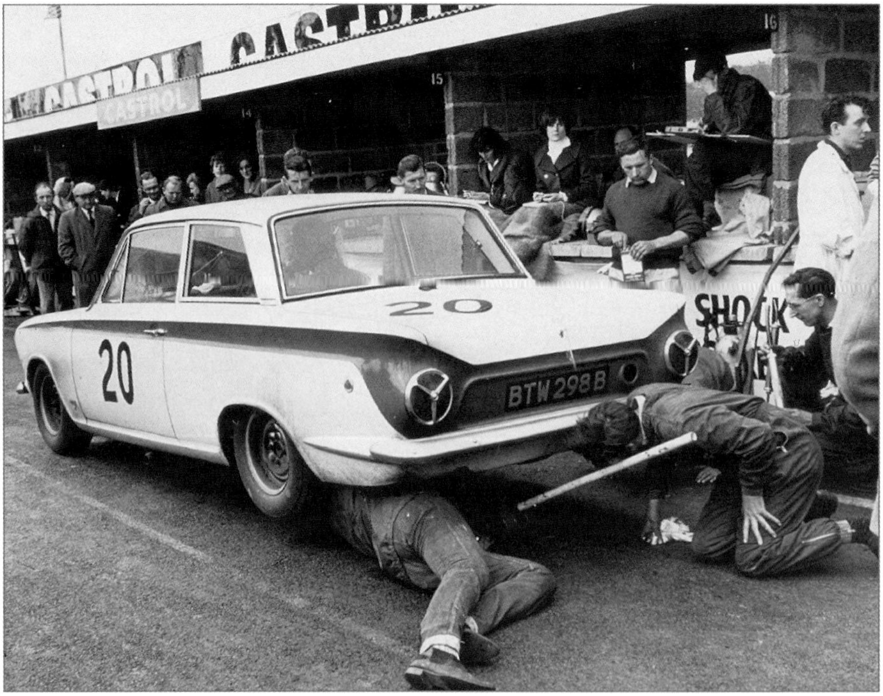

The Henry Taylor/Peter Harper Lotus-Cortina in the Brands Hatch pits during the 1964 European Touring Car 6-Hours. The A-Frame rear suspension stressed the differential to such an extent that it had to be regularly topped up. It finished second to the sister car of Whitmore/Procter but Sir John remarked that the diffs wouldn't have lasted another half an hour

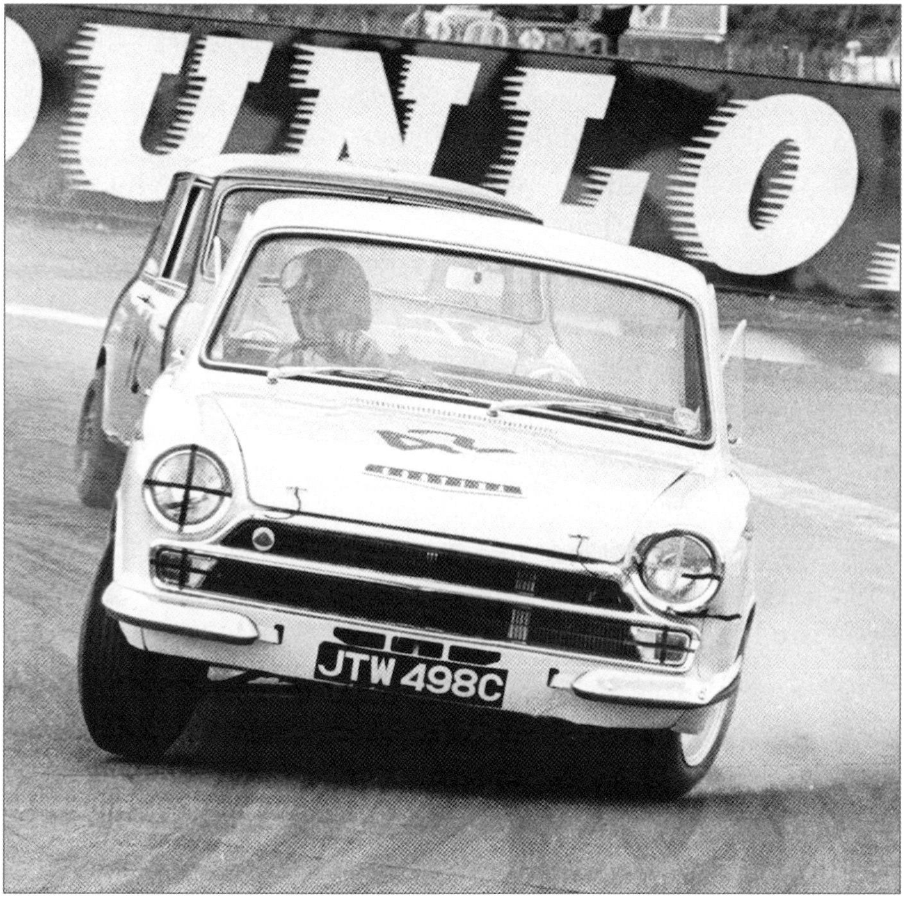

Jack Sears gets the tyres smoking on his Team Lotus-Cortina at Snetterton on his way to third in class at the Autocar International Trophy Formula 2 meeting on 10 April 1965. (This is the tight right-hander at the end of the old Norwich straight, now the site of the flourishing Sunday market.)

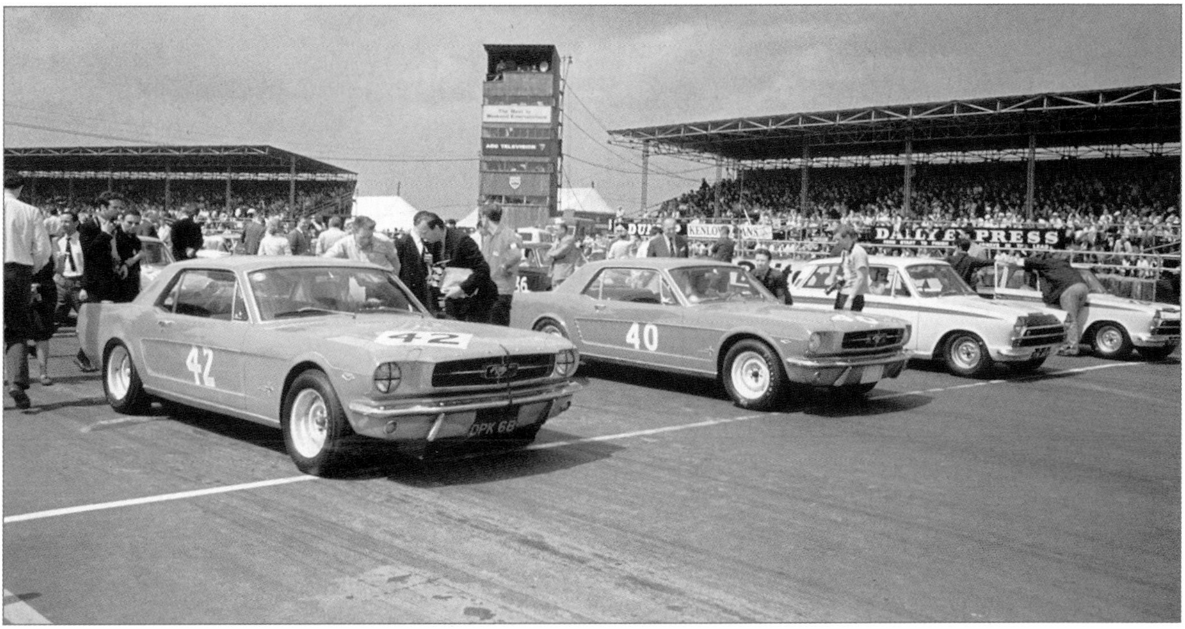

By 1965, big American metal was coming to the fore in British saloon car racing. This is the grid at Silverstone on 15 May 1965, with the Mustangs of Roy Pierpoint (on pole) and Sir Gawaine Baillie flanked by the Team Lotus-Cortinas of Mike Spence and Jack Sears. Pierpoint won from Baillie, with Sears third and Spence fourth, and Frank Gardner was fifth in the Willment Lotus-Cortina. The works Lotuses were first and second in class.

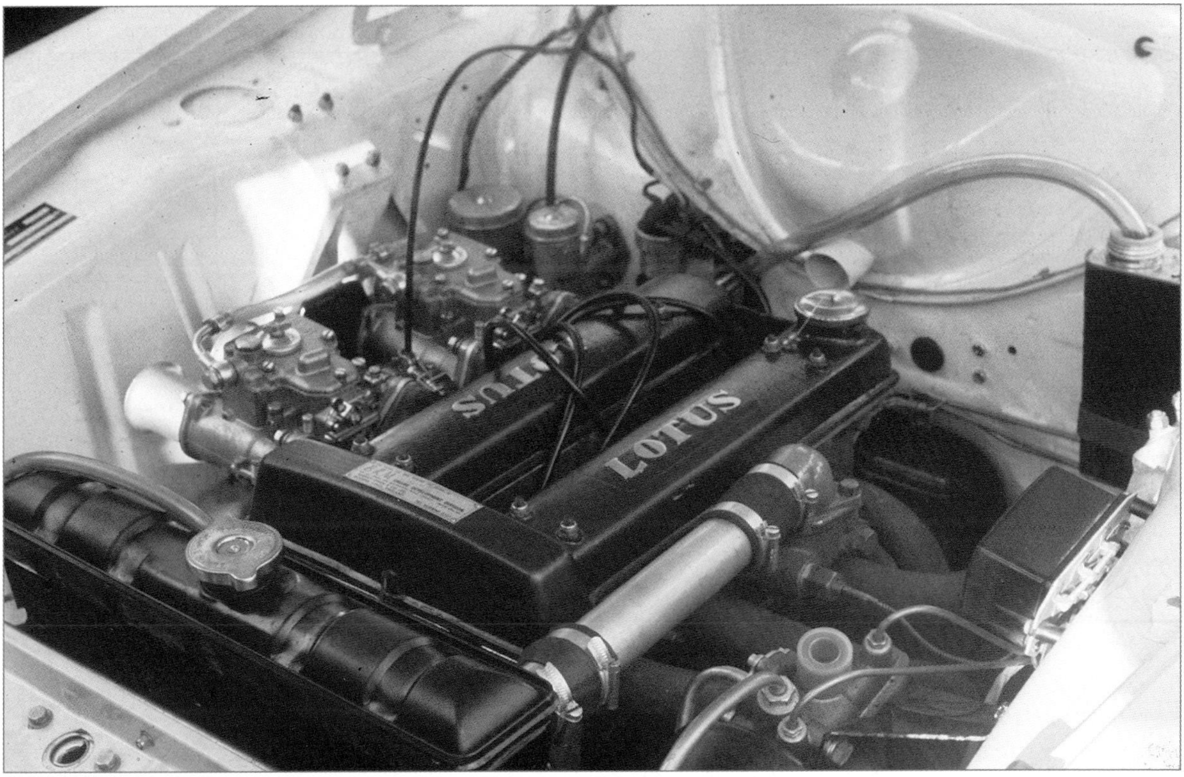

This is a 1965-spec BRM-built Lotus twin-cam with side-draught Webers, fitted in a 1966 Lotus-Cortina shell. Note the 1-quart oil can used as a catch tank on the right and the brake servo at bottom right. It has modified front strut mountings, as raced in the 1966 Group 5 regulations, with the old mounting point blanked off at top left.

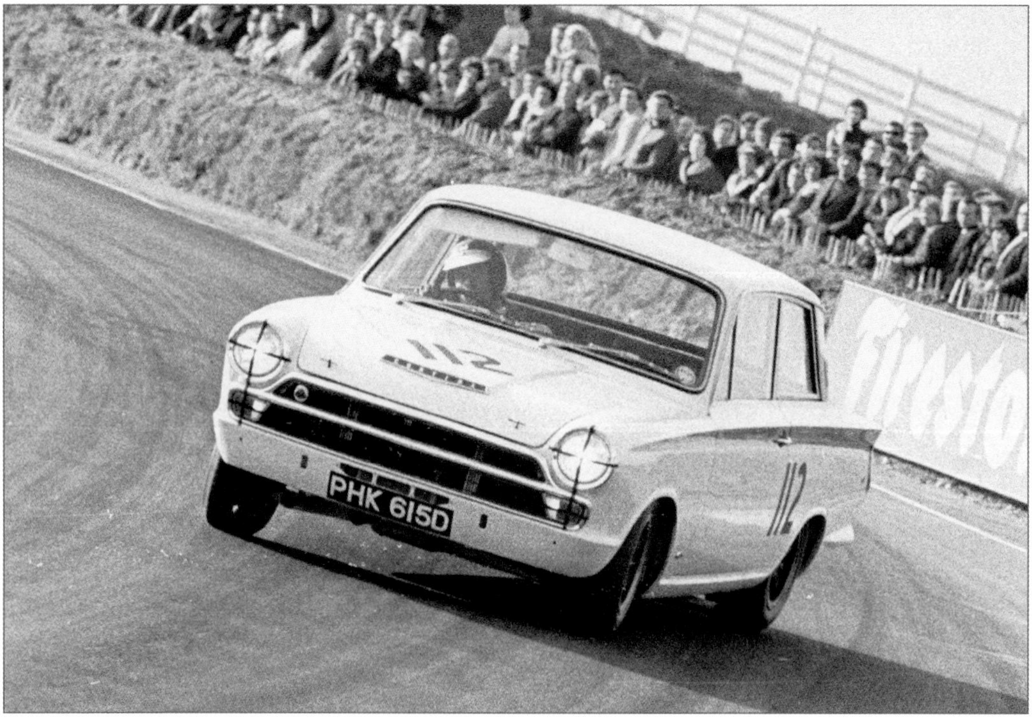

Team Lotus drivers almost always drove the same car, and this is Jimmy Clark's brand new model, PHK 615D, headlights ablaze and drifting nicely through Snetterton's Esses. The date is 8 April 1966, the meeting is the BRSCC International Scott-Brown Trophy race, and Jimmy won his class. The car was also driven by Arundell and Ickx during 1966 and by Chapman in testing. Its major success was Clark's win at the Guards Trophy race meeting. Clark was British Saloon Car Champion in 1964.

1967 Bill Cowe, who had been a Formula 1 mechanic, took over as Lotus-Cortina chief mechanic on the Mk 2 model when Bob went off to run the Lotus Formula 2 cars. By this time the cars were extremely sophisticated, and in 1966 Jim Clark remarked to Bob Dance that 'You've taken all the fun out of driving them!' The suspension was far removed from the 1964 cars, and he wasn't able to do his two-wheeled stunts around Brands Hatch any more. Such antics were manifest in a snowball fight at Ford's post-season party at Cortina d'Ampezzo to celebrate Sir John Whitmore's victory in the European Touring Car Championship, where Jimmy slipped a disc and put himself out of the forthcoming Rand Grand Prix. And as if to demonstrate that racing drivers' eyes are not connected to their brains, Whitmore took a Lotus-Cortina down the vertically banked bob sleigh run.

I asked Bob Dance why Team Lotus abandoned the A-bracket rear suspension set up. 'We didn't have as much trouble with the A-bracket breaking or mountings pulling off the diff as some teams did. But ours were stronger and the fabrication welding was probably higher quality. Meanwhile, Alan Mann worked away on the leaf-spring set up. We weren't in direct competition with him – he was racing in the European Touring car challenge, and we were both working for Ford at the end of the day. We raced against the Willment cars though.' So in 1965 Team Lotus followed the Alan Mann route and the cars were fitted with the less complex leaf spring rear end, while the engines were built by BRM, identified by their dark green cam covers.

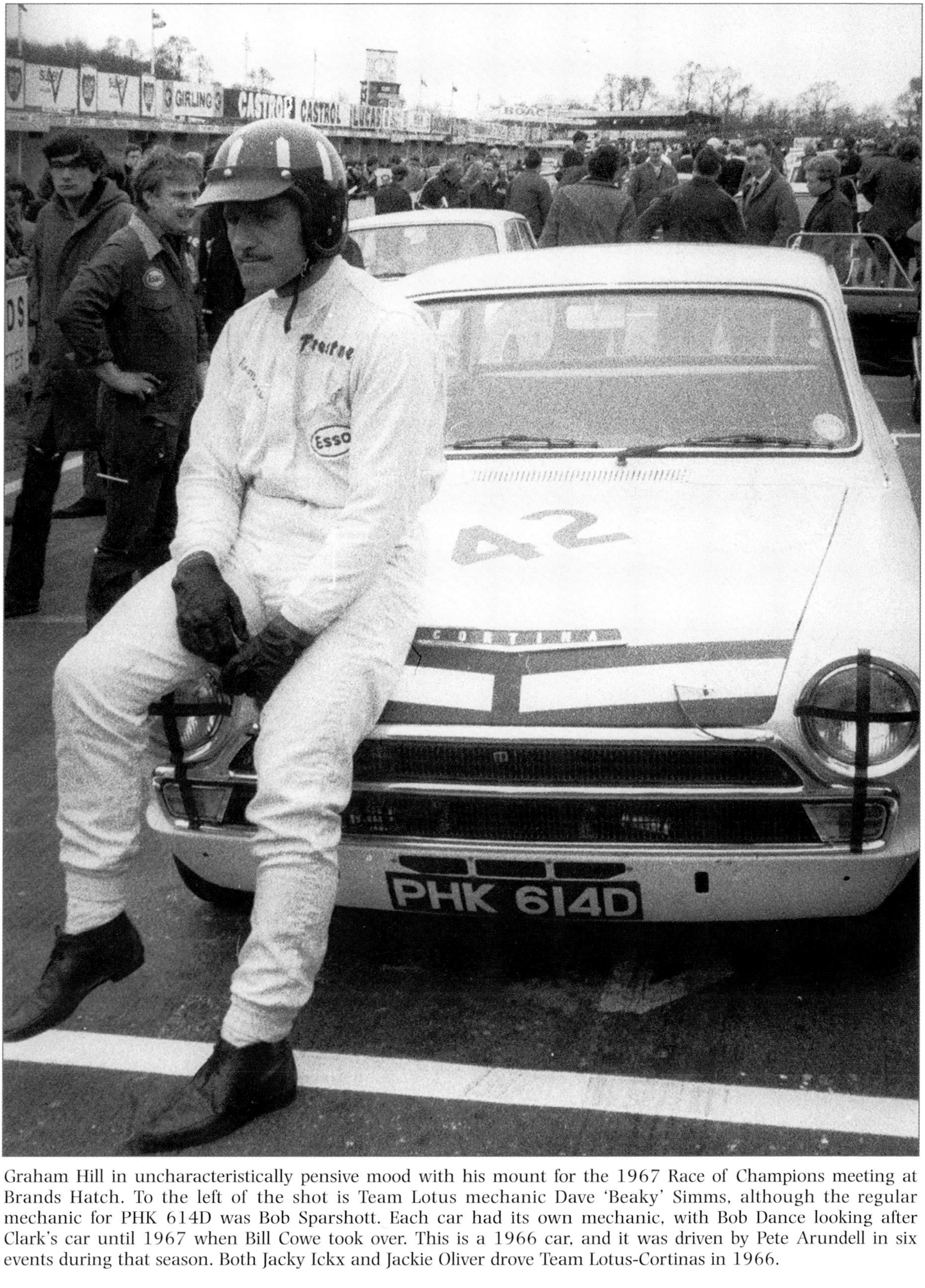

Graham Hill in uncharacteristically pensive mood with his mount for the 1967 Race of Champions meeting at Brands Hatch. To the left of the shot is Team Lotus mechanic Dave 'Beaky' Simms, although the regular mechanic for PHK 614D was Bob Sparshott. Each car had its own mechanic, with Bob Dance looking after Clark's car until 1967 when Bill Cowe took over. This is a 1966 car, and it was driven by Pete Arundell in six events during that season. Both Jacky Ickx and Jackie Oliver drove Team Lotus-Cortinas in 1966.

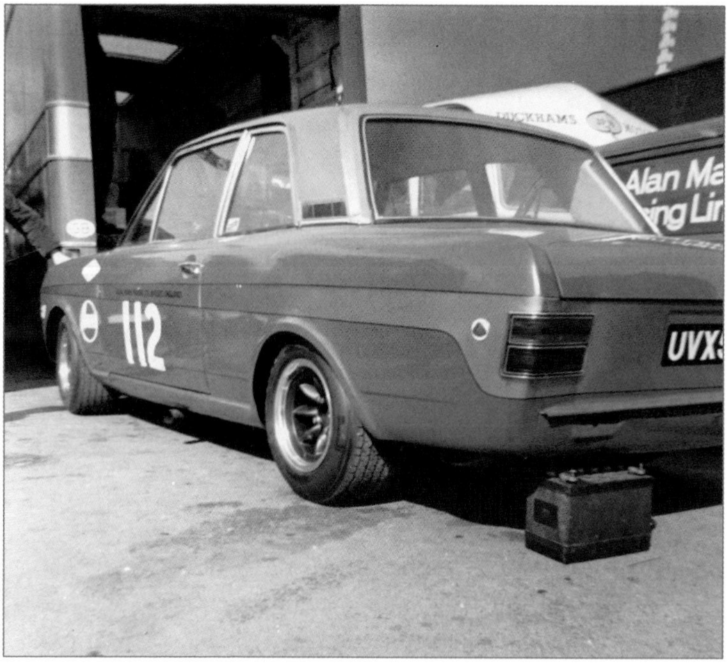

During 1967 the Mk 2 Lotus-Cortina was in the hands of the other Ford-favoured racing teams, and this is the red and gold Alan Mann Racing model driven with spectacular success by Frank Gardner to win the British Saloon Car Driver's Championship. I snapped it in the Brands Hatch paddock on Boxing Day 1967. The following year, the twin-cam Escort would supplant the Lotus-Cortina.

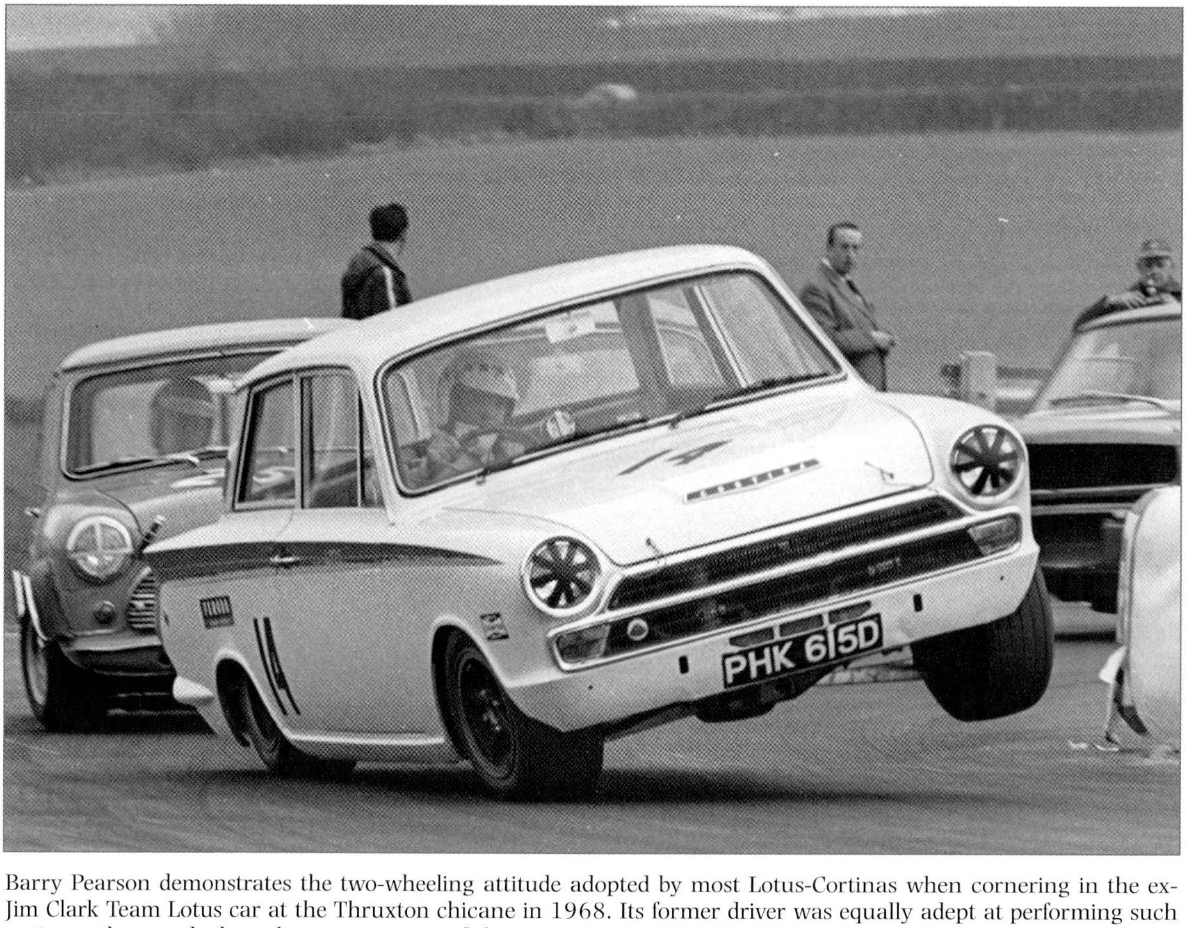

Barry Pearson demonstrates the two-wheeling attitude adopted by most Lotus-Cortinas when cornering in the ex-Jim Clark Team Lotus car at the Thruxton chicane in 1968. Its former driver was equally adept at performing such antics as these and when alarm was expressed, his response was 'You should see what it's like from the inside!'

The Team Lotus cars were driven by Jim Clark and Jack Sears that year, and finished fourth overall in the Championship, with Sears winning the class. Sir John Whitmore won the European Touring Car Challenge driving an Alan Mann car, with wins at Zolder, Snetterton, the Mont Ventoux hillclimb, and sharing a win with Sears at the Nürburgring 6-Hours. The Lotus-Cortina ousted the Alfa Romeo Giulia T.I. Super, but its principal opposition in Touring car racing then came from the Alfa Romeo Giulia GTA coupés. One of the first races I saw involved an intense dice between Whitmore and Jochen Rindt in a GTA in the Snetterton 500km in 1966 – a race so wet that a speedboat could have won it. The two Alan Mann cars were driven by Stewart and Whitmore, and both drivers waved to us spectators as they came out of Russell bend in the heat of the action.

Team Lotus raced the cars in 1966 in Group 5 form. Powered by BRM-built dry sump engines and Lucas fuel injection, they developed 180bhp at 7750rpm, and by this time they were capable of reaching 150mph. Other changes included modified suspension geometry and coil springs, and damper units replaced the standard MacPherson struts. At club racing level, Lotus-Cortinas had become almost too numerous to be interesting, and more outrageous examples like Doc Merfield's V8-powered custard yellow Fraud Cortina seemed more exciting to young race fans. Then the Mk 2 Lotus-Cortina came out in 1967. It was a Dagenham-built car.

Both Team Lotus and Alan Mann Racing ran examples, powered by Cosworth FVA Formula 2 engines. Graham Hill, Paul Hawkins, Lucien Bianchi, Jacky Ickx, John Miles and Frank Gardner were the main protagonists. The Team Lotus race cars were sold off towards the end of the season to Brian Robinson who ran them for himself and Tony Dean. Another of the Mk 1 cars was campaigned effectively by Barry Pearson in 1968.

BIG BANGERS

The quest for more power is often most easily satisfied by fitting a bigger engine, and this practice became something of a growth industry during the 1960s. As is frequently said, there's no substitute for cubic centimetres. The FIA's Group 7 rules catered for just this kind of sports car. These big bangers were originally based on Cooper Monacos with stock-block American V8s shoehorned in. Among the first to fit such an engine were Chris Summers, who ran a Chevy V8 in his Monaco, and Roy Salvadori in Tommy Atkins' 5.0-litre Maserati V8-engined Monaco T61P. Carol Shelby created the 4.7-litre King Cobras in 1963 using Cooper Monaco Mk IVs, and Roger Penske built the Zerex Duralite Special around a Lowline Formula 1 Cooper T61, powered by the 2.7-litre Climax engine out of Brabham's Indy car. This vehicle was uprated to Oldsmobile V8 power and sold to Bruce McLaren, who went on to evolve the definitive range of Group 7 CanAm cars as the decade progressed. Contemporary with McLaren, Lola was developing its own challenger, the T70, and it was inevitable that Lotus would enter the fray. Chapman was a trifle miffed that Ford chose Lola to build its GT car that became the legendary GT40, and the Lotus 30 might have been turned into a GT car to prove a point. It wasn't though, and it goes down as another rare Lotus failure.

In 1964, Lotus introduced the 30 as a ground-hugging open-top Group 7 sports racing car. Its chassis design was based on the Elan-type backbone frame. Against Len Terry's advice, Chapman pushed the project through, and the 30 went ahead as he visualised it. Up front, its transverse box-section carried the front suspension, and at the back end, two diverging arms carried the rear suspension pick-up points. The

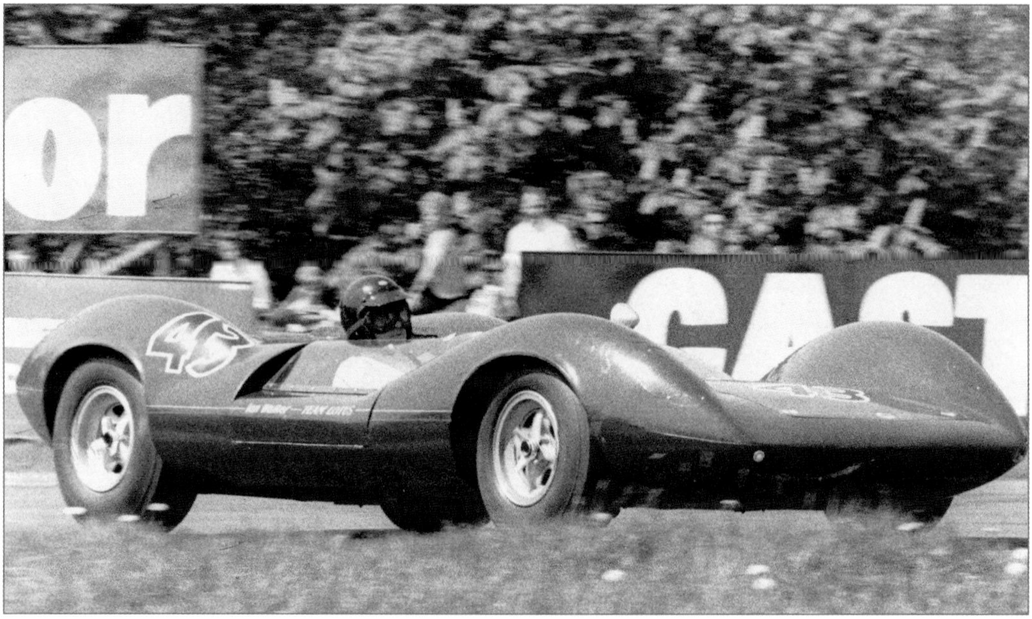

The 30 was Lotus's shot at producing a Group 7 'big banger', a rival to the V8-engined Lola and McLaren sports racers that would come to dominate the North American CanAm series. Based on a box-section backbone chassis that was conceptually similar to that of the Elan, the 30's main drawback was that it was powered by the heavyweight cast-iron 4.7-litre Ford V8, while rivals used aluminium alloy engines. Even the genius of Clark couldn't make it a success.

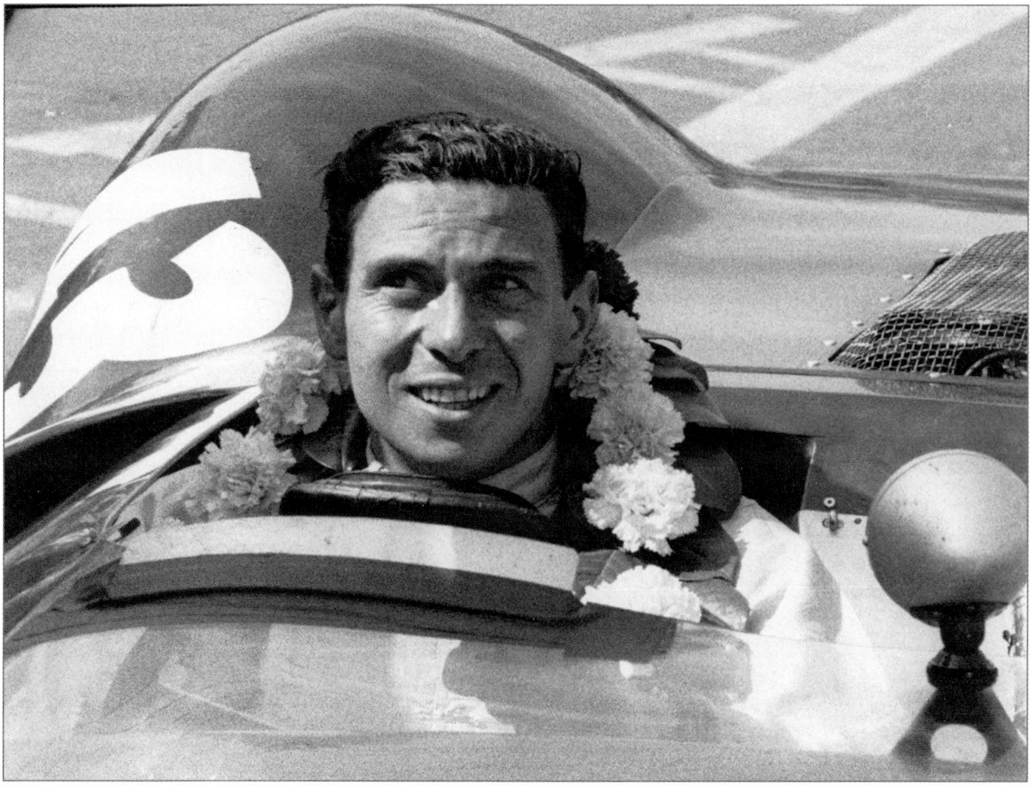

Garlanded after his Mallory Park success, Jim Clark scored the 30's only significant victory against minimal opposition at Mallory Park in June 1964.

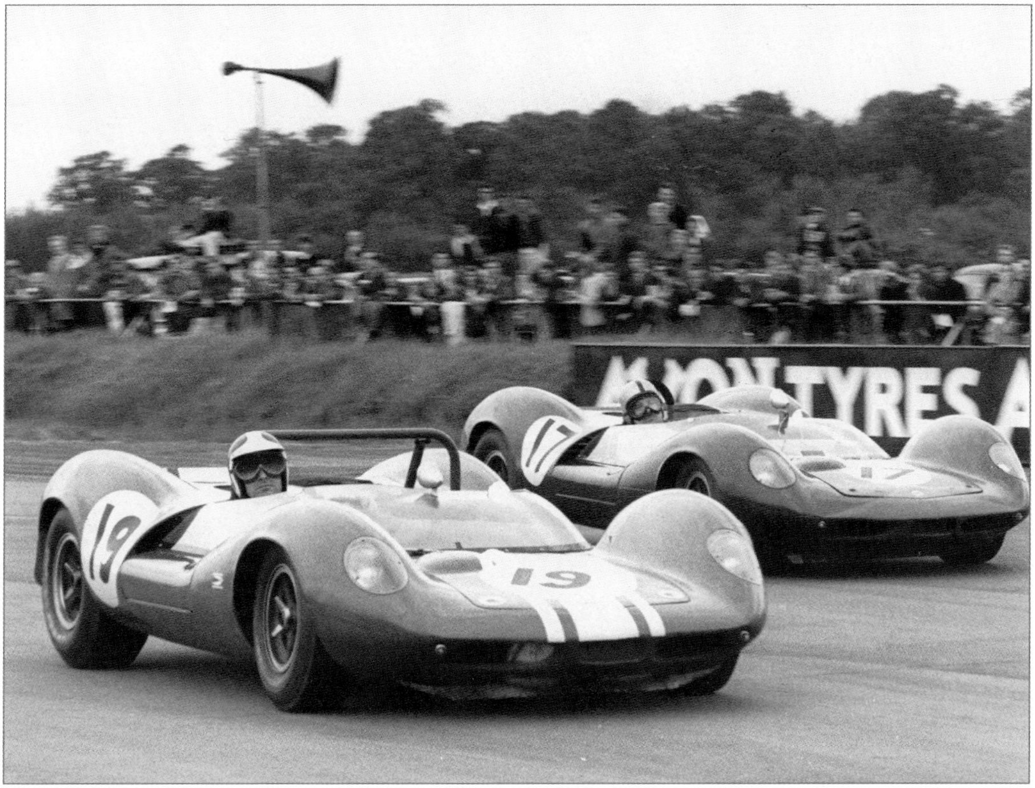

Lotus Components built 33 units of the 30 in S1 and S2 guise. Here, privateers John Dean and Bob Bondurant give their thundering monsters an airing at Silverstone in May 1965.

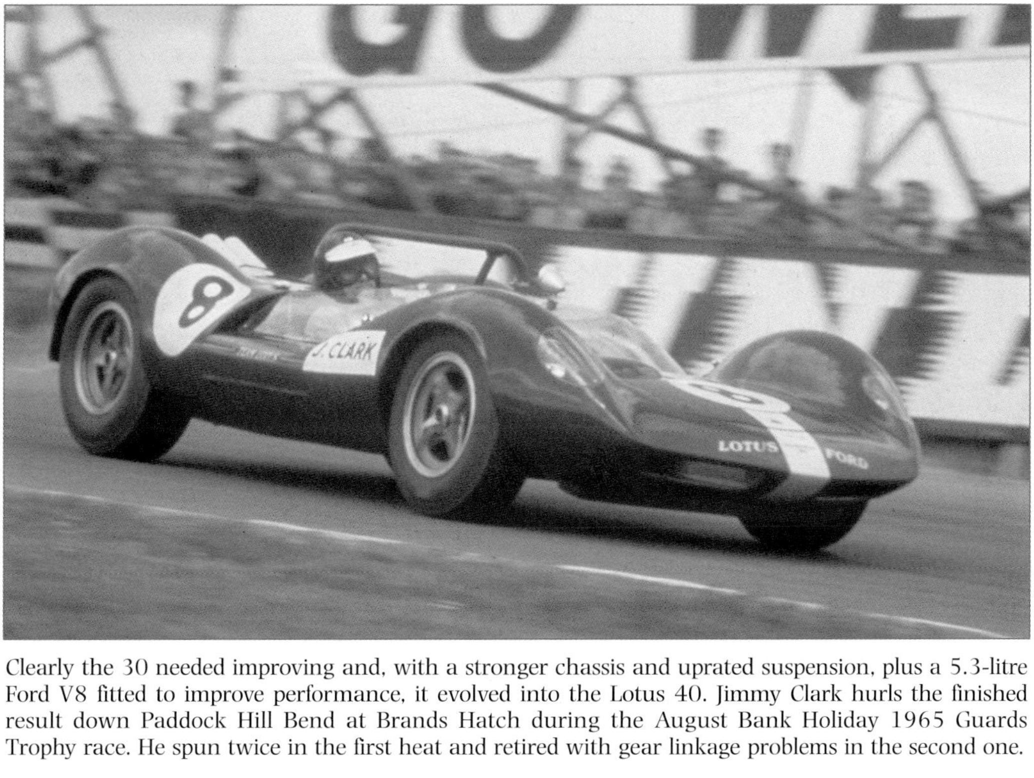

Clearly the 30 needed improving and, with a stronger chassis and uprated suspension, plus a 5.3-litre Ford V8 fitted to improve performance, it evolved into the Lotus 40. Jimmy Clark hurls the finished result down Paddock Hill Bend at Brands Hatch during the August Bank Holiday 1965 Guards Trophy race. He spun twice in the first heat and retired with gear linkage problems in the second one.

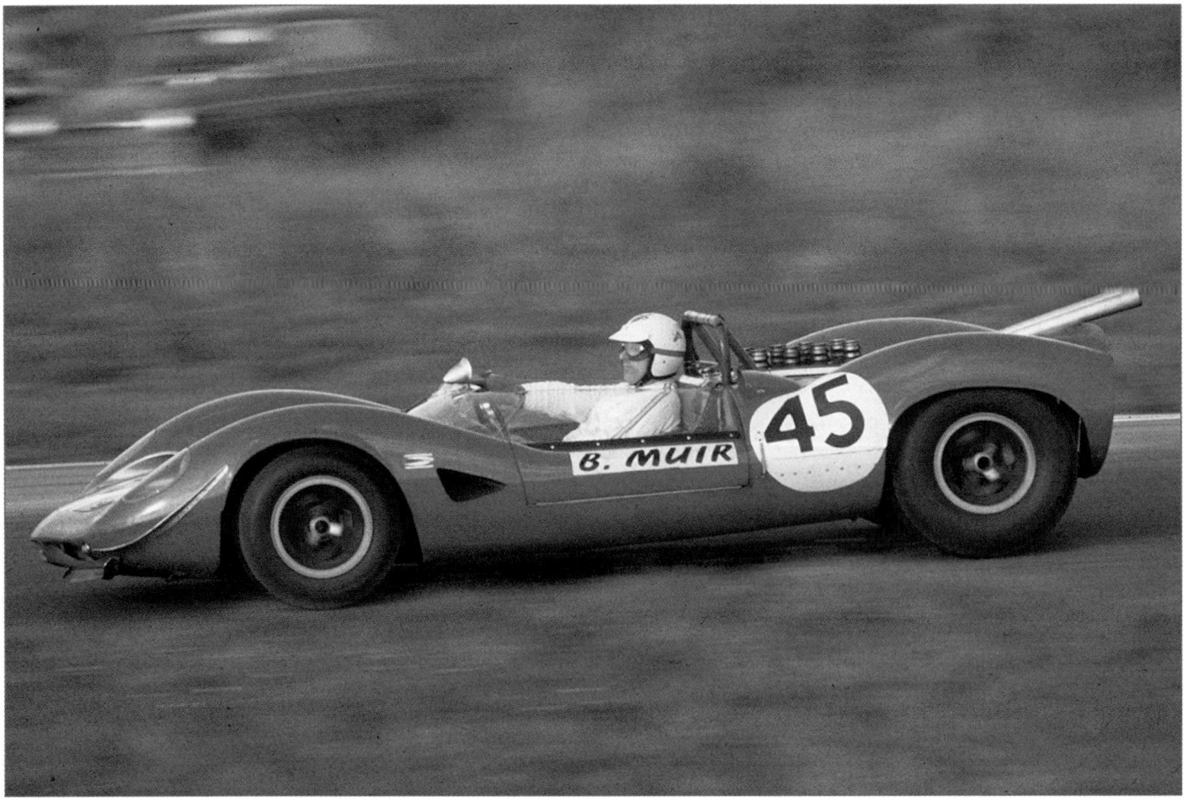

Brian 'Yogi' Muir drives the Willment 40 in a round of the 1966 Guards Championship, in which he scored a few wins. The 40 features a North American SCCA rollover hoop, and its induction trumpets and upward slanting exhausts are also prominent.

Ford V8 engine was mounted amidships, and the central box-section was created from 20-gauge sheet steel, within which was the 13-gallon rubber bag tank. Auxiliary tanks were carried in the sills. The 30's front suspension was by unequal-length double wishbones and coil springs and dampers, while at the rear were upper wishbones, reversed lower wishbones, lower radius rods and coil springs and dampers. The 271bhp 4.7-litre Ford Hi-Performance engine featured four downdraught Webers, and was linked to a five-speed ZF all synchromesh gearbox. With a 3.00:1 final drive, it was potentially capable of 200mph. The 30 ran on 13 inch four-spoke wheels similar to the Halibrands used at Indianapolis, with Girling 11 inch solid disc brakes.

After numerous development problems, the first 30 was supplied to Ian Walker Racing which ran it as a quasi-works car. Bearing the legend 'Together we chose a Lotus 30' on the rear, Jim Clark drove it to second place in the 1964 Aintree 200. McLaren's Cooper was the sounder car, and even the Lotus 19 club racers were more reliable. Jim took an easy win at Mallory Park, but the prototype was wrecked by Tony Hegbourne at Brands Hatch. Clark also drove the 30 in North America, retiring at Mosport Park, but finishing third at Riverside Raceway, where the big-engined 19s were also running. Jim's 30 also led the Tourist Trophy at Goodwood until brake and suspension problems got the better of him. Twenty-one units of the

30 were made in Series 1 form, and the majority used thicker 18-gauge steel to create a stiffer chassis.

The Series 2 version came out in 1965. It had a detachable centre-section to the bridge carrying the rear suspension so gear ratios could be changed more easily. The bodywork featured an upswept tail and rear spoiler, and roll-over bars were now fitted, along with bigger brakes and 15 inch diameter wheels. Tecalemit Jackson fuel injection replaced the Webers, and power rose to 360bhp. At £3,500, it was not an expensive racing car. Lotus Components production manager Roy Badcock calculated that each one took 600 hours to construct, but only twelve were built that year. Successes were few, mostly because Lotus was stretched in other directions. Clark won the curtailed Senior Service 200 at Silverstone, halted due to heavy rain, and also earned a victory at Goodwood on Easter Monday. Privateers to compete in 30s were David Prophet, who raced extensively in South Africa, Simon Lautour, Vic Wilson, and Tony Dean, while JCB ran cars for Trevor Taylor and Peter Sadler, and Willment ran a car for Frank Gardner. Taylor's caught fire at Silverstone, but Gardner won at Mallory. However, the cast-iron Ford engine meant the Lotus 30 weighed as much as the aluminium-engined Olds and Chevy-powered McLarens and Lolas, and it was never as developed as these rivals. The answer could lie in fitting a bigger engine and that is what the Lotus 40 was all about.

What most people remember about the Lotus 40 is Richie Ginther's quip that 'the 40 was the same as the 30 but with ten more mistakes'. The intention was to upgrade the Group 7 sports racing 30 by building it with a stronger chassis and fitting it with a 5.3-litre Ford engine. It would have a Hewland LG500 transmission instead of the ZF unit. The transformation into the 40 was carried out by Colin Chapman and Lotus development engineer John Joyce, with Dave Lazenby in charge of the racing team. Externally it differed by having modified bodywork and twin upswept exhausts. The 40 debuted in 'Sound of Music' country in the Austrian Grand Prix at Zeltweg, where Mike Spence retired. Jim Clark drove a 40 in the Guard's Trophy race at Brands Hatch, spinning into retirement in the first heat and crashing out of the second. A pair of works cars were driven by Clark and Richie Ginther in the Californian Group 7 Professional Series, and Jim managed a second place at Riverside. Another 40 was sold to Holman & Moody for A.J. Foyt but it was wrecked by Bob Tattersall. Foyt later drove it at the Nassau Speedweek, but it literally fell apart. The 40 carried more engine weight and more power than the backbone chassis could tolerate. A joint venture with De Tomaso in 1966 could have produced a light alloy 5.5-litre V8, capable of turning out 500bhp, to power the 40 but nothing came of the project.

Back at the Brickyard

Meanwhile, there continued to be much interaction with the USA, and a full account of the Team's activities can be found in the fascinating *Team Lotus: the Indianapolis Years* by its then competitions manager Andrew Ferguson. Back at Indianapolis for the 1964 race, Team Lotus and its mentor Ford entered Clark and Gurney to drive the Lotus 34, which was a slightly modified version of the previous year's 29. It was now powered by the 410bhp four-cam fuel-injected Ford V8 engine, allied to a ZF two-speed transmission rather than the Colotti gearbox. There were new

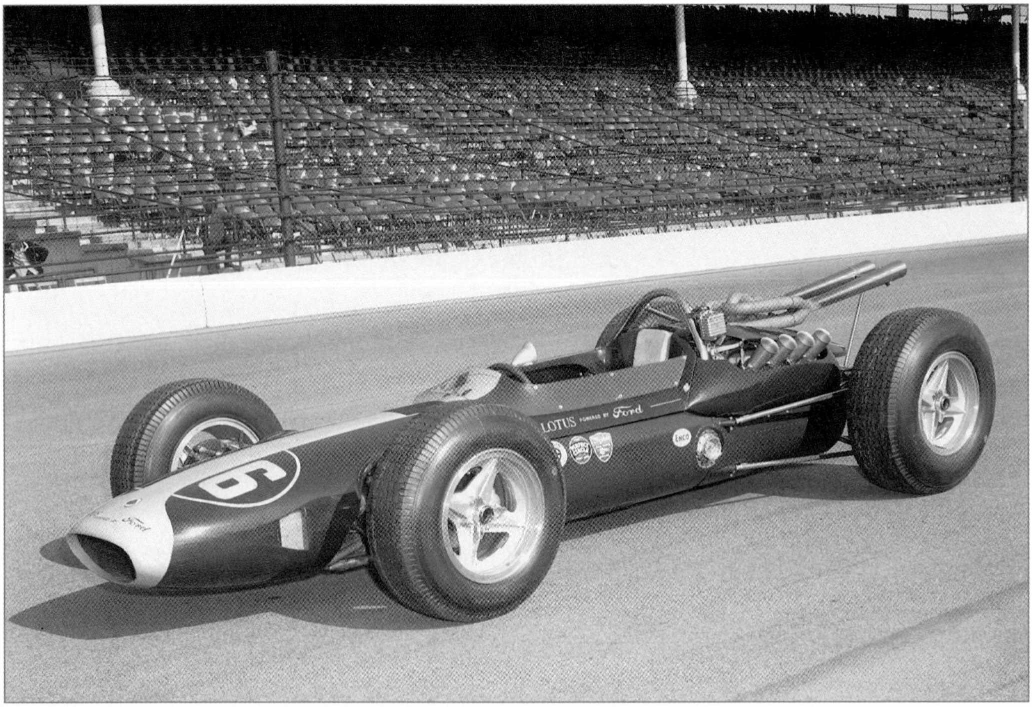

The Lotus-Ford 34 was a development of the 29, and was Team Lotus's weapon for Clark and Gurney to contest the 1964 Indianapolis 500. It was powered by the double overhead cam version of the 4.2-litre Ford V8, now putting out around 410bhp. Spoked Lotus wheels replaced the previous year's Halibrands and are shod with Dunlop tyres. Those used in the race were specially made at the last minute after treads had been thrown during qualifying, but they weren't up to the job. One of Clark's tyres threw a tread while he was leading, and although he didn't have an accident, rather than have Gurney's car suffer a similar fate, it was brought in – much to Ford's dismay.

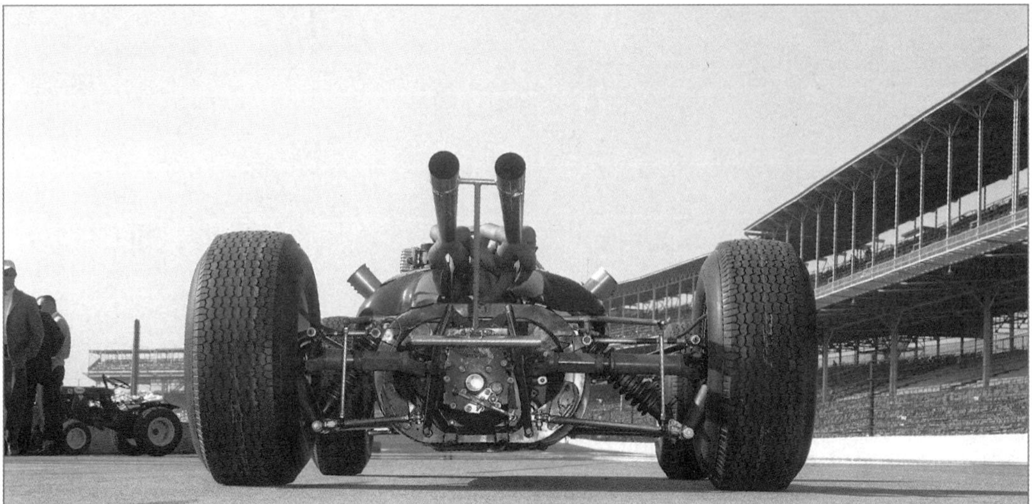

Rear view of Lotus 34/3 Indycar, showing the complex exhaust manifolding of the Hilborn fuel-injected engine and its ZF two-speed gearbox. The suspension on the left-hand side of the car is clearly shorter than that on the right because of the absence of right-hand bends on oval circuits.

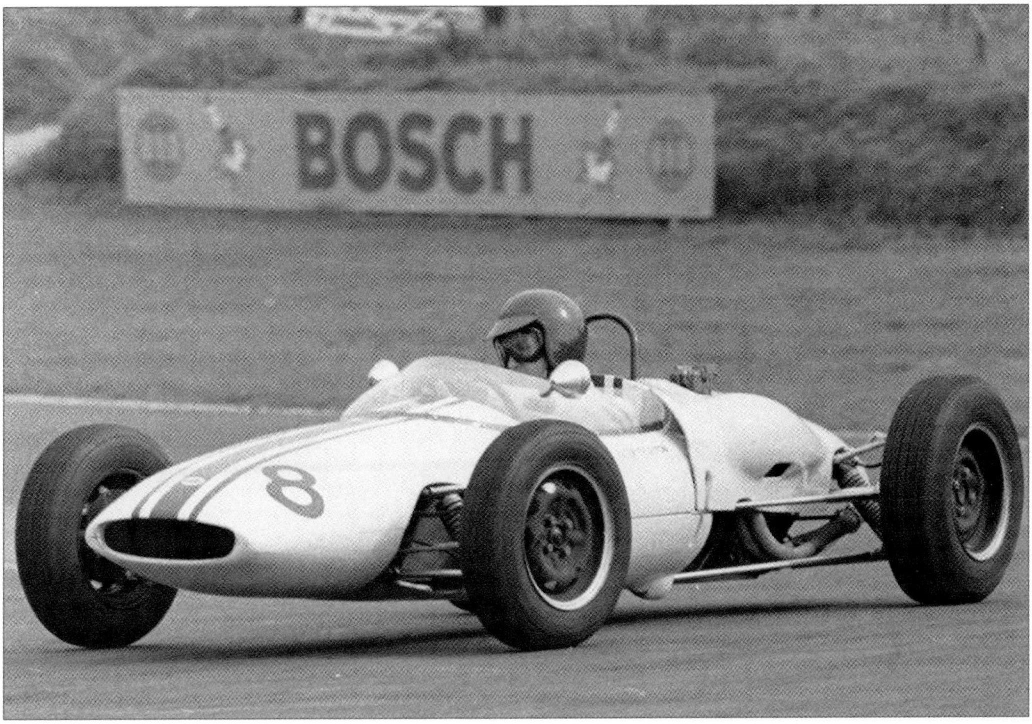

For the 1964 Formula 3 season, Lotus Components introduced the 51. It was developed from the 22, and powered by the 85bhp 997cc Ford 105E-based unit. Cosworth having turned its attentions to developing an F2 engine, the cars to beat in F3 that year were the Tyrrell Cooper-BMCs of Jackie Stewart and Warwick Banks. Top Lotus runners in F3 were Piers Courage, Jonathan Williams and Harry Stiller. This is P. Thurston's 31 at Brands Hatch in 1967.

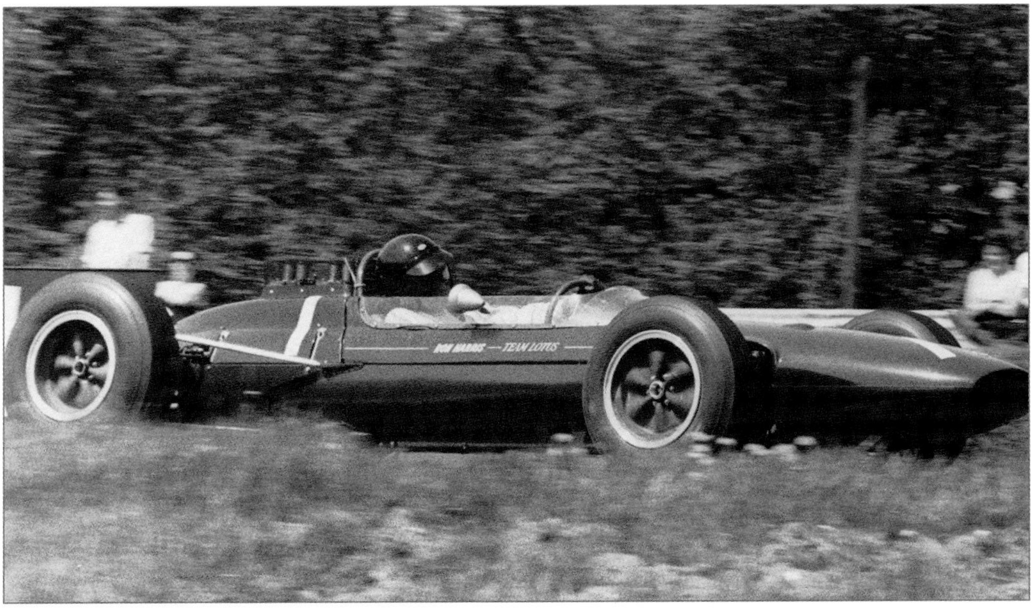

The 32 was a monocoque Formula 2 car, powered by the 997cc Ford-Cosworth SCA engine, which superseded the redundant 27 Formula Junior car. Lotus Components produced twelve of the 32, and the works' cars were run for Jim Clark – seen here – Peter Procter, Mike Spence and Peter Arundell under the Ron Harris-Team Lotus banner. Others to have occasional outings were Jackie Stewart, John Fenning and Brian Hart, and the Team scored seven victories in 1964.

magnesium alloy wheels shod with specially made Dunlop tyres, although these were chunking badly and a fresh consignment was made with just hours to spare before the race. Clark had used the sticky rubber to good effect, taking pole position. A fiery and fatal seven-car accident caused the race to be halted, but Clark retook the lead, although he was briefly headed by Bobby Marshman in the old 29, running on Firestone rubber. But serious vibrations set up as the Dunlop tyres threw their treads caused Clark's left rear suspension to collapse and, mindful of the earlier fatalities, Chapman withdrew Gurney's car as a precaution. Ford was not impressed, but sponsored the 34s to run in certain USAC races later in the season. A 34 driven by Parnelli Jones won at Milwaukee and Trenton, where Clark retired, and in the 1965 Indianapolis, the previous year's winner, A.J. Foyt, drove a 34 to second place.

FORMULA TWO FORTUNES

It was a busy time for Lotus. A fresh set of rules came out for Formula 2 in 1964, calling for 1.0-litre unsupercharged four-cylinder engines, running on commercial fuel. Although he was ambivalent about the new Formula 3, Chapman was serious about F2 and developed the 32 accordingly.

Formula 3 was dominated by Jackie Stewart and Warwick Banks in the Tyrrell Cooper-BMC T72s in 1964, Stewart winning ten out of twelve races. While the Lotus 31 Formula 3 car was based on the Mk 22 Formula Junior car, the 32 took as

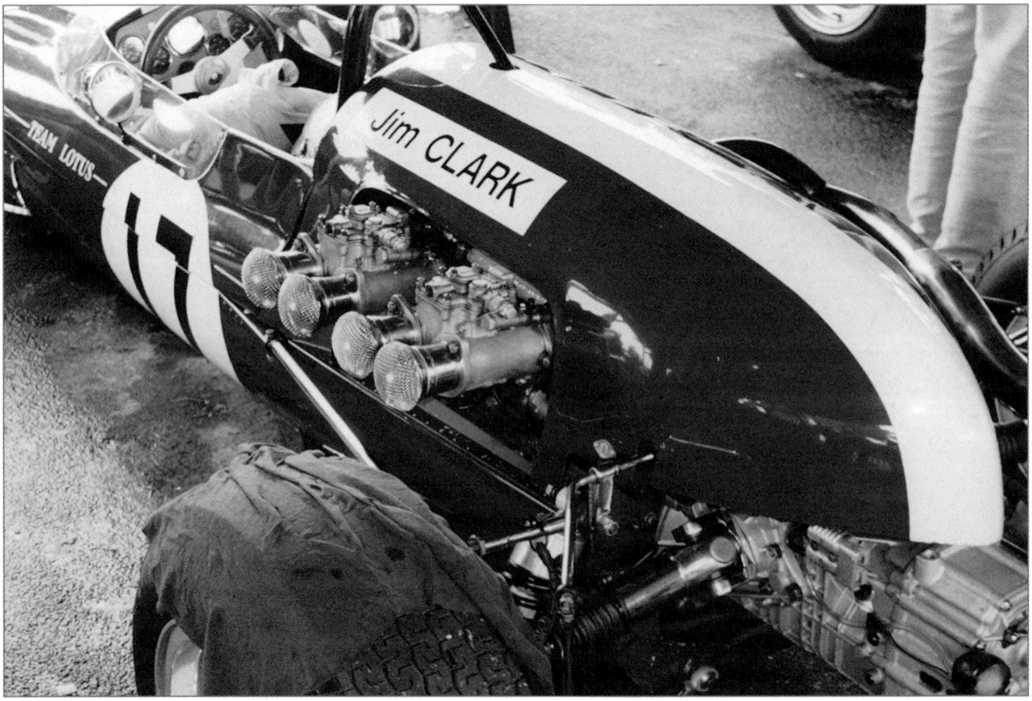

Jim Clark's victorious and hard-worked car for the Tasman Series that began in January 1965 was the one-off 32B. The regular F2 car was modified to accept the four-cylinder 2.5-litre Climax FPF engine, hence the 32B's hump-back engine cover. It was built largely by the Lotus-Cortina mechanics of the day, who included Ray Parsons, Bob Dance and Bob Sparshott. The wheel came full circle at the Coy's International meeting at Silverstone in 1999 when it was looked after by Bob Dance on behalf of driver Malcolm Ricketts

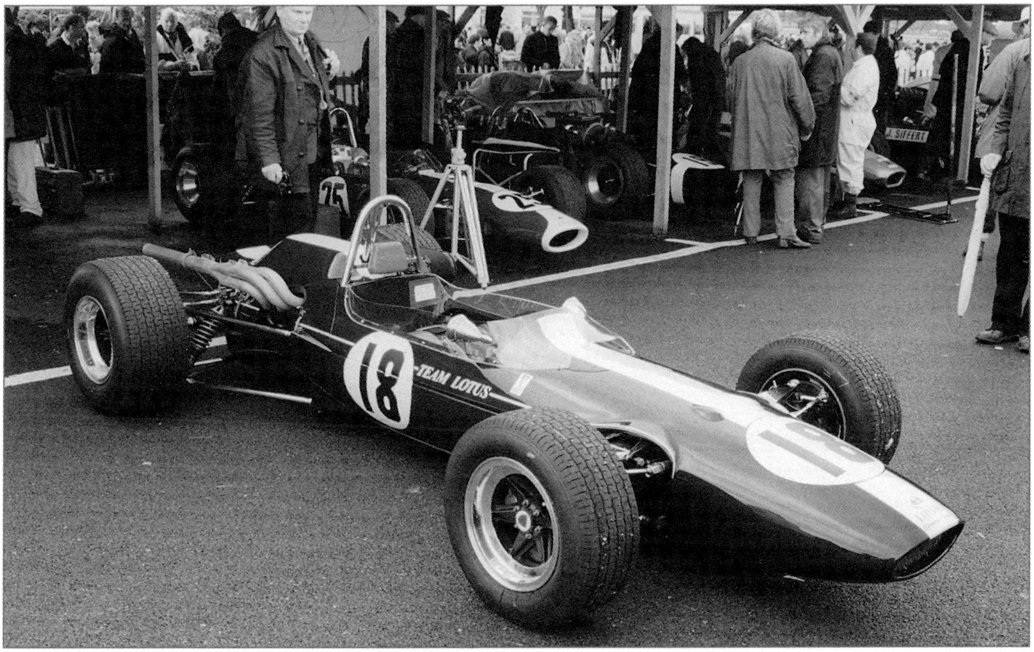

The Lotus 35 was a 1965 development of the 32 and consisted of a monocoque tub that could be used for Formula 2, Formula 3 or the North American Formula B series. By Lotus standards it was not that quick in F2, losing out on the straights, but Jim Clark still scored five wins to take the 1965 European F2 title (billed as the Grands Prix de France series). He also won the UK Autocar F2 Championship. Pictured in the Goodwood paddock at the 1999 Revival meeting is the John Dimmer car.

its inspiration the aluminium monocoque of the 27 FJ car. It was constructed of sturdier 16-gauge steel sheet panelling, however, with a pair of 9-gallon bag tanks housed in the sides of the body. The 32's front suspension of upper rocking arms, lower wishbones, coil springs and dampers was sourced from the 24 and 25 Formula 1 cars, and at the rear the top radius arms were mounted on tall uprights, with reversed lower wishbones, coil springs and dampers.

The 32 F2 car was powered by one of the best engines of the era, the single overhead cam 115bhp 997cc Cosworth-Ford SCA unit, leaning over at 25 degrees, and fed by two Weber 40DCM2 downdraught carburettors. As was regular Chapman practice, the operation was farmed out to Ron Harris Racing and the cars were entered as Ron Harris-Team Lotus. Regular drivers were Jim Clark, Peter Arundell, Mike Spence and Peter Procter, with Jackie Stewart, John Fenning and Brian Hart also getting drives when the top men were committed to F1. Wins for Clark came at Pau, the Nürburgring, Mallory Park and Brands Hatch, and Spence took the Formula 2 class at the Aintree 200 meeting. Brian Hart won at Enna Pergusa and Jackie Stewart scored at Snetterton in the Vanwall Trophy race. At Reims, Arundell collided with Ginther and crashed badly. Despite a full recovery, he never regained his previous form. Most F2 races were won by Brabhams – nine out of eighteen – and Lotus took seven, with one each to Cooper and Lola. A total of twelve 32s were built, and the 32B derivative was a single car made to contest the Tasman series. It was powered by a 2495cc four-cylinder Coventry-Climax engine, and Jim Clark was victorious in five out of eight races. Leo Geoghegan was also successful down under in his private 32.

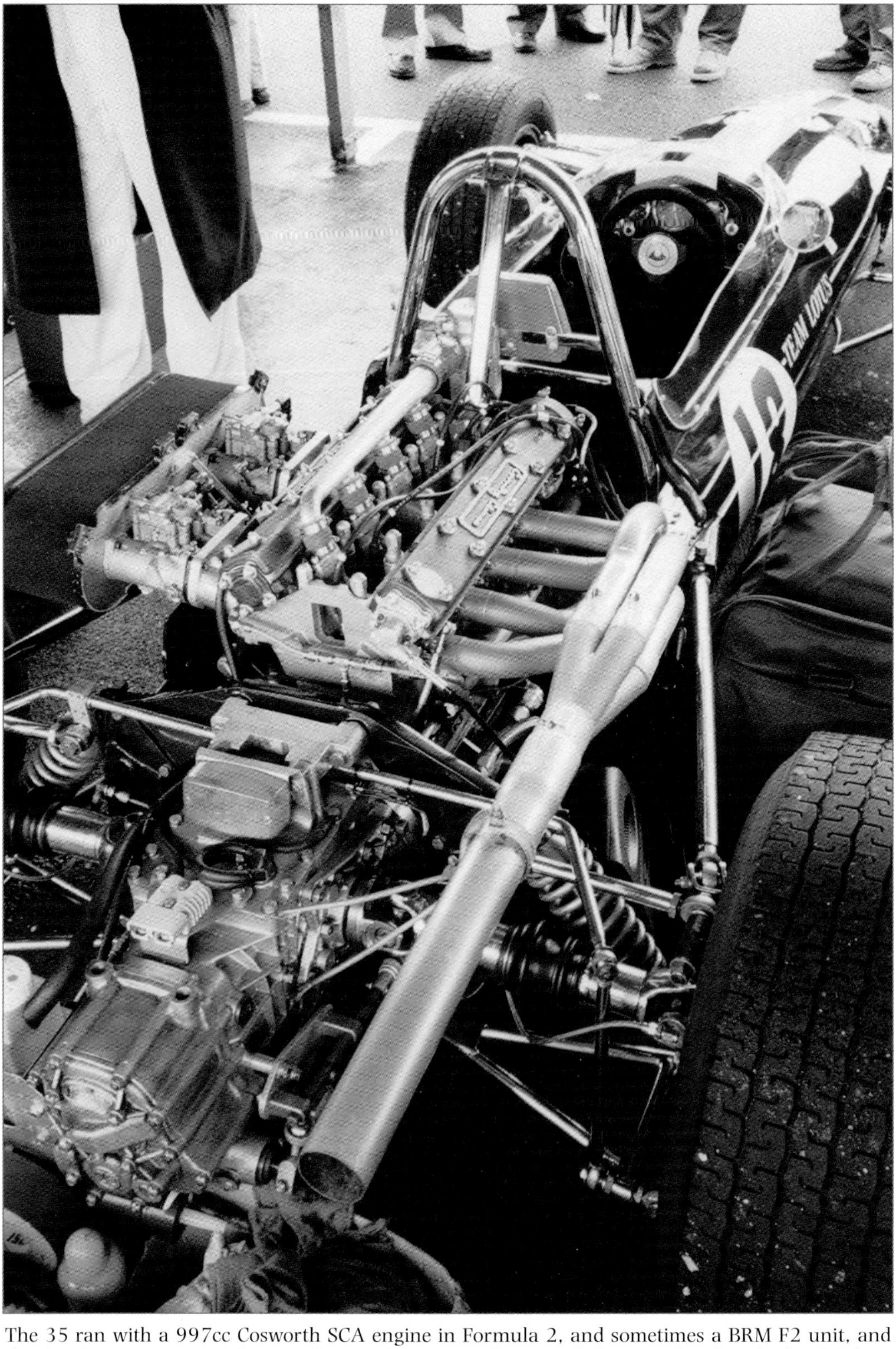

The 35 ran with a 997cc Cosworth SCA engine in Formula 2, and sometimes a BRM F2 unit, and the 1600 twin-cam in US Formula B. John Dimmer's example here is fitted with the 2.5-litre Climax FPF engine and Hewland transaxle as for the Tasman series. Twenty-two units of the 35 were built by Lotus Components in 1965.

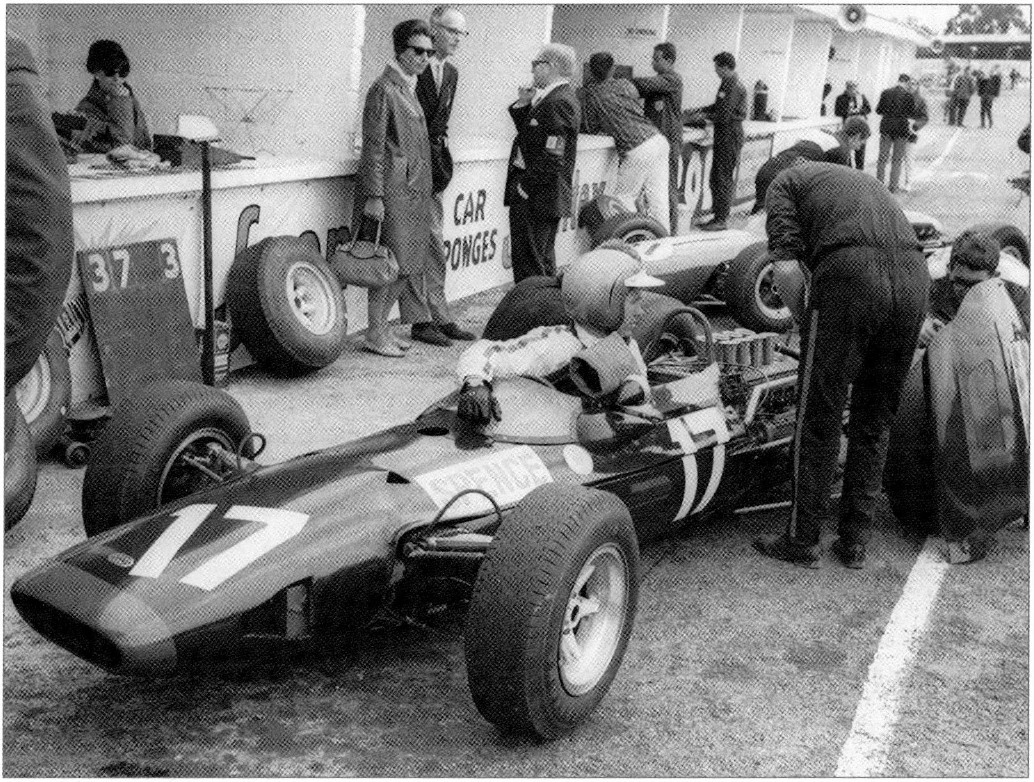

Mike Spence moved to the Tim Parnell team for 1966, and his 25B R13 was fitted with the 2.0-litre BRM V8 engine. Success was very limited, and Spence, here in the Brands Hatch pitlane during practice, retired after fifteen laps of the British Grand Prix with an oil leak. He managed fifth at Zandvoort and Monza, however.

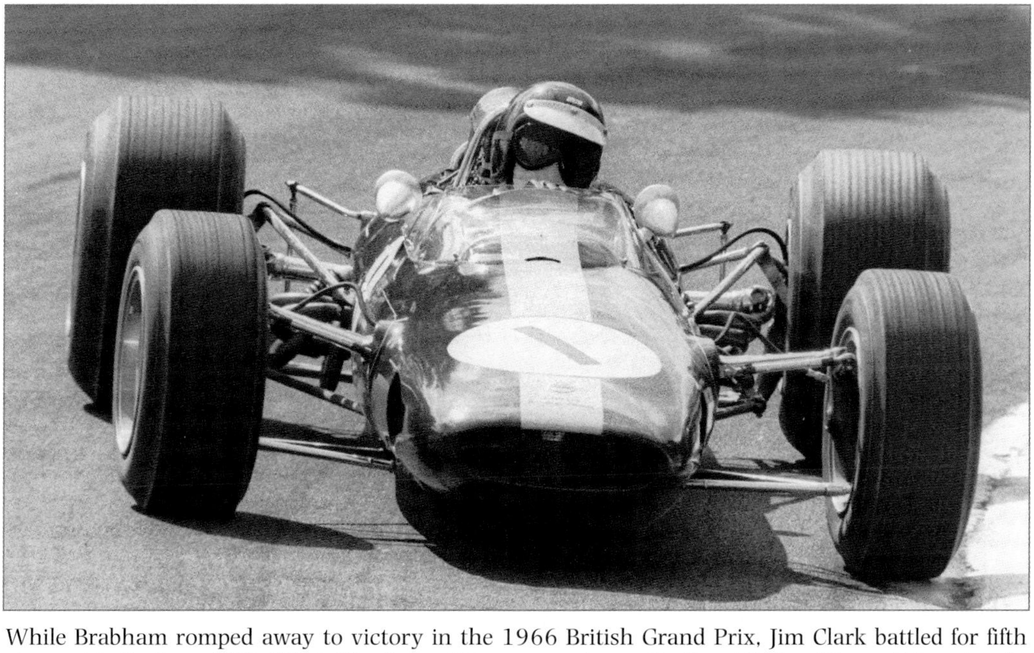

While Brabham romped away to victory in the 1966 British Grand Prix, Jim Clark battled for fifth place with Graham Hill's BRM. The Lotus 33 was running with the 2.0-litre Climax V8 engine in the first year of the 3.0-litre Formula 1, and he did well to finish fourth with ailing brakes.

NEW MONOCOQUE

Racing car design evolves when data acquired during the development of one car are fed into its successor. It can be the case that the new model is closer to the final evolution of its predecessor than that car is to its original version. The new 33 benefited from lessons learned from racing the 25, and the suspension modifications that were incorporated into the 25B also appeared on the 33. Len Terry saw to it that the latter's monocoque was both straighter and stiffer, without the 25's inner kink at the pedal box, and with a fractionally longer wheelbase to incorporate the latest flat-crank Climax V8 engine. It had heavy duty driveshafts, solid steering arms and additional rivets in areas found suspect in the 25. Such was the affinity between the two cars that the 33's chassis numbering followed on from that of the 25.

It was not an instant success, however. A promising debut in the 1964 Aintree 200 found Clark battling with Brabham for the lead, only to be pushed off by a back-marker. The car was destroyed. The 33 didn't appear again until the non-Championship Solitude Grand Prix in July. It was unseasonably cold and wet, with much aquaplaning, but Jimmy won a tussle with Surtees' Ferrari, setting fastest lap on the way. It was Solitude's last F1 event, and it left a trail of broken cars in ditches. Clark and Spence drove 33s at the Nürburgring, but the Scot had engine failure and Spence was eighth. At Zeltweg for the Austrian GP, now a championship qualifier, Clark was overwhelmed by the pack as he started in second gear, and demonstrated his talent by moving swiftly from twelfth place to fourth in just five laps. However, both 33s broke their driveshafts. These were replaced for Monza with Mercedes-Benz type items, but on this quick circuit the Lotuses were no match for the straight-line speed of BRM and Ferrari. Mechanical glitches largely put paid to any further success in 1964.

The following year was the last season of the 1.5-litre Formula 1, and changes within Team Lotus saw Jim Endruweit become Racing Manager and Dick Scammell take over as Chief Mechanic. After Indianapolis, Len Terry joined Gurney and Shelby at All American Racing to work on the Eagles. At the wheel of a 33, Jim Clark's total domination of the season began with victory in the South African Grand Prix on New Year's Day, followed by success in a heat of the Race of Champions, victory in the Syracuse Grand Prix and the *Sunday Mirror* Trophy race at Goodwood. He had started from pole position at each venue except Goodwood, and set fastest lap in every race. Chapman and Clark were contracted to Ford to race at Indianapolis and the schedule was just too tight for Jim to race at Monaco. Back in Europe, he won at a totally drenched Spa-Francorchamps using the 32-valve-engined car, setting fastest lap. Driving a 25 at the roller-coaster Clermont-Ferrand circuit in the Auvergne he took victory from pole and set fastest lap there too. It was four British Grand Prix wins in a row for Clark, although Hill saw to it that it wasn't a Lotus whitewash. At the end of a thrilling Silverstone chase, Hill's BRM was just 3.2 seconds adrift. Paul Hawkins was to drive Dickie Stoop's 33, but crashed in practice.

Although Ginther's Honda showed its potential yet again at Zandvoort, it was Clark's race, from Stewart's BRM and Gurney's Brabham this time. This was the event where Chapman decked a Dutch auxiliary policeman who tried to prevent him from going to his pit at the start of the race. Notwithstanding Jimmy's win, the police came to arrest Colin after the race, and a scuffle erupted with the mechanics protecting Chapman. The intercession of BRM owner Louis Stanley ensured that

Chapman had to spend just the one night in gaol, while Jabby Crombac was deputed to receive the trophy on his behalf.

Team Lotus took three 33s to the Nürburgring, with Clark and Spence driving R11 and R9. Jimmy won, but Spence, together with all the privately entered Lotuses, retired. Jimmy then took second place in the Mediterranean Grand Prix at Enna Pergusa, where 33 R6 caught fire during practice refuelling. Although he retired from the three remaining Grands Prix, Clark had done enough to clinch the 1965 World title. Team Lotus had been the most successful entrant by far during the five years of the 1.5-litre Formula 1. But as the company prepared to move its entire operation up to Norfolk, 1966 was to prove a lean year for race results.

For the new 3.0-litre Formula 1 instituted in 1966, Lotus continued to use the 33, powered by a large-bore Climax V8 engine fitted with a small-bore crank, giving dimensions of 1974cc and developing 240bhp – way off the 400bhp that could be expected from a full 3.0-litre engine. However, it was better than nothing, and had been built specially for Lotus under the direction of Wally Hassan at Coventry-Climax. This firm had recently been taken over by Jaguar, and that spelled the end for Climax as a race engine builder, despite the creation of a one-off flat-16 engine that never left the test-bed. Brabham, on the other hand, had stolen a march on everyone else, using 3.0-litre Repco engines, and it was to be Denny Hulme's year.

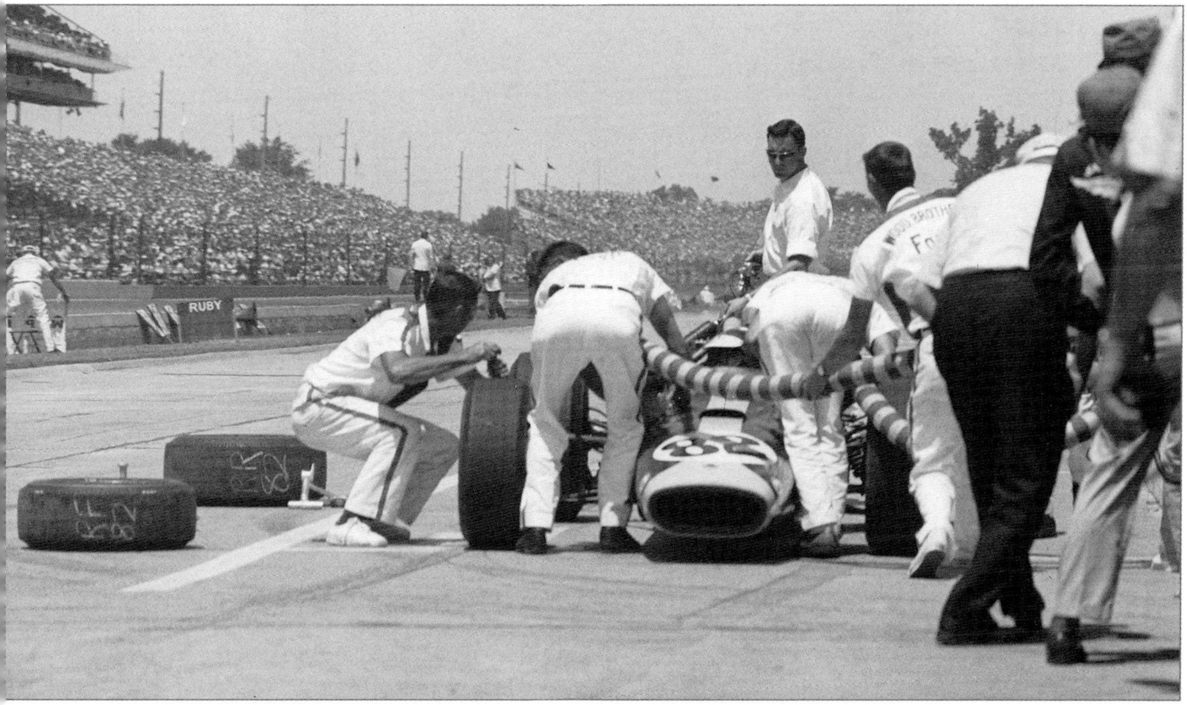

Pitlane drama in the 1965 Indy 500 as Jim Clark stops for fresh tyres and fuel, administered by the Wood Brothers pit crew who were hired by Ford on account of their legendary prowess in NASCAR stock car racing. They took 16 seconds to fill the Lotus-Ford 38's 50-gallon fuel cells, and Clark's two stops were accomplished in 44.5 seconds, against Parnelli Jones taking 98.5 seconds for three stops. Jimmy went on to win at an average 150.686mph having built a two-minute advantage over Parnelli Jones. Bobby Johns in the second Lotus-Ford 38 drove a steady race to finish seventh.

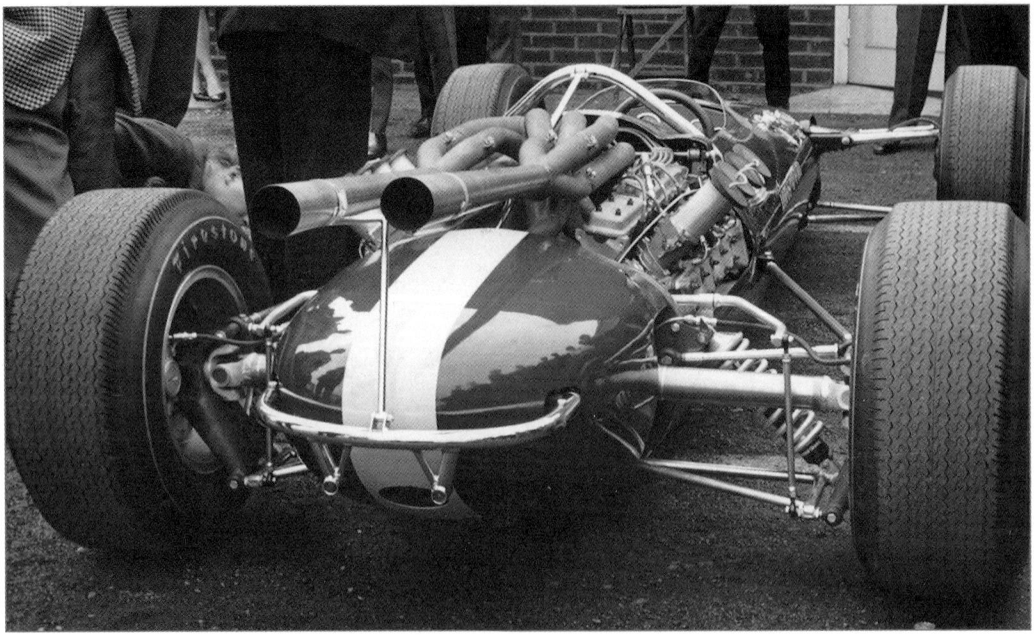

The tail end of the 38 Indycar shows the mandatory 'bumper' used for push-starts conveniently placed to support the exhaust pipes, which were finished in yellow. The quad-cam centre-exhaust V8's side engine covers that would clad the injection trumpets are removed here. The shot also emphasises the short left-hand suspension and massive drive shaft.

At Monaco and Spa, Jim drove 33-R14, non-starting at Reims after being hit in the face by a bird and sustaining an eye injury. By now the 43 with its BRM H16 engine was up and running, but totally lacking in development. The 33s of Spence, Arundell and one-off outings for Pedro Rodriguez and Geki Russo used BRM 2.0-litre V8 engines but were outclassed by the Brabhams and Ferraris. Clark came fourth at Brands Hatch in the British Grand Prix, was third at Zandvoort and third in the Oulton Park Gold Cup. Mike Spence won the non-Championship South African Grand Prix, with Arundell third, and Spence came fifth at Monza. The 33 sustained Clark into 1967, as he took five victories and three second places driving R14 to win the Tasman Series in New Zealand and Australia during January and February.

Another unique car that served in the Tasman series was the Lotus 39. It was a one-off, originally conceived as a Formula 1 car to use the Coventry-Climax flat-16 engine. This unit never left the test-bed, and the 39's chassis was surplus to requirements. One of Maurice Phillippe's first tasks as Chief Designer was to adapt the 39 to take the four-cylinder 2.5-litre Climax FPF engine for Jim Clark to use in the 1966 Tasman series.

The most successful car of the Antipodean season was the 2.0-litre V8 BRM of Jackie Stewart, but Jimmy won at Warwick Farm, and was second at Levin, Melbourne's Sandown Park. He was also third in the Lady Wigram Trophy at Christchurch and in the Australian Grand Prix at Brisbane's Lakeside. The 39 was then sold off to Leo Geoghegan, who ran the car in the Tasman series until 1970, powered by a 2.5-litre Repco V8 engine after 1968.

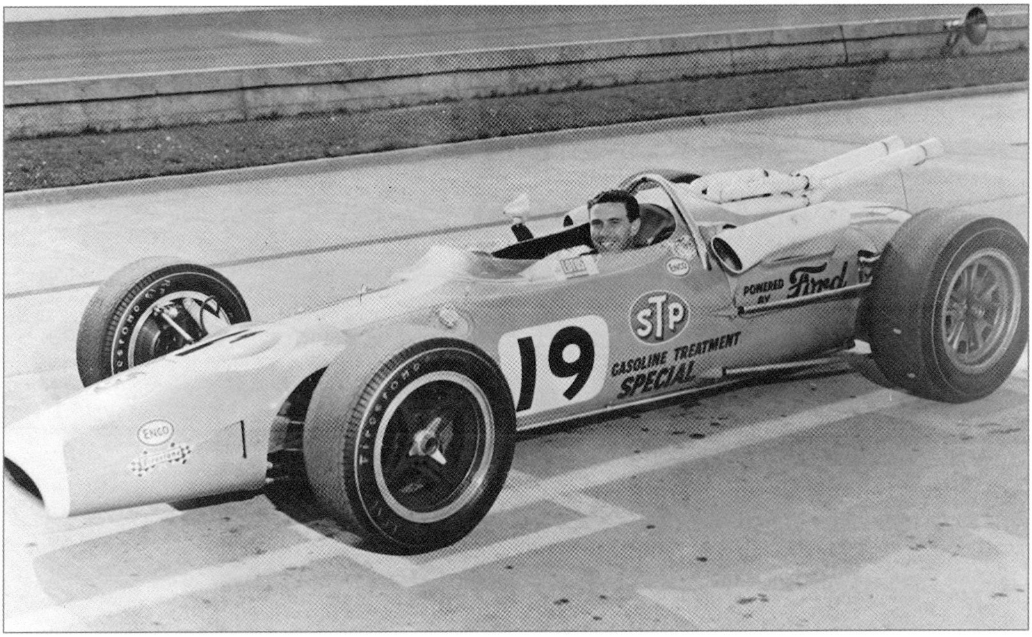

When it became obvious that the 4.2-litre H16 BRM engine wasn't going to be ready for the 1966 Indianapolis 500, the previous year's Team Lotus Ford V8-powered Lotus 38s were brought out again. Now with sponsorship from STP gasoline treatment, the cars of Jim Clark and Al Unser were painted day-glow pink. A third 38 was driven by A.J. Foyt but eliminated in the eleven-car start-line pile up.

Graham Hill returned to his old team to partner Clark for 1967. The Scot continued to race the 33 with its 2.0-litre Climax engine, while Hill's 33 was fitted with a 2.1-litre BRM V8 engine. In this car Graham came fourth in the Daily Express race at Silverstone and was second at Monaco, but they wouldn't have to wait long to get their hands on the brilliant new 49.

Stressed Skin

Prompted by the fatal accident in the 1964 Indianapolis 500, the rules for 1965 called for changes to the fabrication of the cars' structure. To this end, Len Terry designed a completely new vehicle, the 38. It had a stressed-skin construction, and this superseded the bath-tub monocoque of the 25 and its derivatives. The new body incorporated three fuel tanks, one on either side and a third behind the seat. Still smarting from the previous year's debacle, Ford offered Chapman only a flat fee to do the race, instead of the lavish terms he'd enjoyed before. The 38 was equipped with more powerful 500bhp Ford V8 four-cam engines mated to the ZF two-speed transmission, featuring a ram-air system and a distinctive central exhaust system. The suspension was offset 3 inches to the left, and a longer nose cone was fitted. With Dave Lazenby as Team Lotus's Indy racing manager, two cars were entered for Clark and US stock car driver Roger McCluskey, as well as a third car for Gurney, which was sponsored by Yamaha and run by All American Racers. However, McCluskey crashed one of the cars in testing at Trenton when the throttles jammed open, and an argument ensued over whether it was the fault of Team Lotus or Ford's engine workshop. For the Indy 500, the two Team Lotus 38s were driven by Jim

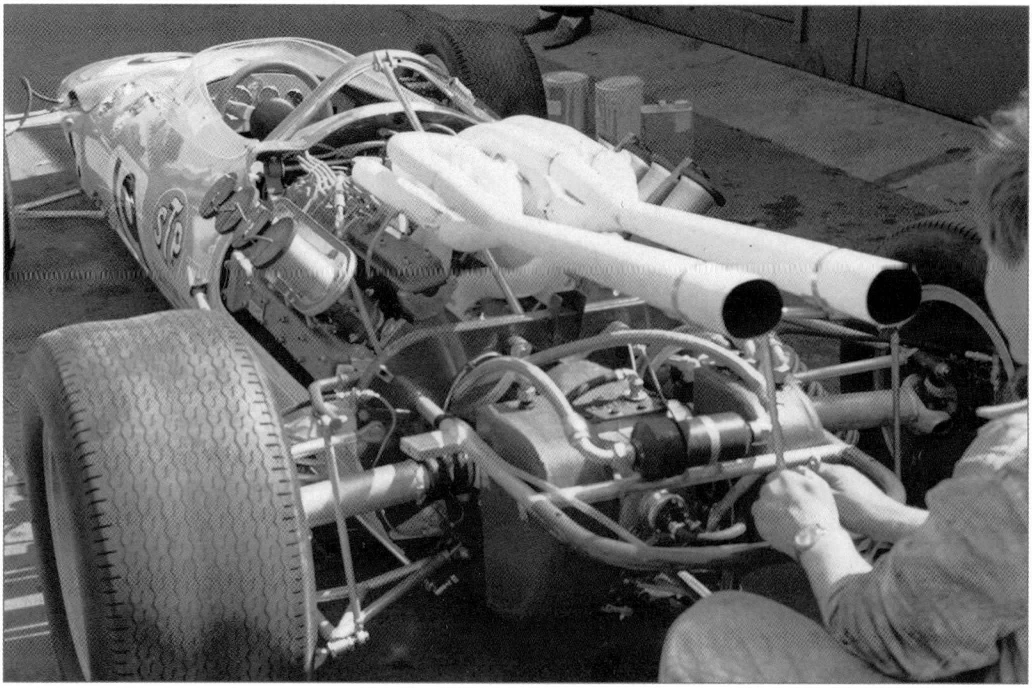

The brackets supporting the exhaust pipes are tightened up on Jim Clark's 1966 Lotus 38 Indianapolis car. The dohc Ford V8 engines were supplied by the Meyer-Drake company belonging to three-time Indy winner (and purveyor of Offenhauser engines) Louis Meyer, and no fewer than twenty-four of the thirty-three starters ran these engines in '66. Jim finished second to Graham Hill after Jackie Stewart had dropped out, earning almost $77,000 in the process.

Clark and journeyman US driver Bobby Johns, and at the end of lap one, the first five cars were Lotuses. Although pole sitter Foyt was running on nitro-methane, which gave a slight power advantage, Jimmy still beat him. Aided by a couple of well-managed pit stops by Ford's highly professional NASCAR stock car specialists (the Wood brothers), Clark emerged the winner at 149.334mph. Parnelli Jones's Lotus 34 was second, and Al Miller's 29 was fourth, with Johns in seventh place.

The following year's Indianapolis 500 was noted for an appalling start line crash that sidelined eleven cars and saw A.J. Foyt escape from his wrecked 38 by scrambling up the catch fencing. Miraculously, he was the only person injured, and that was a cut finger from the chain-link fencing. Lotus had intended racing the new 42 but since it wasn't ready, the Team returned with the 38s for Clark and Al Unser – liveried in STP red. Two further 38 shells were built by Abbey Panels but they needed strengthening before delivery to their new owners. Mario Andretti elected not to run his new 38, but Foyt drove the other, albeit briefly. After the race restarted, there was much uncertainty over laps completed, due to many spins occasioned by the oily track surface. This was caused by the enthusiastic over use of the sponsored STP lubrication treatment and thinner engine oil, which was prone to leaking. Clark spun his 38 twice during the race, finishing second to Graham Hill's Lola. Andretti retired with engine failure while Clark's team-mate Al Unser crashed out.

Team Lotus had intended to run the 4.2-litre version of the BRM H16 engine in the 42 at the Indy 500 in 1967. But the engine proved uncompetitive and for Clark

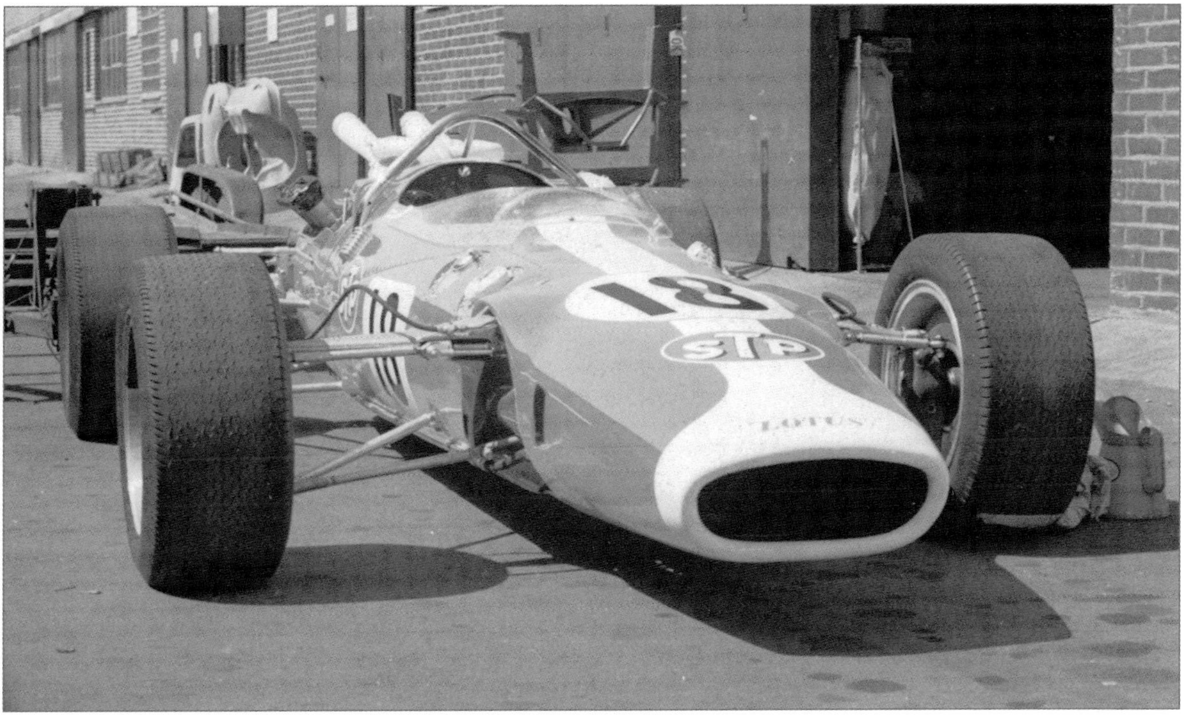

USAC legend Al Unser drove the second Team Lotus 38 (number 18) in 1966, qualifying on the seventh row at 162.27mph, and retiring after 161 laps after hitting the north-west wall.

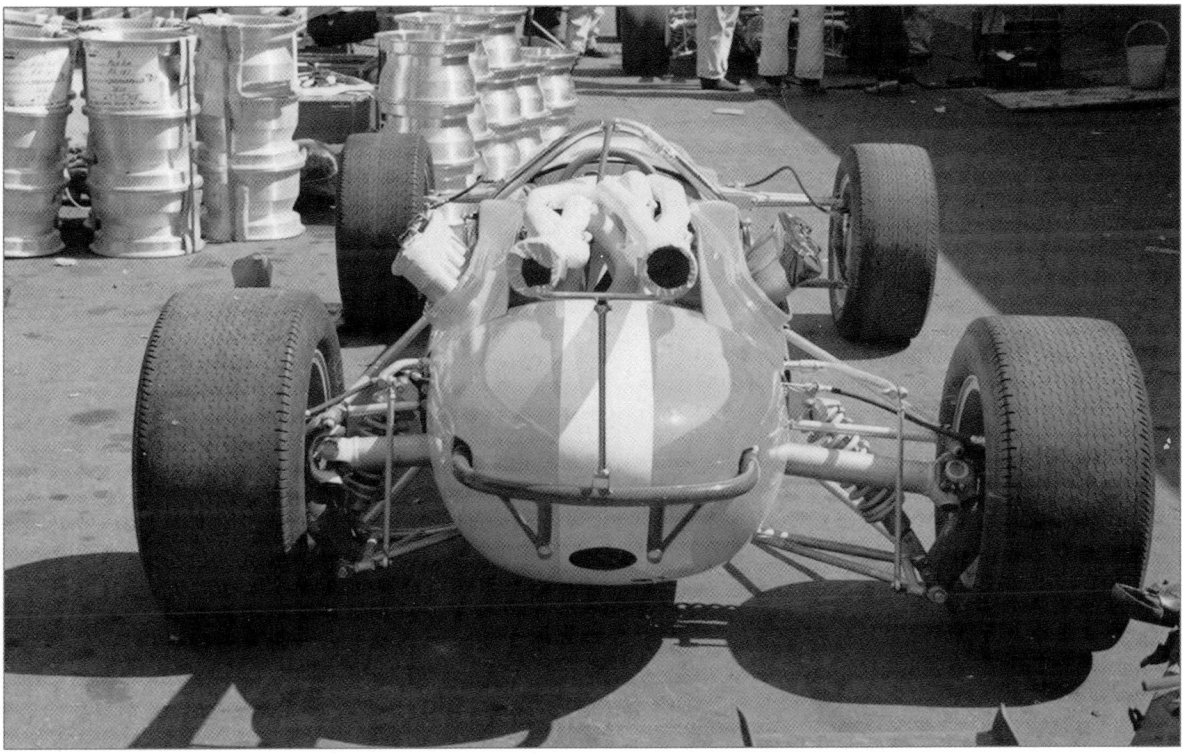

Here's what the business end of the 38 looked like with its day-glow pink STP liveried engine cover in place. Colin Chapman referred to this particular hue as 'Granatelli green', a jest referring to the Indy establishment's consternation at Lotus's regular BRG paint scheme. Green was 'unlucky' at the Brickyard.

they relied yet again on the 38 with Ford V8 power. In the event it was a low-key affair, stopped by rain when Clark was running in eighteenth place. His 38 succumbed to a broken piston when the race was restarted the following day.

The Lotus 42 was Graham Hill's mount for the 1967 Indianapolis 500. The monocoque chassis was originally made to accommodate the H16 BRM engine, but was converted with a tubular sub-frame behind the cockpit to accept the Ford V8 unit. Although Clark was a lowly sixteenth on the grid in his 38, Hill was thirty-first out of thirty-three runners, despite having qualified faster than Jimmy – but only on the second weekend of the practice period. Like Clark, Hill retired with a broken piston. The car was later sold off – as were all ex-Team cars – and ended its days in ruins after owner-driver Bill Brack shunted hugely at Lime Rock in 1969. By this time its BRM engine was long gone back to Bourne and it ran with a 5.0-litre Chevrolet unit in F5000 and Formula A events. Brack was the Toronto-based Canadian concessionaire for Lotus, and the 42 was known as the Lotus GTX, which was also appropriate as he had sponsorship from Castrol. The GTX letters were embossed on the castings for the suspension uprights.

Roots Racer

Lotus never forgot its club racing roots, and a number of quite senior employees drove Lotuses of one kind or another. New designer, Maurice Phillippe, raced a Series 1 Seven. For 1965, a single Seven was produced for the Clubman's Formula category that was essentially the province of Arthur Mallock's U2s, which looked

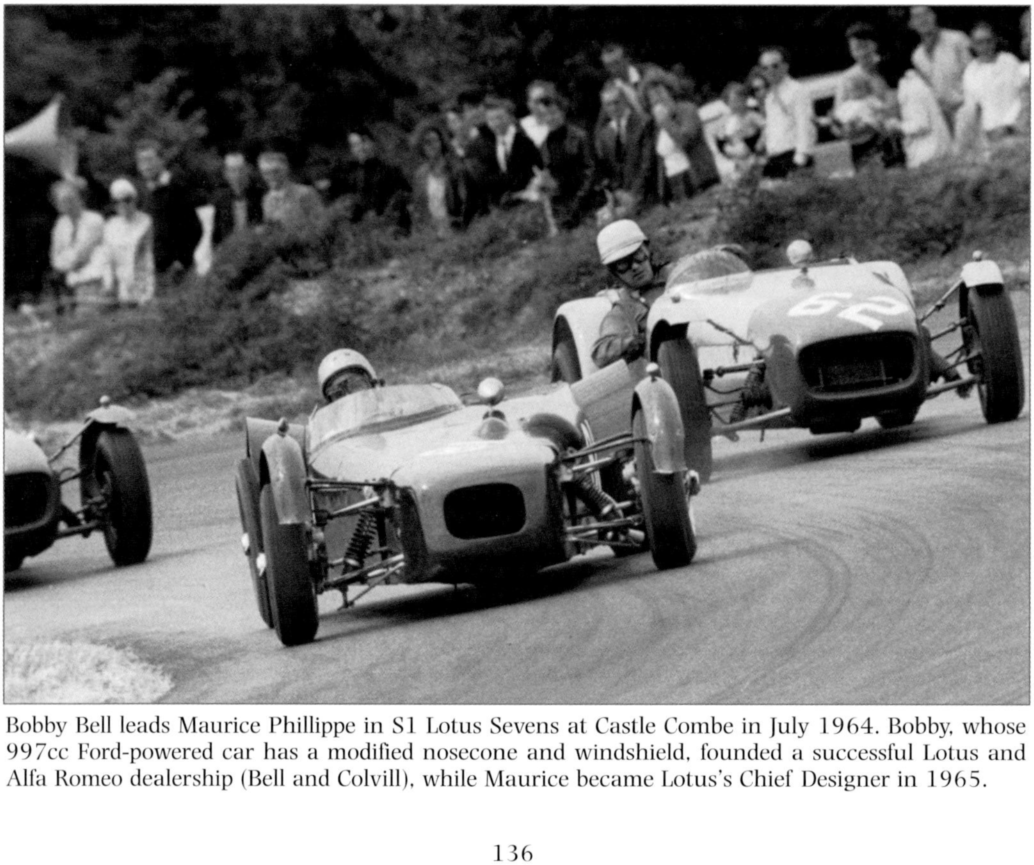

Bobby Bell leads Maurice Phillippe in S1 Lotus Sevens at Castle Combe in July 1964. Bobby, whose 997cc Ford-powered car has a modified nosecone and windshield, founded a successful Lotus and Alfa Romeo dealership (Bell and Colvill), while Maurice became Lotus's Chief Designer in 1965.

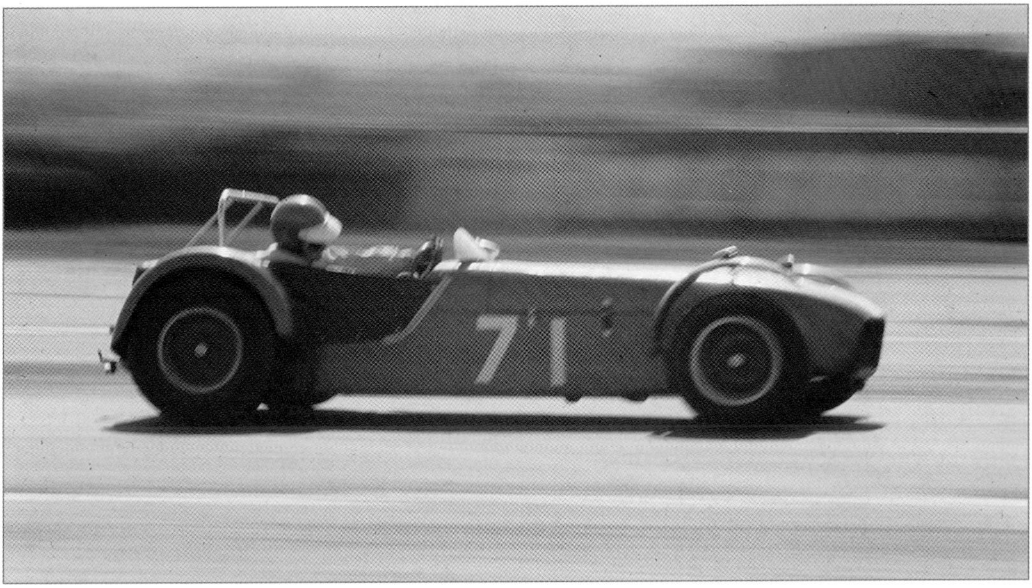

The Seven in Series 3 guise continued to be a force to be reckoned with, seen here at speed at Silverstone in 1968. It was available with 1300 Escort GT engine, or 1600 Ford 225E Cortina unit, and 13 SS models were fitted with the Lotus-Ford Holbay-tuned twin-cam. A single car, known as the Holbay S, used a highly tuned 225E engine with R120 high-lift cam and dry sump lubrication.

like Sevens that had been squashed. The 'Three Seven' was Lotus type number 37 and was exhibited at the 1965 Racing Car Show. It was basically a heavily modified Series 2 Seven powered by a 125bhp Cosworth-Ford 116E 1499cc engine, with independent rear suspension, an Elite differential and inboard Girling discs at the back. It ran on 13 inch magnesium-alloy wobbly-web wheels, and featured with considerable success in Clubmans' races in the hands of John Berry. Five years later, another ground-breaking Seven appeared. It was designated the 7X and was raced in club events by Tim Goss.

Versatile Single Seater

The 41 was launched by Lotus Components late in 1965 as a Formula 3 contender. It had a tubular spaceframe chassis similar to that of the 22 and 31 rather than a monocoque. This design was chosen because it was simpler to build and easier for customers to repair. The weight factor was of little consequence, since the cars had to carry ballast in any case to meet the minimum weight limit for F3. Additional strengthening was provided by sheet steel around the pedal box and the undertray, and there were stressed steel bulkheads fore and aft of the cockpit. The chassis tubes also transported oil and water between the engine and radiators. Suspension was by double wishbones and outboard coil springs and damper units at the front, with reversed lower wishbones, single top links, radius rods and outboard coil springs and dampers at the rear. The 41 in F3 form was powered by the Cosworth MAE66, canted at 30 degrees to the right. It developed 100bhp at 9500rpm, driving through a Hewland Mk 4 four-speed transmission. The vehicle ran on six-spoke cast magnesium 13 inch wheels, with 10½ inch Girling disc brakes mounted outboard front and back.

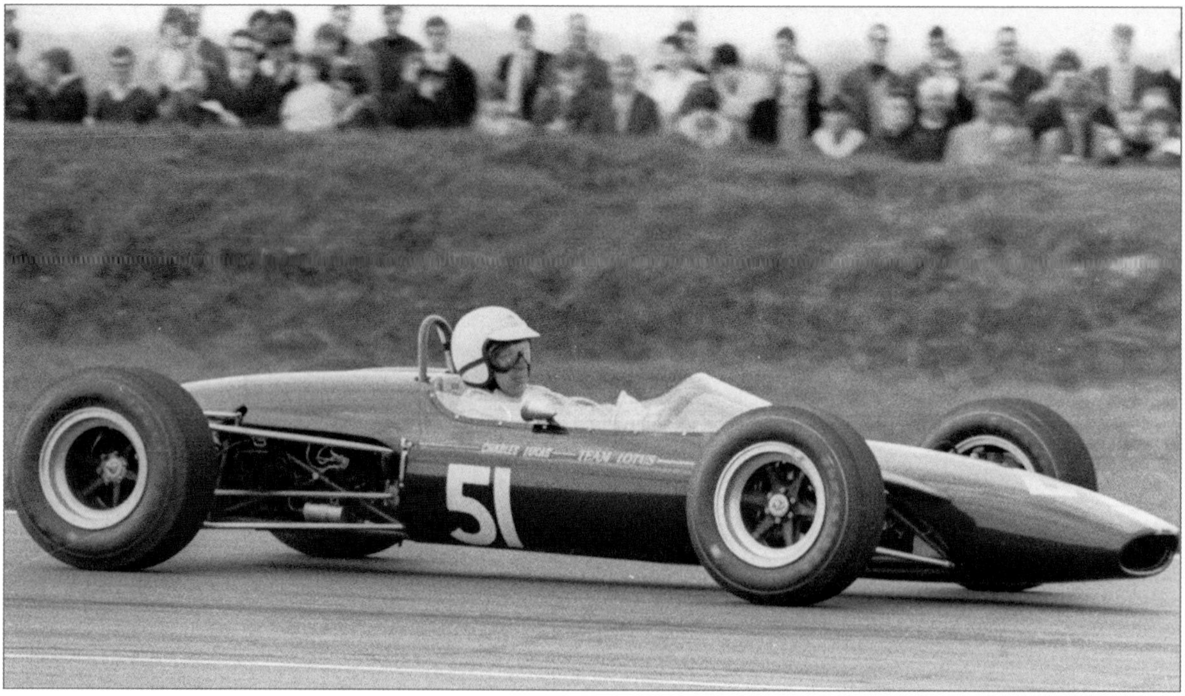

The works' cars were run in Formula 3 by Charles Lucas-Team Lotus for Roy Pike and Piers Courage, seen here at Snetterton in 1966. Piers drives the 41, a spaceframe car powered by the mandatory 997cc Cosworth SCA screamer. Although the spaceframe was at first seen as regressive, it was a reliable basis for a customer F3 car, and far less costly to produce than the monocoques. It was similar to the previous 22/31 spaceframe, and Lotus Components produced sixty-one of the 41.

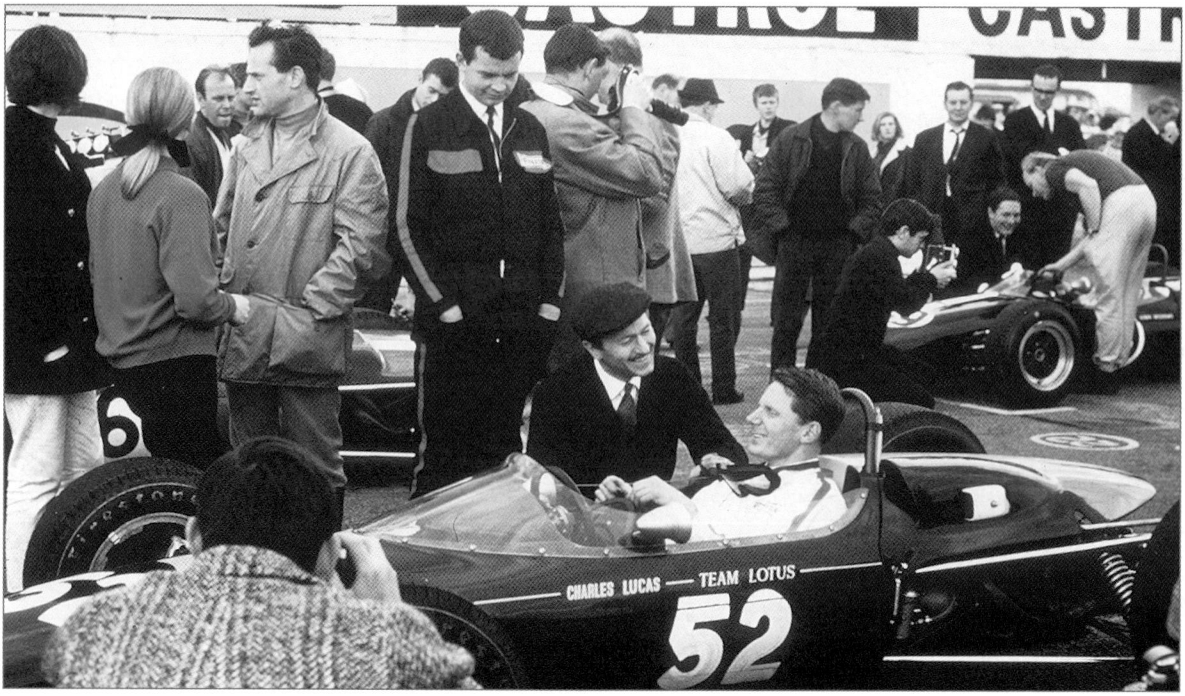

Colin Chapman jokes with Roy Pike on the Snetterton grid in 1966. The Charles Lucas-Team Lotus 41 was the car to beat in Formula 3, and Pike won two major F3 races to Courage's five. These events were notable for incredibly close slipstreaming battles along the straights, in which all the main protagonists excelled.

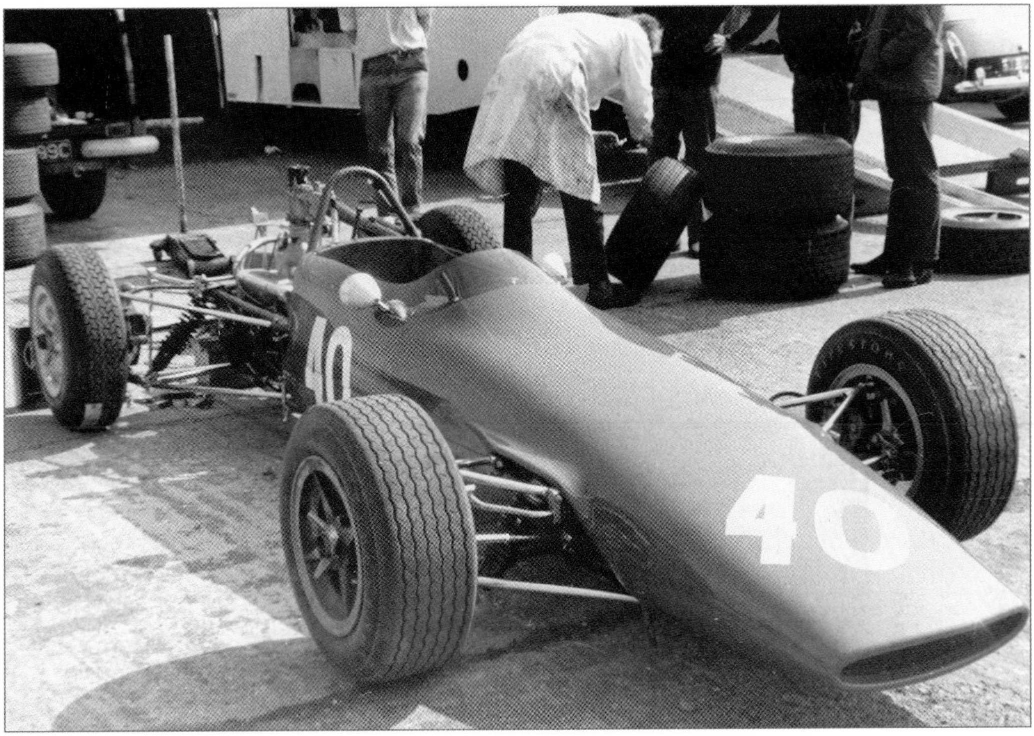

Unquestionably one of the sleekest racing car bodies, the 41X was the works Formula 3 development car for 1986, based on the 41C of 1967. It used Lotus 47 front suspension uprights and wheels that made it more stable under braking. The wedge-shaped engine cover is missing here and diminutive mule rear wheels are temporarily fitted. John Miles debuted the car at the 1968 Silverstone International Trophy meeting, winning here, plus at Brands Hatch, Croft and Zandvoort. The 41X was driven by Mo Nunn when Miles was busy with the GT cars, and if productionised it would have been the Lotus 55. But the car was shelved.

The Lotus works cars were run by Charles Lucas Racing under the banner Charles Lucas-Team Lotus. Lucas was no mean F3 protagonist himself, and was surrounded by a coterie of up-and-coming drivers who enjoyed a party. His drivers were Piers Courage and the American Roy Pike. Courage took five significant victories to Pike's two, Courage winning the French 'Craven A' series, while the Lotus Components 41 was driven by Jackie Oliver with some successes in minor F3 races. Three further cars were run by Ron Harris Racing for Peter Revson, John Cardwell and Rob Slotemaker, while privateers to show well in 41s included Derek Bell, John Hine and Mo Nunn.

The 41 was sufficiently adaptable to be used as a Formula B and F2 car when fitted with a heftier chassis and Cosworth FVA engine, and it was sold to the States as the 41B – Canadian Lotus dealer Bill Brack was a notable customer. An example was entered by Lotus Components in Formula 2 for Jackie Oliver. The 41C was the 1967 version of the Formula 3 car, entered by Lotus Components for Mo Nunn and John Miles. The company produced a total of sixty-one units of the 41. In 1968 there was a postscript to the 41, and this was the works' built 41X, equipped with suspension uprights and wheels from the Lotus 47 GT car, and clad in a wedge-like fibreglass body reminiscent of the '56 Indianapolis car. In the hands of John Miles, and entered by Gold Leaf Team Lotus, the 41X notched up wins at Silverstone, Brands Hatch, Croft and Zandvoort. Mo Nunn drove the car when Miles was otherwise occupied. However, Tecno was the team to beat in F3, and Lotus won just seven out of sixty-five European Formula 3 races. Plans to productionise the 41X as the 55 were shelved.

In 1966, the monocoque construction 44 was developed from the 35 for Formula 2. It was powered by the Cosworth SCA engine and endowed with the wider track suspension of the 41 Formula 3 car. Two cars were built and run by Ron Harris-Team Lotus but the vehicles to beat were the Honda-powered Brabhams. The 44 made its first appearance at Barcelona, where Jim Clark retired with engine trouble. While Clark scored a few podium places, none of the eight drivers who drove the second 44 achieved any notable success.

A Multitude of Cylinders

During the gestation period of the 49, Team Lotus raced the 33 with Climax and BRM V8 power during 1966 and in the 1967 Monaco Grand Prix. They also built the 43 to accommodate the highly complicated BRM 3.0-litre H16 engine and six-speed BRM gearbox. It prefigured the 49 in so far as the BRM unit was mounted on the rear bulkhead of the stressed-section monocoque and acted as a chassis member. In fact from certain angles the 43 looked like a bulbous version of the 49. The 43's front suspension was similar to that of the 33 with upper rocking arms, lower wishbones and inboard coil springs and damper units, while the rear end featured reversed lower wishbones, lateral top links and coil springs and damper units.

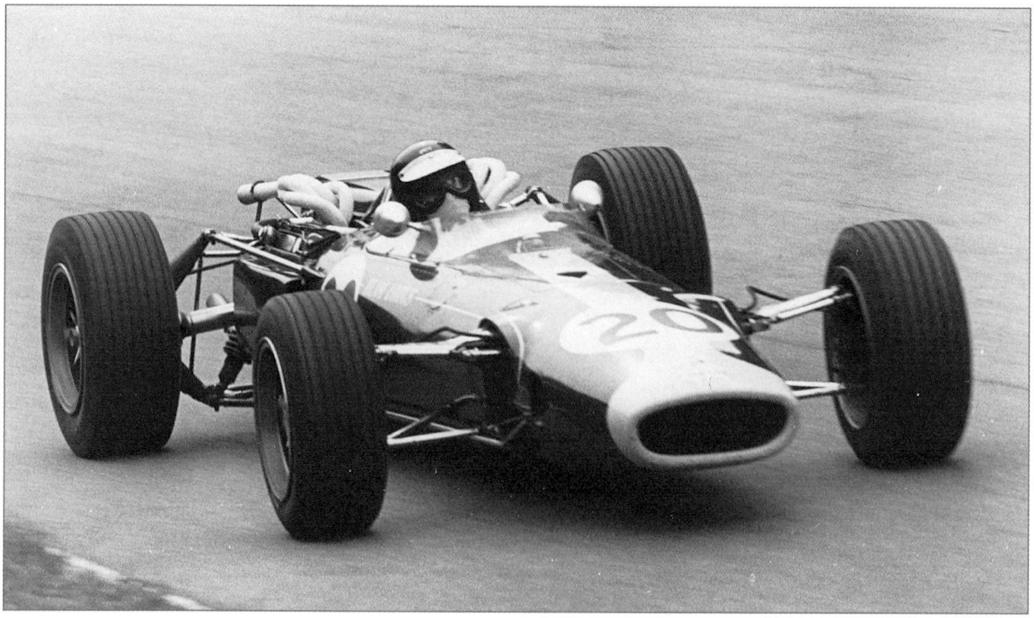

The Lotus 43 was designed by Maurice Phillippe and built up around a full monocoque tub that ended at the rear of the cockpit. The engine was bolted to it as a fully stressed chassis member. The power unit was the Tony Rudd-designed 2996cc sixteen-BRM H16, which developed 405bhp. The 43 was debuted at Reims in 1966 by Peter Arundell, retiring after four laps with transmission trouble. Two 43s were made, but the combined engine and gearbox assembly weighed as much as the F1 minimum weight limit, and the car's only success was Jim Clark's victory in the US GP at Watkins Glen.

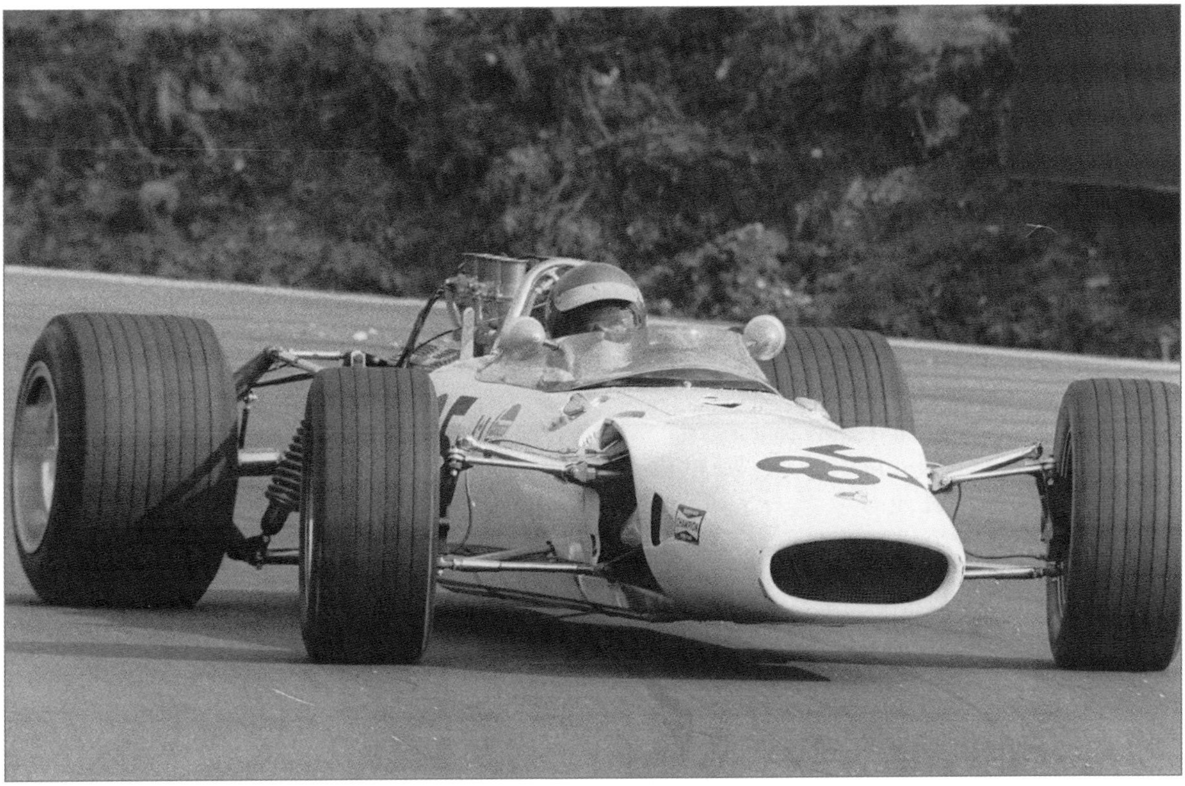

What happened to pensioned-off Formula 1 cars, or even the odd Indianapolis racer? In the late 1960s, their destiny could be Formula 5000, and this is Lotus's Canadian distributor Bill Brack in an ex-Team Lotus 1967 Indianapolis 42. Brack acquired both 42B/1 and B/2, and fitted a 5.0-litre Ford V8 lump to go racing in F5000 and Formula B. He's pictured at Brands Hatch in April 1969, but by August that year one of the cars was in ruins, having been somersaulted by Brack at Lime Rock, Connecticut.

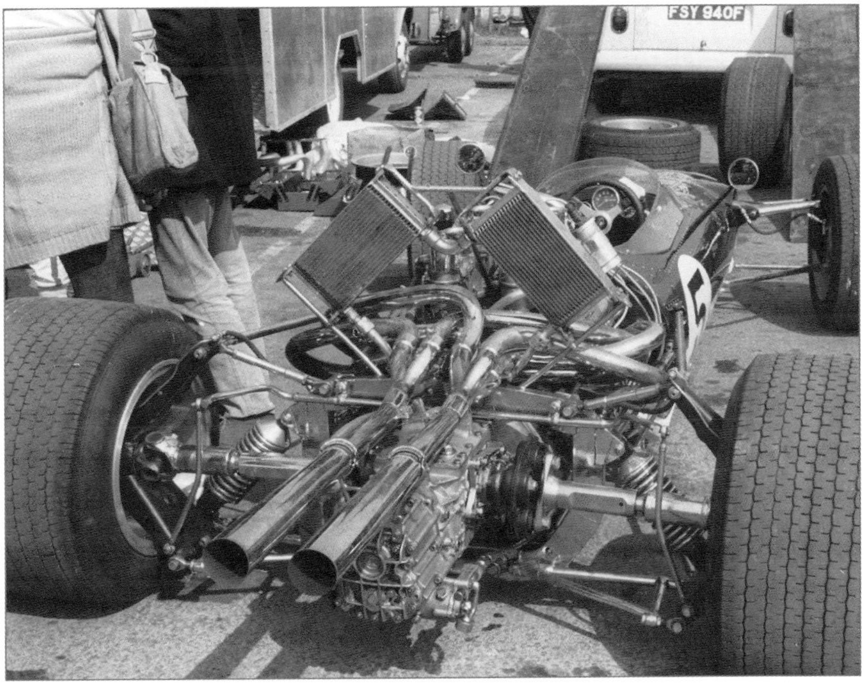

One of the ex-Team Lotus 43s, relieved of its H16 engine, was also equipped with a Ford 4.7-litre unit – plus curious oil cooler arrangement – by Robs Lamplough in 1968, and was acquired by Jock Russell in 1968. It was severely damaged at Brands Hatch the following year during a Formula 5000 race.

The sixteen cylinder H16 unit was designed by BRM's Chief Engineer Tony Rudd, and he eventually joined Lotus in 1969. The unusual layout of this power unit was based on the 1.5-litre V8 engine, with the angle widened to 180 degrees, making it a flat-eight, and a second bank of eight cylinders placed on top of it. Not surprisingly, there were initial difficulties in mating the separate crankshafts, but it howled magnificently on full song. The BRM engine wasn't ready until the second race of the season at Spa, and Arundell in the single 43 failed to make the start. He retired at Reims, and the 43 didn't appear again until Monza, when Clark drove it. Many were astonished when it survived the entire US Grand Prix at Watkins Glen to give Clark the win. It was Lotus's only World Championship victory of 1966 and the only World Championship win that the H16 engine achieved. As the Team waited for the 49, the 43 was pressed into service for both Clark and Hill for the first race of 1967, the South African Grand Prix at Kyalami, but both retired. The two 43s were pensioned off and featured in Formula 5000 events.

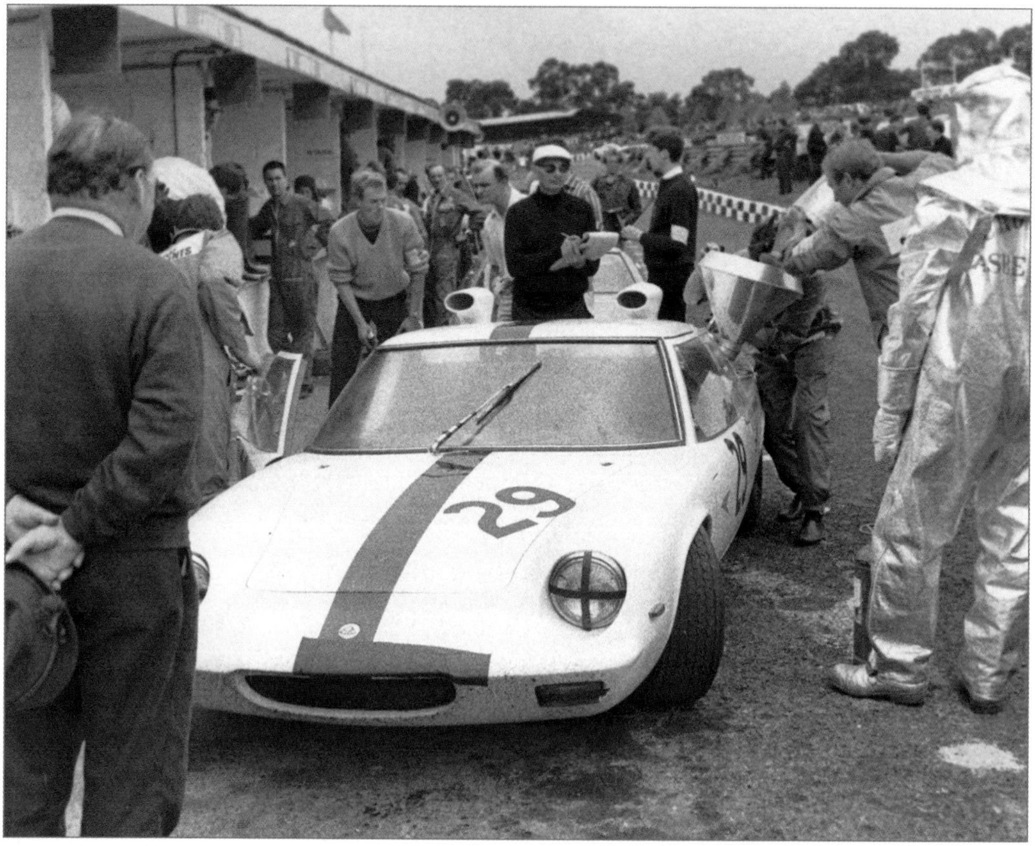

The 47 was the competition derivative of the mid-engined Europa, and in Group 4 trim it was powered by the 165bhp 1594cc Lotus-Cosworth twin-cam engine. The curious 'hair-dryers' protruding from the engine cover are air intakes for the T-J fuel injection. The event is the 1967 BOAC 500 6-Hour race, in which Jackie Oliver and John Miles finished ninth overall and first in the under 2.0-litre class in the Lotus Components car. The Taylor/Preston Team Elite 47 came nineteenth and last, while the Chris Barber 47 of Hine/Greene retired. The race was won by the winged Chaparral 2F of Mike Spence/Phil Hill, with masses of interest from Ferrari P4s, Mirages, GT40s, Lola T70s and works Porsche 907s.

A Car for Europe

Colin Chapman had originally perceived that the mid-engined John Frayling designed Europa would be a replacement for the Seven and would be produced by Lotus Components. In January 1965 he entered negotiations with Renault as a supplier of 1470cc engines from the R16 for the new Europa. The car was launched at the end of 1966. At first, all were exported to France.

By this time Lotus had moved lock, stock and barrel to Hethel, a move accomplished amazingly over one weekend in November. The majority of the workforce moved with it, and cars in build were transported in trucks. By the afternoon of the first Monday, finished cars were rolling off the line. Eventually the Europa proved too complex to serve as a Seven replacement, and it was built as a main-line vehicle.

The racing version of the Europa was constructed by Lotus Components, fitted with the twin-cam engine, and designated the 47. Its debut came at the 1966 Boxing Day Brands Hatch meeting – an ideal occasion for sustaining the festivities of the previous day – and John Miles won the prototype class in the sports car race. The 47 couldn't be homologated as a Group 4 racer until fifty units had been built, however, and some cars got plain Cortina engines to speed up the process. It's even possible that only thirty units were ever completed.

The 47's chassis was similar to the Europa's, but somewhat lighter, and this caused flexing and cracking of the bodywork. More fundamentally, its proportions differed in that the T-shape front cross-member was deeper than the Europa's, and the rear Y-section arms were shorter and buttressed by a box-section cross-piece in order to accommodate the engine. This was the 165bhp dry sump Lotus-Cosworth 1594cc twin-cam, with Tecalemit-Jackson fuel injection on the works cars, and two twin-choke Weber 45 DCOE carburettors on most private owners' cars. Transmission was via a Hewland FT 200 five-speed gearbox and final drive unit. The 47's suspension was similar in layout to its road-going sibling, but there were light alloy rear uprights with separate upper tubular links, lower reversed wishbones, and the tubular upper and lower trailing arms were longer. The 47 had disc brakes all round, located inside the rear suspension uprights, and it was fitted with centre-lock magnesium-alloy wheels. A pair of aluminium fuel tanks situated behind the rear bulkhead carried 20 gallons.

The 47 made little impact abroad, although John Miles won eight races in the UK during 1967. More notable was his ninth overall and 2.0-litre class win partnered by Jackie Oliver in the BOAC 500 Six Hour race at Brands Hatch, pitched in with the cream of the World Championship for Makes. The following year the works fielded a pair of cars for Miles and Oliver, liveried in the red, gold and white of the new sponsor John Player's Gold Leaf cigarette brand. Two other top names drove 47s – Trevor Taylor and John Hine – but they were less successful. By this time the 47 had a stronger chassis, better brakes and a detachable body.

The race record for 1968 shows that Miles was sixth overall. He won the 2.0-litre class in the John Player Trophy at Silverstone, was second in the 2.0-litre class of the sports car race supporting the *Daily Express* International Trophy meeting, and took third at the Vila Real sports car race in northern Portugal. In the Guards Trophy at Brands Hatch, Oliver came sixth overall and won the 2.0-litre class. A few fair results, then, but nothing spectacular. Team Lotus moved on to experiment with the bigger 62 in 1969, while privateers carried on with the 47. One of the most successful exponents was Chris Meek.

New for F2

A new set of regulations for Formula 2 came out in 1967, with a limit of 1600cc, and Team Lotus built the 48 to contest the series. The 48 was an attractive looking car consisting of a monocoque tub with a spaceframe engine bay, and its suspension was similar to its immediate predecessor the 1.0-litre Lotus 44. At the front were upper rocking arms, lower wishbones and inboard coil springs and damper units, with reversed lower wishbones, top links, twin radius rods and coil springs and dampers at the rear. The 48 was powered by the Cosworth FVA 1599cc engine, mated to the ZF 5DS12 transmission after the old Hewland Mk V gearbox failed on Hill's car in the Australian Grand Prix at Warwick Farm. It was to prove a more successful car than the 44, winning four of the twenty-five international F2 events in 1967, and placing well in others.

Jimmy Clark won at Barcelona, Jarama and Keimola, and was second at Zolder, and third in Scandinavia at Karlskoga and Hameenlinna and at Albi in France. Jackie Oliver's regular mount was the 41B, but he drove the 48 when Clark was busy elsewhere. He won the Formula 2 class of the German Grand Prix in the 48 and was placed fifth overall. Graham Hill scored second places in F2 events at Snetterton, Reims and the Oulton Park Gold Cup.

The 1968 Hockenheim tragedy is well documented and, to use the old President Kennedy analogy, enthusiasts of a certain age well remember where they were when

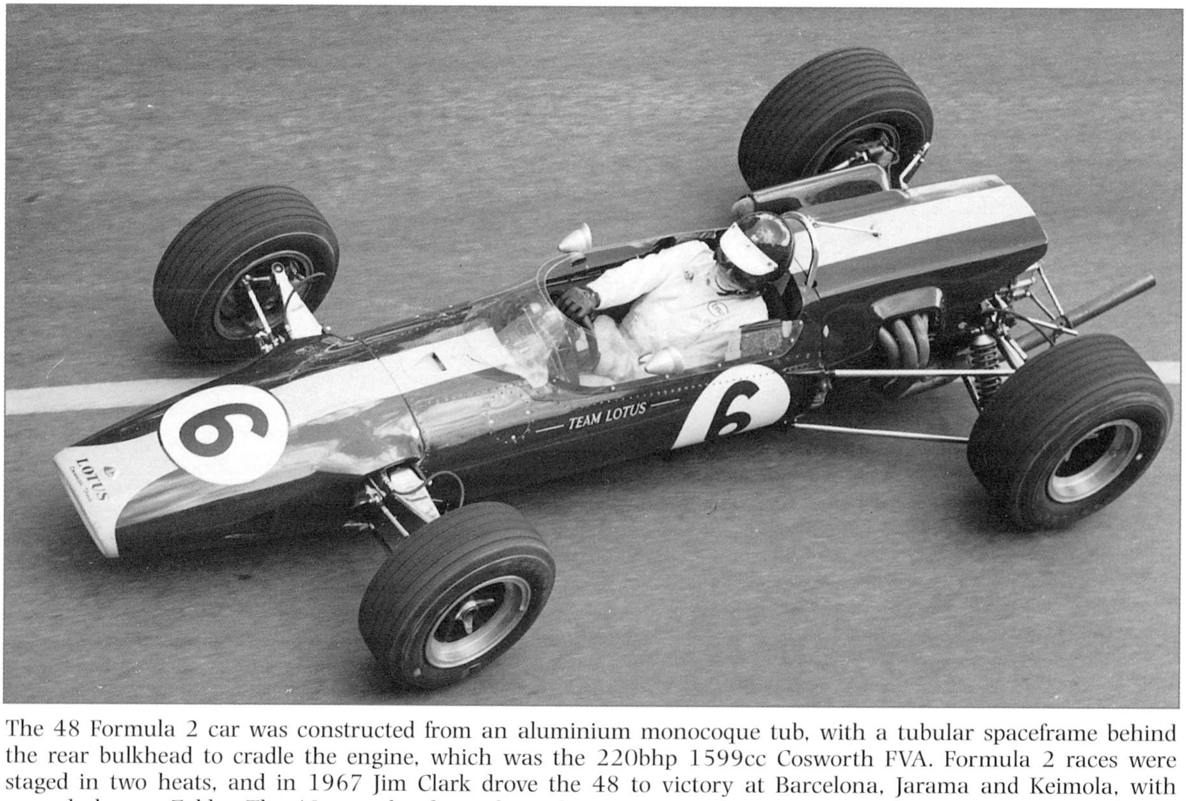

The 48 Formula 2 car was constructed from an aluminium monocoque tub, with a tubular spaceframe behind the rear bulkhead to cradle the engine, which was the 220bhp 1599cc Cosworth FVA. Formula 2 races were staged in two heats, and in 1967 Jim Clark drove the 48 to victory at Barcelona, Jarama and Keimola, with second place at Zolder. The 48 was also driven by Jackie Oliver when Clark and Hill were away, and he won the F2 division of the German Grand Prix in 1967.

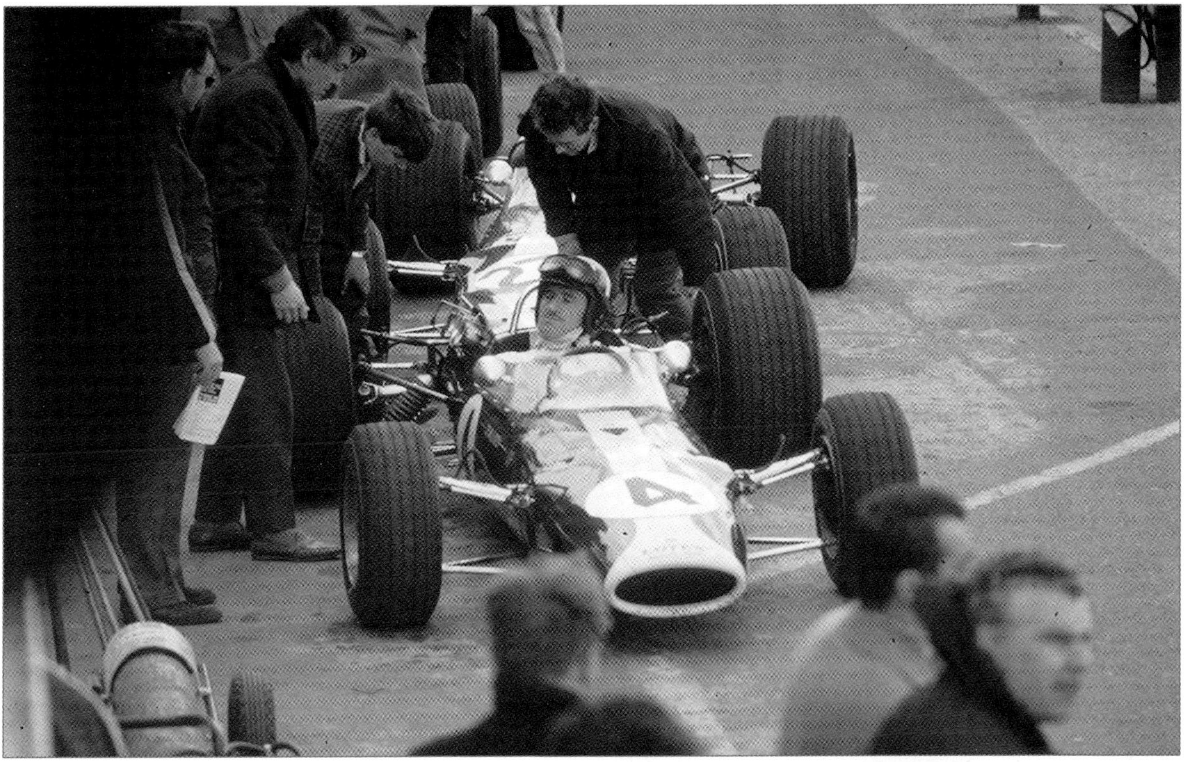

Graham Hill in the Team Lotus 48 F2 car in the Oulton Park pits during the Gold Cup combined Formulae 1 and 2 race on 16 September 1967. Jim Endruweit is to the left, and mechanics Bob Sparshott and Dick Scammell examine the back of the car. Graham came third behind the F1 car of Brabham (Brabham-Repco BT 24) and Stewart's F2 Matra MS7, while Jackie Oliver drove the other Lotus 48 but retired.

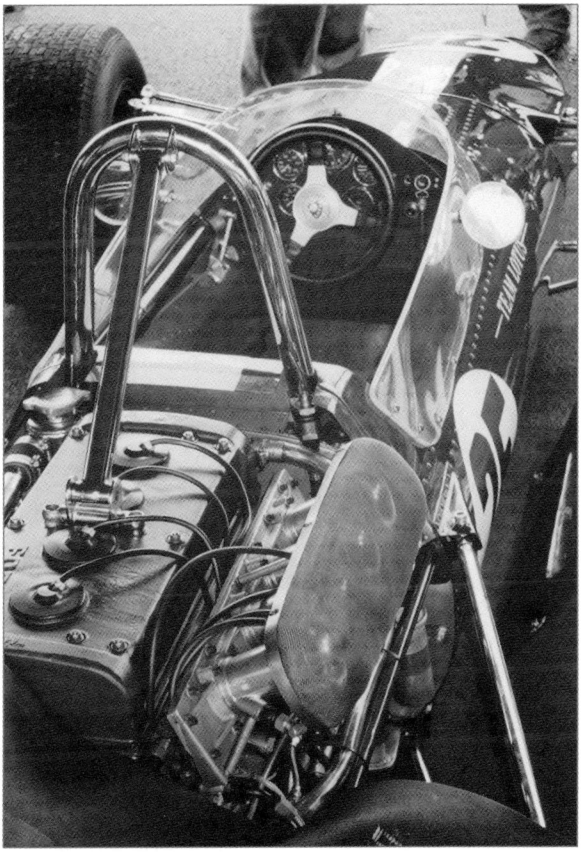

Close up of the 1599cc Ford-Cosworth FVA, minus engine cover, of Bob Tabor's 48, which was driven by Simon Hadfield at the Goodwood Revival meeting in 1999. The roll-over hoop is braced from the cam cover.

they heard Jim Clark had been killed. Many of us expected to see him racing an Alan Mann-run Ford P65 prototype in the BOAC 500km at Brands Hatch, and indeed he had provisionally agreed to do so since the new sports racer had the full backing of Ford's hierarchy. But the drive was never officially confirmed, and it appears that he assumed his services weren't required and opted instead to race at Hockenheim. By the time Alan Mann Racing confirmed his Ford prototype drive, it was too late. Clark was never one to break his word, so he stood by the entry for the Lotus 48 F2 car in Germany. Jimmy Clark could instinctively adapt to any deficiency in a car and drive around it, to the extent that he could overlook a slow puncture until the tyre was completely flat. But at Hockenheim, when a deflating tyre suddenly left the wheel rim as Clark negotiated a high-speed bend, his 48 was pitched off the Armco-less circuit and into the trees. Ten or twenty years on, it was the kind of accident that he'd have simply walked away from, or maybe never had in the first place. A postscript to the crash was that later wheels were fitted with a bead that prevented damaged tyres from leaving the rim in such an abrupt fashion.

The 48 can be seen as a transitional car because, like the 47 and 49, it was raced in Lotus's traditional colours as well as those of its new sponsor in 1968. Nowhere is this fact more poignantly recalled than in the statement by RAC scrutineer Peter Jowett called in by Chapman to investigate the wreck of Jim's 48. As recorded in the Graham Gauld book *Jim Clark Remembered*, the 48's original British Racing Green was visible beneath the red, gold and white of the Player's livery where the tub was damaged.

It was left largely to Graham Hill to remotivate the team, and a distraught Chapman was on the verge of stopping racing altogether. Indeed, Jowett thought that had Clark's accident been caused by a failure of the car, rather than a

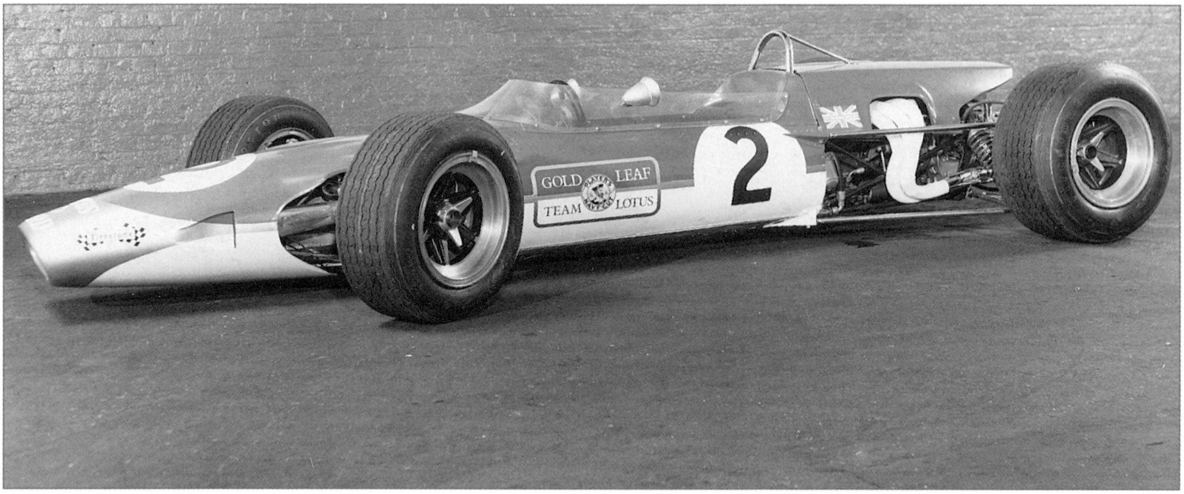

The 48 F2 car is painted in the Gold Leaf cigarette livery of Team Lotus's new sponsors John Player & Son. Jim Clark drove a 48 at Barcelona with an extra inch in the tub to accommodate his taller team-mate, but was sidelined in a shunt from Ickx's Ferrari. Then at Hockenheim the sudden decompression of a deflating tyre caused him to crash with fatal consequences. Meanwhile, Graham Hill had eight races with his 48 during the 1968 season and finished just two, with fourth at Reims his best result. Jackie Oliver's 48 was painted in the British Racing Green of the Herts and Essex Aero Club, and he finished eleven out of eleven F2 races with second place his best result at the ill-starred Hockenheim event.

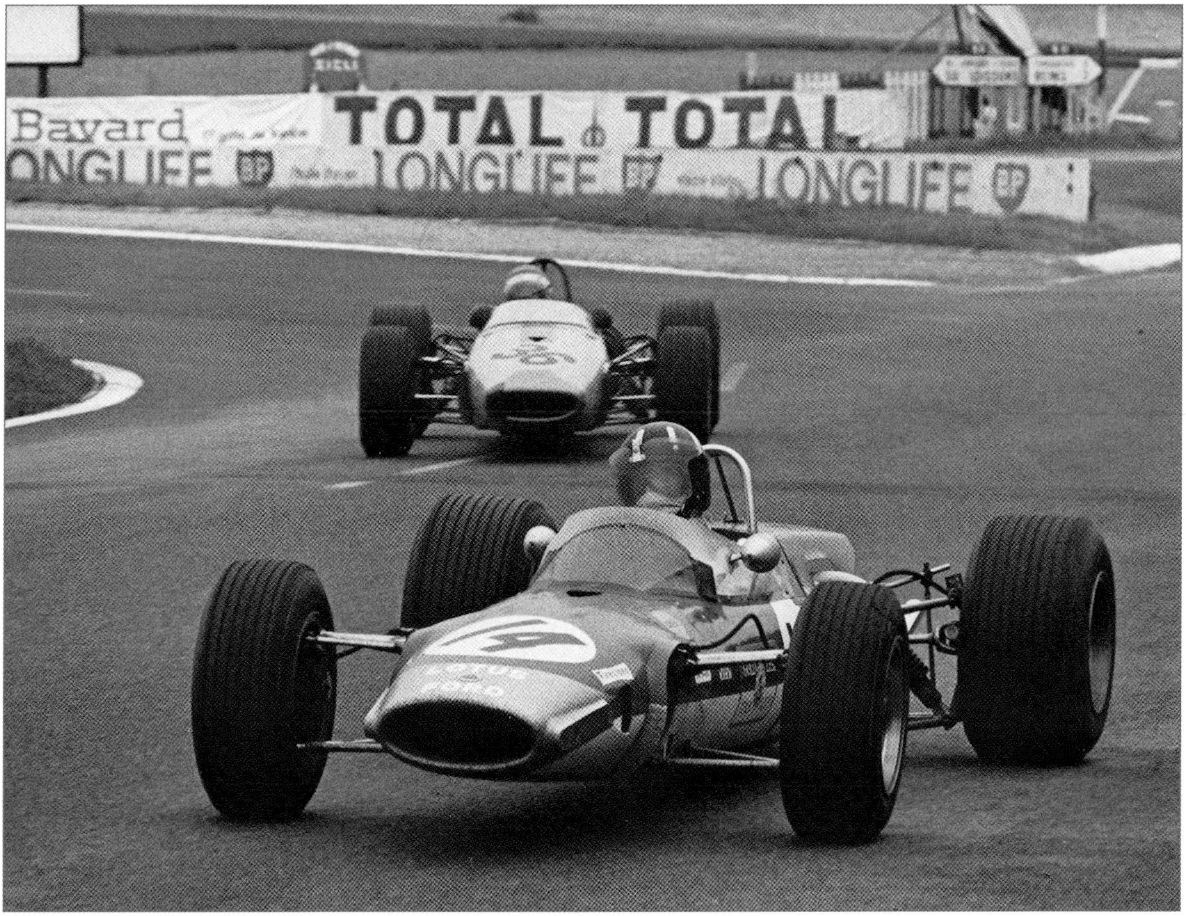

Graham Hill drove the 48 to fourth place in the F2 race at Reims on 15 September 1968, heading home Brian Hart's Brabham-Cosworth by just over a second. The 35-lap race lasted just over 1 hour 23 minutes, and Jackie Stewart's winning average was 207.869kph.

component outside his control, Chapman would indeed have quit completely. The situation was compounded only a month after Clark's death, when Mike Spence was killed in qualifying at Indianapolis when he was struck on the head by a front wheel off his STP-Lotus 56 gas turbine as it glanced off the wall. The impact of Clark's loss was starkly clear to race fans as well as to the Team. Suddenly, everyone else seemed doubly vulnerable in a racing car. If Clark, the absolute maestro, could get killed, what chance did the rest of them stand? Meanwhile, Graham Hill soldiered on, and drove a 48 in eight F2 races that year, achieving nothing at all in the way of results.

Just four examples of the 48 were built, and the car that Graham had first driven in Australia in early 1967 was raced by Jackie Oliver during 1968 with sponsorship from the Herts and Essex Aero Club. Oliver finished second at Hockenheim and when I saw him drive it to fourth place in the F2 race at Crystal Palace on Whit Monday, in the wake of Clark's death, Jackie, with his dark helmet in a dark green Lotus, looked uncannily like Jim.

Chapter Seven

Another Milestone Car

Without question, the Lotus 49 was one of the greatest racing cars ever. It was a car of its time and an example of the ingenuity and audacity of Colin Chapman. Back in 1965, Chapman had already been negotiating for an engine that would power his cars in the 3.0-litre Formula 1, and it seemed his best hope lay with Aston Martin. Almost resigned to the potential of having a dominant David Brown on board, Chapman managed to convince Ford's Walter Hayes of the potential publicity inherent in sponsoring the development of an F1 engine that would be built by Cosworth. The resulting DFV – double four-valve – engine was developed within the extremely modest budget of £100,000, and Team Lotus had exclusive use of it for one season. Key figures in the development of the 49 were Chapman, Phillippe, Duckworth, Costin, and of course Hill and Clark. Chief mechanic Dick Scammell – now a Cosworth director – had the task of strengthening anything that broke. Mechanics involved with the 49 included Allan McCall, Dale Porteous, Leo Wybrott, Dougie Bridge, Gordon Huckle, Bob Dance, Eddie Dennis, Trevor Seaman, and Bob Sparshott. Wybrott, Dance and Dennis went on to be Team Lotus chief mechanics.

A full description of the genesis of the DFV engine, and indeed the story of the car itself, can be found in the thoroughly excellent *Lotus 49* book by Michael Oliver. Suffice to say that the 90 degree V8 grew out of the sixteen-valve four-cylinder FVA 1600 twin-cam that was in service in Formula 2 and sundry other applications. It had the advantage of being compact, with the water and oil pumps, distributor and ancillaries located down beside the crankcase. Indeed, the DFV's width was greater than its length, and like the H16 in the 43, it was mounted to the tub's rear bulkhead to form a stressed unit with the monocoque. Power output was rated at 408bhp at 9200rpm, and fuel consumption was around 5mpg.

The 49's tub was made of 18-gauge L72 Alclad aluminium-alloy sheet, with a fabricated front bulkhead and a sub-frame carrying the oil and water radiators. Up front, the suspension was by upper rocking arms, inboard coil spring and damper units, with a fabricated lower link and radius rod on each side. The rear suspension comprised single top links, reversed lower wishbones, twin radius rods and inclined outboard coil springs and damper units. These components were attached to a tubular frame that was located at pick-up points on either cylinder head, and a cross-frame, which supported the ZF 5DS12 five-speed gearbox and final drive unit. Chapman remained steadfastly loyal to the German make even though Duckworth recommended the new Hewland box for ease of changing ratios. This was an important factor because of the DFV's power delivery. It was desirable to have exactly the right gear ratio for every corner on any given circuit as the power came in quite suddenly at 6500rpm. By 1969, Hewland gearboxes were fitted. Sliding-spline joints were originally incorporated within the output shafts, later replaced by solid torsion shafts.

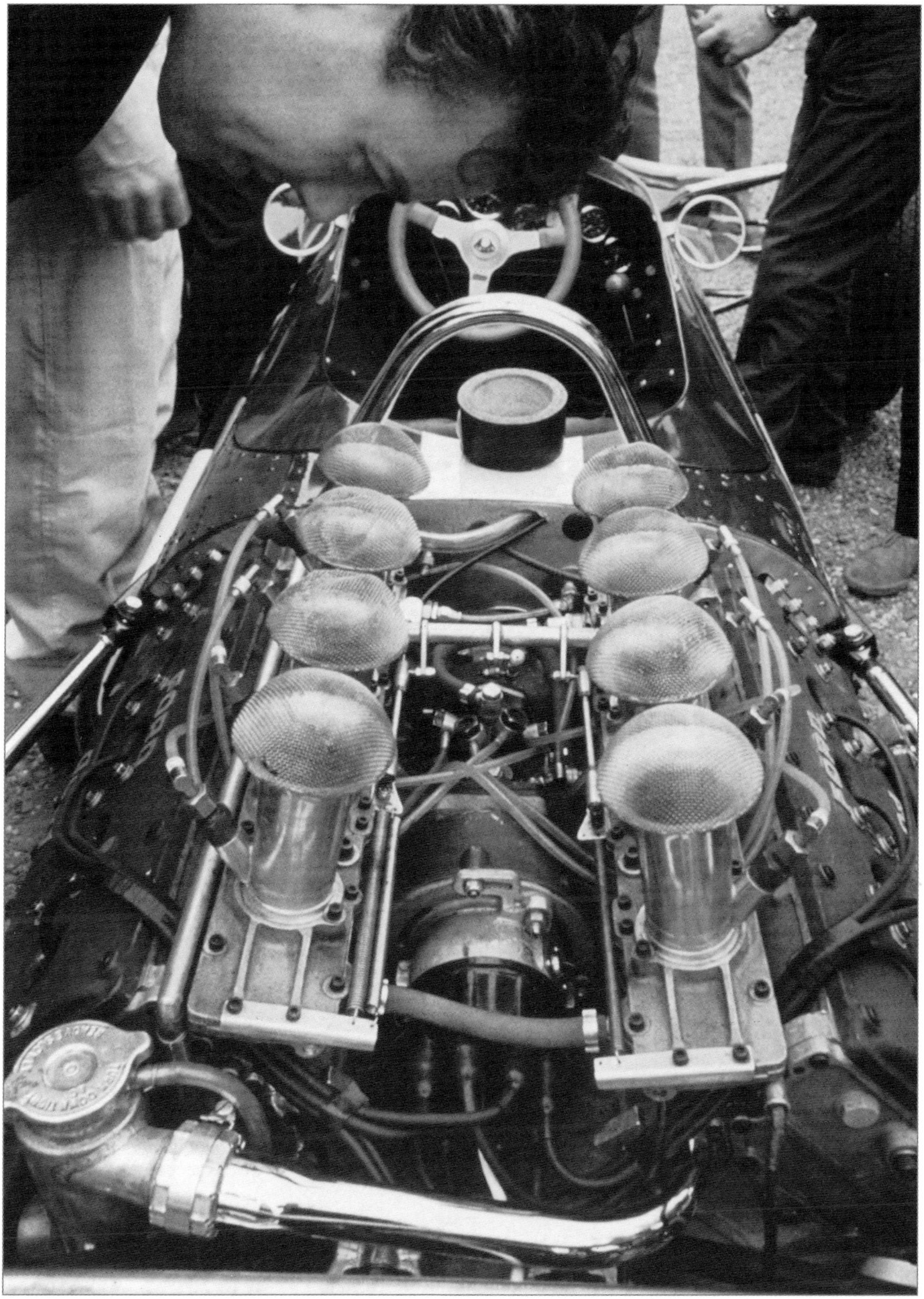

From drawingboard to running engine, production of the Ford-Cosworth DFV took less than a year. Its principal creator Keith Duckworth looks into the 2995cc 90 degree V8. It had a flat-plain crank, pent-roof combustion chamber with four-valves-per-cylinder. A half-speed compound gear within the V of the engine drove the four camshafts, plus belt drive for the water, oil and fuel pumps. It also provided drive for the fuel injection metering, distributor and alternator. Note the roll of black tape!

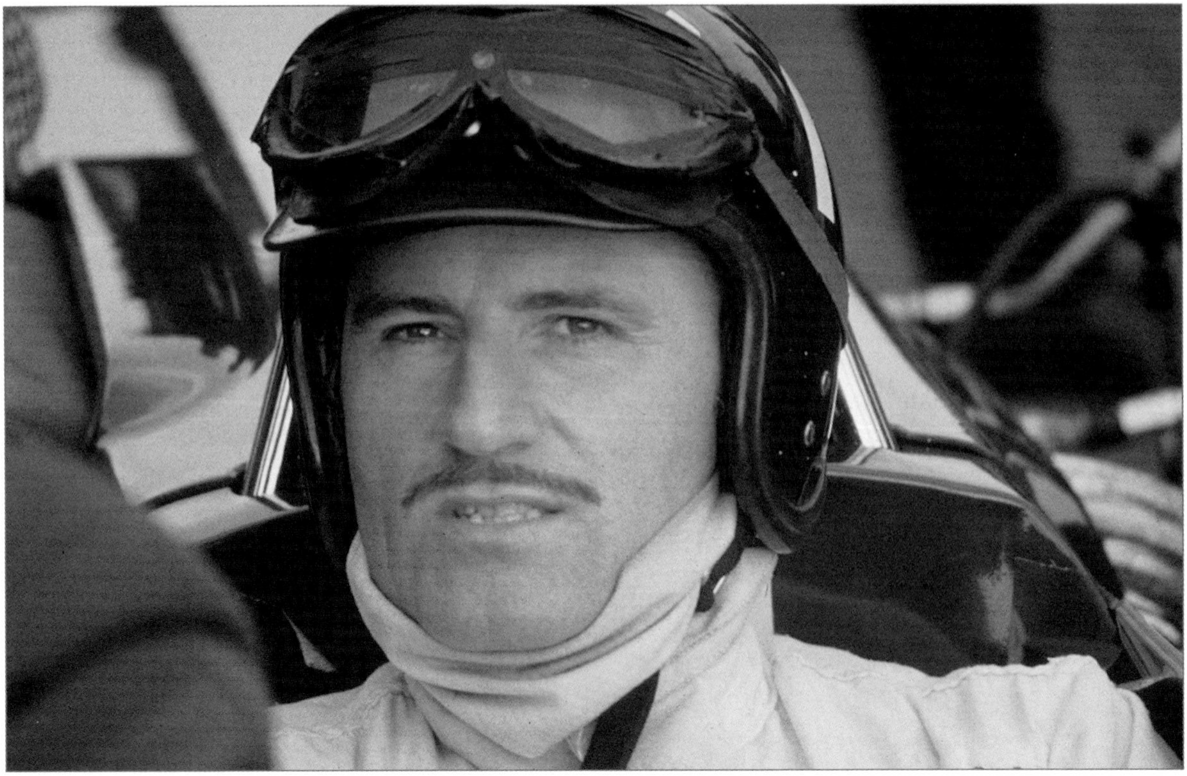

Debonair and determined, Graham Hill rekindled an old alliance when he joined Team Lotus for 1967. On his day he was as quick as Clark, but rather harder on his equipment.

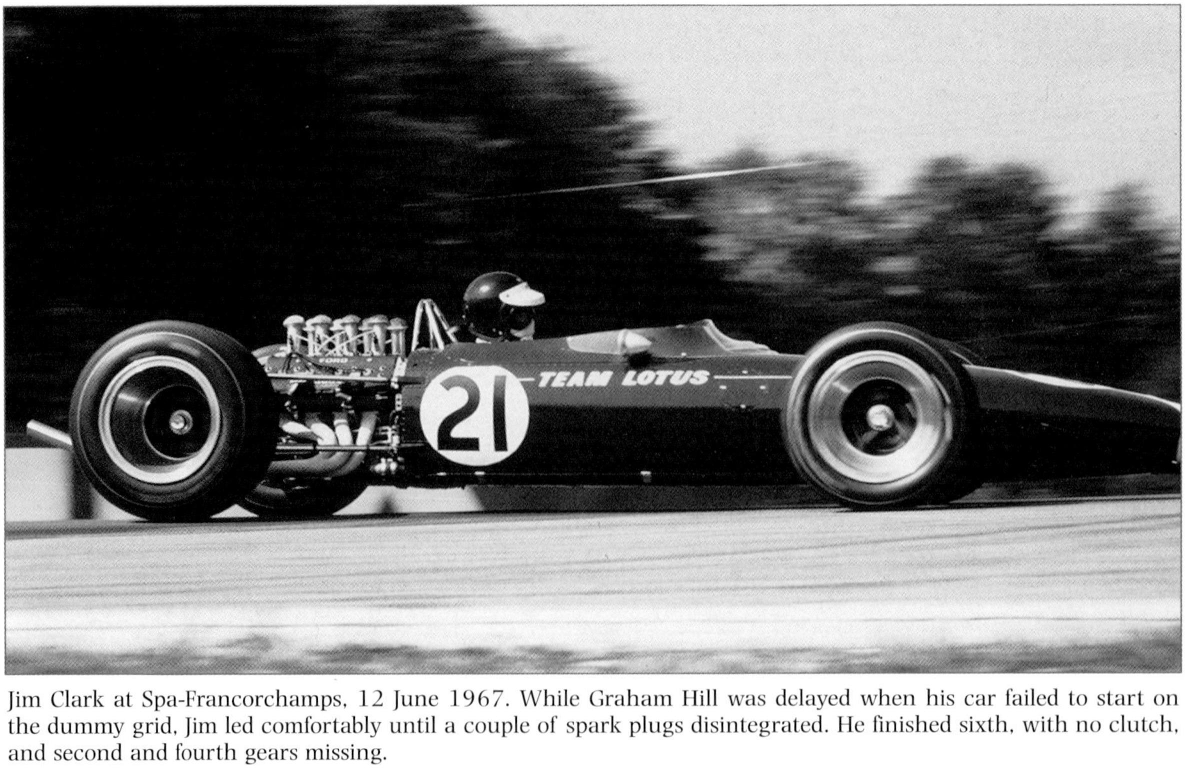

Jim Clark at Spa-Francorchamps, 12 June 1967. While Graham Hill was delayed when his car failed to start on the dummy grid, Jim led comfortably until a couple of spark plugs disintegrated. He finished sixth, with no clutch, and second and fourth gears missing.

The 49 carried its fuel in two 15-gallon bag tanks and a 10-gallon tank mounted behind the seat. The only section of bodywork that the original 49 wore was its detachable fibreglass nosecone, since the engine and transmission were unclad. Aerodynamic theories were still in their infancy, and the nosecone had yet to sprout its twin aerofoils. These didn't emerge until Monaco 1968 on the 49B, although a 'beard' spoiler was tried on a couple of occasions.

The 49 didn't appear until the third race in the 1967 season, which was at Zandvoort. Testing had been minimal and Clark had no knowledge of the car, yet he not only won but set fastest lap. His new team-mate Hill led from pole position only to retire after eleven laps. Clearly the ingredients were promising, even if they needed refining. By the British Grand Prix, the series was still wide open, with Gurney, Brabham, Hulme and Rodriguez all having won races. Clark and Hill were the class of the field at Silverstone, and while Graham was sidelined with suspension failure, Clark took the victory. Hill was fourth in Canada, and Clark third in Italy. Then they went to Watkins Glen for the US GP.

So dominant were the 49s in practice that Clark and Hill tossed a coin to decide who was going to win. Although Hill led from pole, a slipping clutch forced him to let Clark by, and the pair traded places twice more until Clark took the chequered flag first – albeit with a collapsed rear top link.

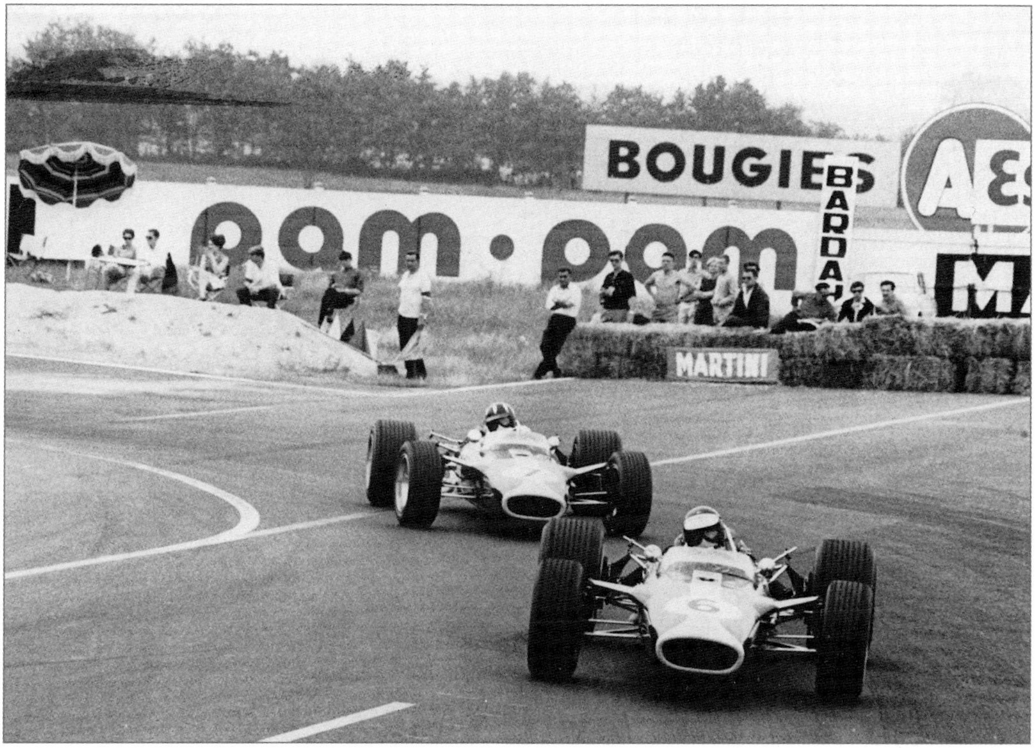

Graham Hill on opposite lock following Jim Clark's 49 at the 1967 French Grand Prix. The venue is the 2.8-mile Le Mans Bugatti circuit (which was possibly the first time the expression 'Mickey Mouse' was used to describe a race track). It centres on the Le Mans pits and grandstand complex. Both 49s were dogged with mysterious misfires, and both retired from the race with sheared crown-wheels and pinions.

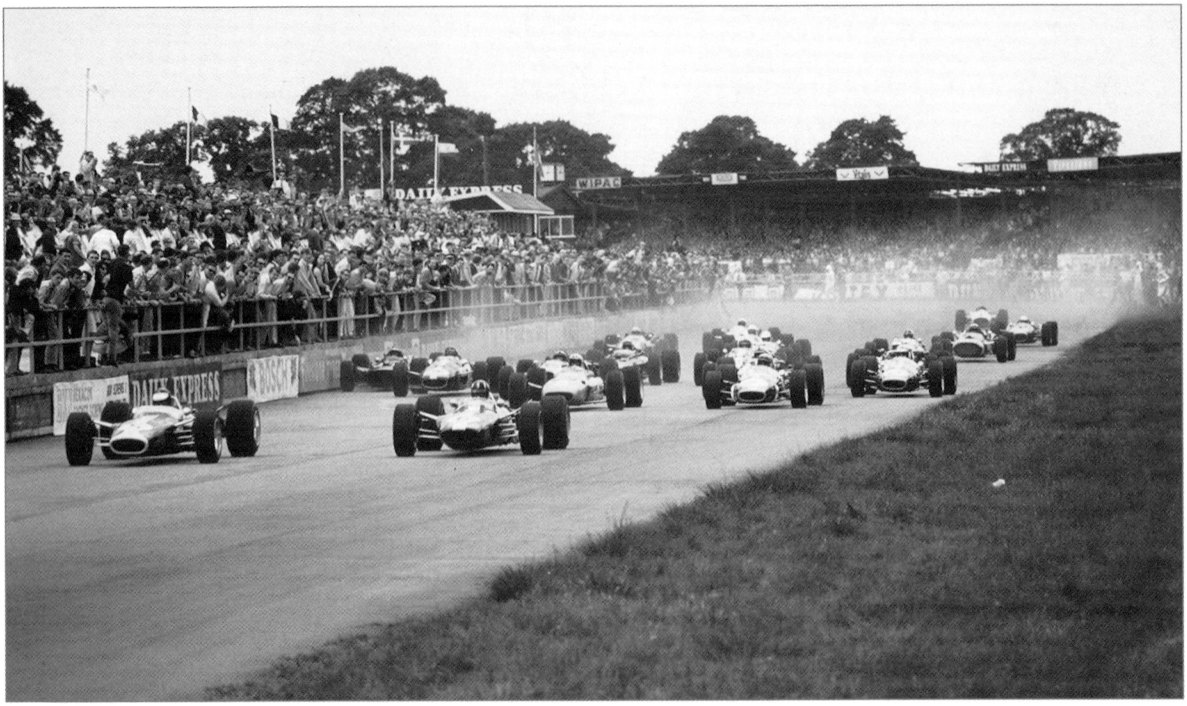

Jim Clark and Graham Hill streak away ahead of the field to lead the 1967 British Grand Prix – an indication of the superiority of the Lotus 49. After his regular car was wrecked in practice, Graham's car was built overnight, amidst much toing and froing between Hethel and Silverstone. It came complete with a 33 nosecone. He led Jim until first the suspension shed a bolt, and then the engine let go. Clark won.

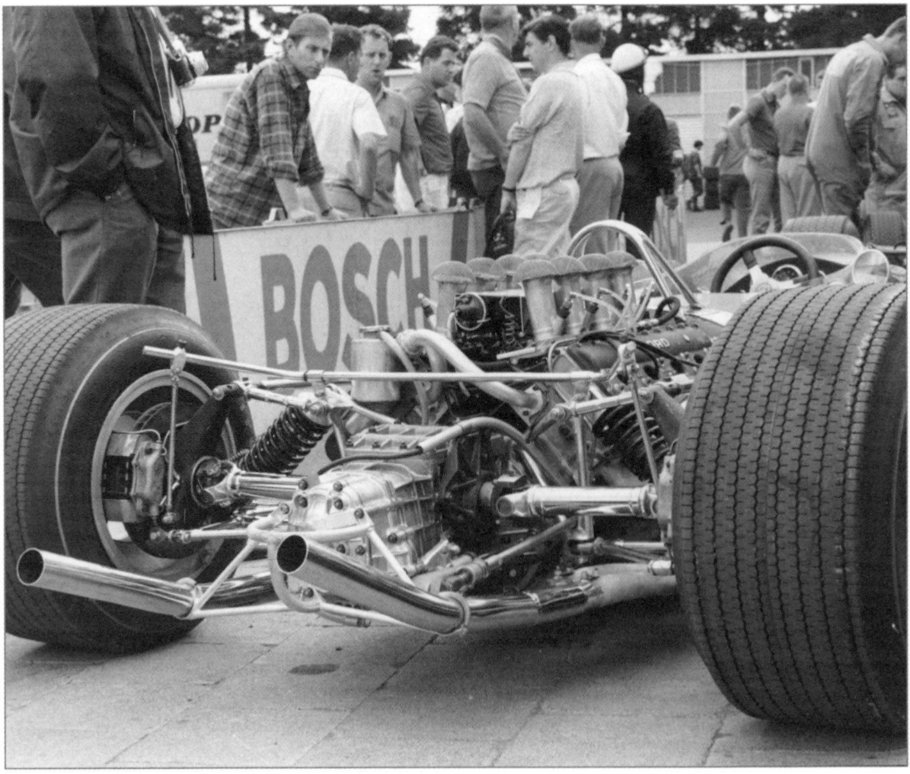

Graham Hill's 49 in the Nürburgring paddock on its way to scrutineering prior to the 1967 German Grand Prix. For this race the ignition ancillaries were relocated within the V of the engine between the injection trumpets, and the Allen screws in the suspension were replaced with fixed studs and lock nuts.

Mechanic Doug Bridge (right) supervises Graham's entry into the Nürburgring pits. In practice for the German Grand Prix, Hill played with suspension settings on his 49/R3. Initially, the brakes were cooling too efficiently and the pads were glazing, so solid discs replaced the ventilated ones. Hill's car also got 16-gauge front rocker arms after Clark's had collapsed in practice. But when a wheel came loose, Hill retired on lap eight with deranged rear suspension.

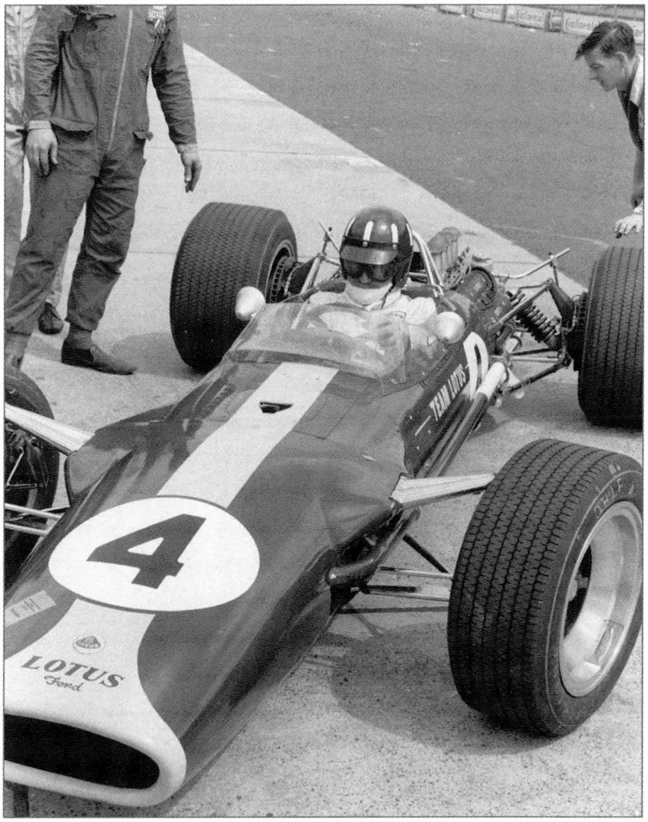

Canadian Eppie Wietzes was the first 'outsider' to drive a 49, competing in his home GP in the third Team Lotus car but retiring with drowned electrics. Jimmy went on to win the Mexican Grand Prix and the non-Championship Madrid Grand Prix staged at Jarama in November, where Graham was second. Moises Solana guested in the third Team Lotus 49 in his home Grand Prix at Mexico City, retiring with suspension failure. In the World Championship standings, it was the more straightforward Repco-Brabhams of Hulme and Brabham that emerged at the top of the table, although Clark had won more races.

With the coming of the 1968 season, Lotus no longer had the monopoly of the DFV engine, and the McLaren, Matra and Rob Walker teams hurried to get Cosworth power for the South African Grand Prix. There was no stopping Clark and Hill, however. The 49s were fastest in practice and first and second in the race. It was Jimmy's 25th World Championship victory, surpassing Fangio's record. It was also his last Grand Prix, if not his last race, and the circus moved down under for the Tasman series. Now fuelled by Shell rather than Esso, who had funded Lotus to a great extent since 1953, the 49s were also painted red, white and gold and entered by Gold Leaf Team Lotus for the majority of the series. The race at Levin in New Zealand was the last event that the works Lotuses ran in their green and yellow stripe colour scheme, and the Lady Wigram Trophy race was the first at which the 49s appeared in their new livery. They were nonetheless required to blank off the tobacco company's sailor's head logo that featured on the Gold Leaf insignia at

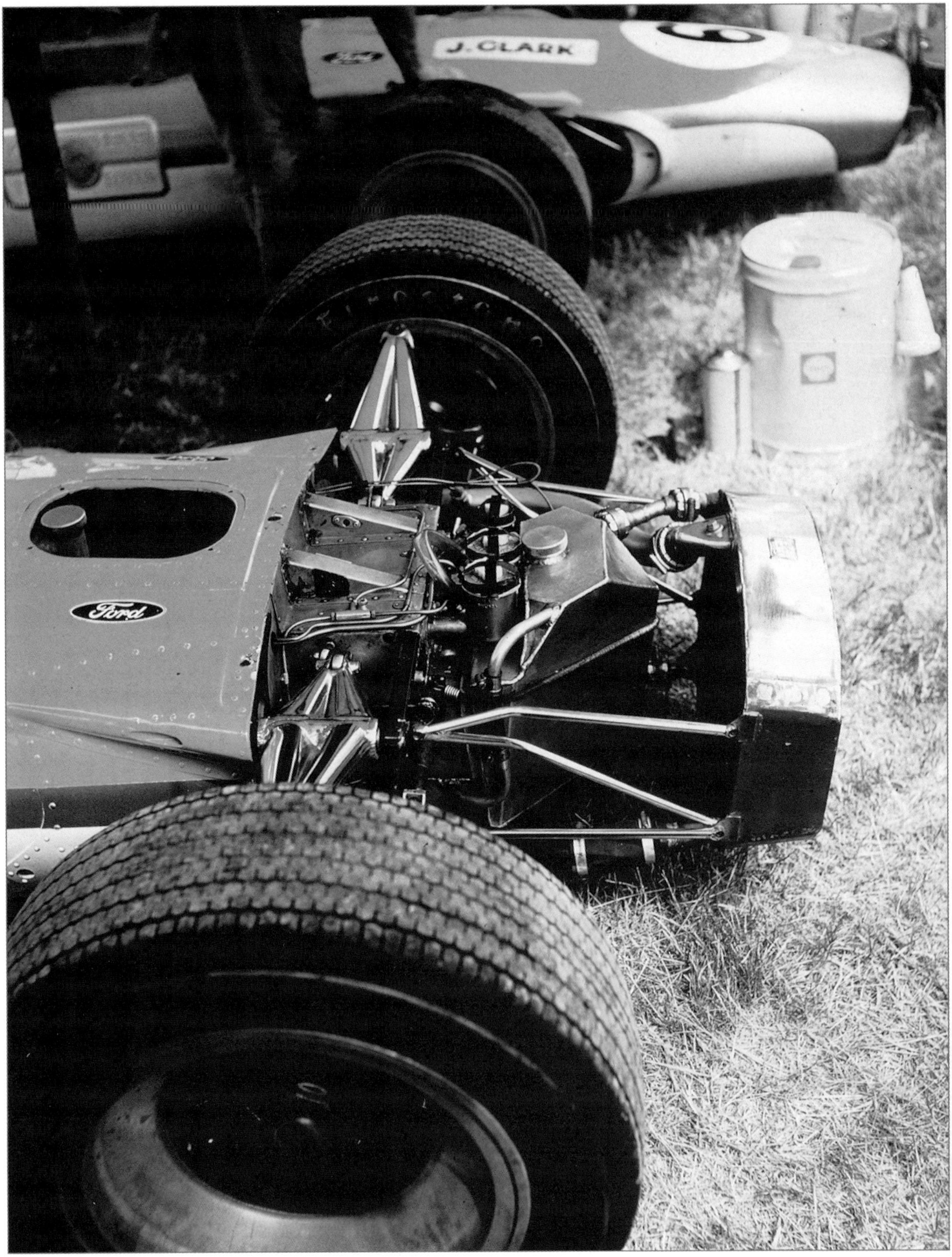

Front subframe of Graham Hill's Tasman 49, showing the triangular oil tank, radiator, and brake and clutch fluid vessels, with front suspension rocker arms prominent.

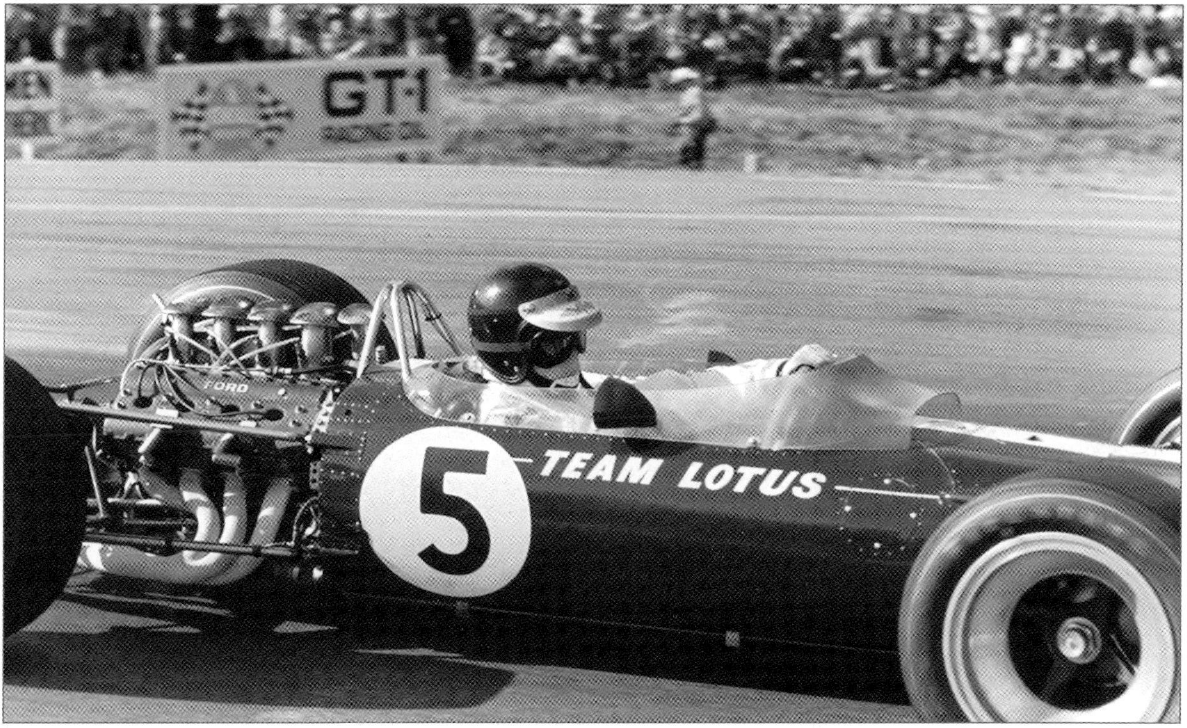

Jimmy Clark on his way to victory at Watkins Glen on 1 October 1967. By the time he crossed the line his right rear suspension had broken and a wheel was almost off. Some said that was the mark of a true racing car – one that holds together till the finish and then falls apart. Graham Hill was second, 6.3 seconds behind with a failing clutch.

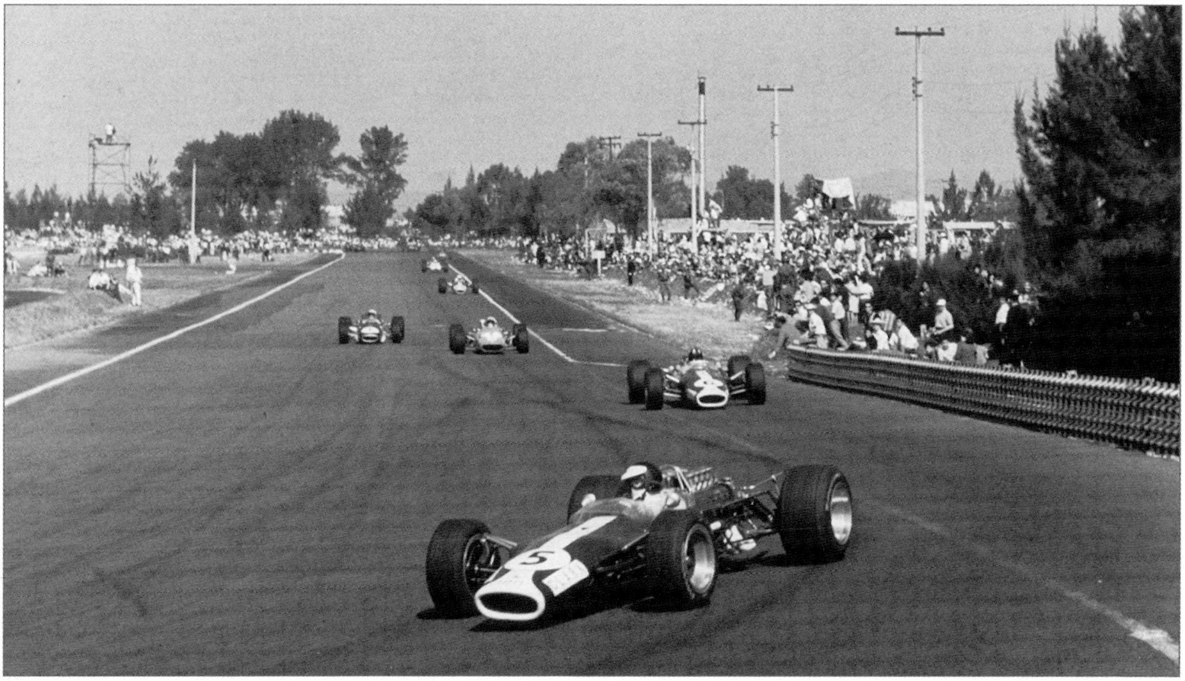

By the last race of the 1967 season in Mexico, the 49's rear suspension had been beefed up to counter breakages earlier in the season. The art of designing a racing car was to go as light as possible to start with, and strengthen anything that broke subsequently. Taking weight out of a car was never the most efficient way. Here, Jimmy takes an easy win after Graham's car snapped a driveshaft.

the Surfer's Paradise meeting. There's a certain irony in the fact that Team Lotus was sponsored by a tobacco company – Colin Chapman didn't smoke.

To comply with the maximum 2.5-litre regulations, the 49s – known as 49T for Tasman – were equipped with sleeved-down DFVs and called DFWs. They developed 350bhp at 9200rpm. A further quirk was that Clark had a tiny aerofoil specially made up and located above the rear anti-roll bar, similar to one he'd seen on a Vollstedt Indycar at Riverside in 1967. Chapman forbade its use, but it was a harbinger of things to come in the aerodynamics department. Clark had won at Christchurch, Surfer's Paradise, Warwick Farm and Sandown Park, taking the Tasman Cup after some close duels with Amon's Ferrari.

The Lotus 56 Turbine car scheduled to race at the 1968 Indianapolis 500 was begun in November 1967. Four cars were to be built and raced under the STP–Lotus banner, powered by the Pratt & Whitney ST6 aircraft engine – located behind the driver rather than alongside in the manner of the 1967 STP-Paxton Indycar. The 56 used Ferguson four-wheel drive transmission, and since gas turbines don't need gearboxes because they develop sufficient torque at all engine speeds, the transmission amounted to two crown-wheel and pinion assemblies, with a differential gear unit each, plus one acting as a torque splitter. The car was tested by Jim Clark at Indianapolis in March 1968, and although its competition career was relatively brief (the 56B was resurrected as a Formula 1 car in 1970), it was also important because of its wedge shape, which demonstrated Chapman's thoughts on aerodynamics at the time that wings were in the offing.

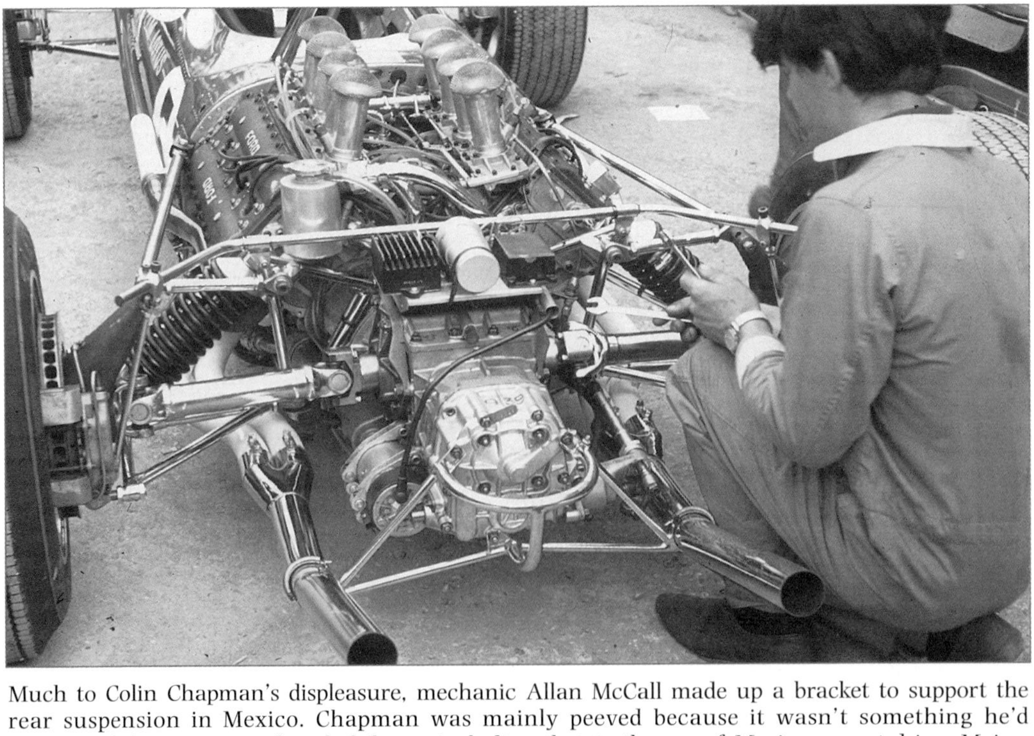

Much to Colin Chapman's displeasure, mechanic Allan McCall made up a bracket to support the rear suspension in Mexico. Chapman was mainly peeved because it wasn't something he'd sanctioned, but a string of such failures, including that to the car of Mexican guest driver Moises Solana, led to McCall quitting Team Lotus for McLaren.

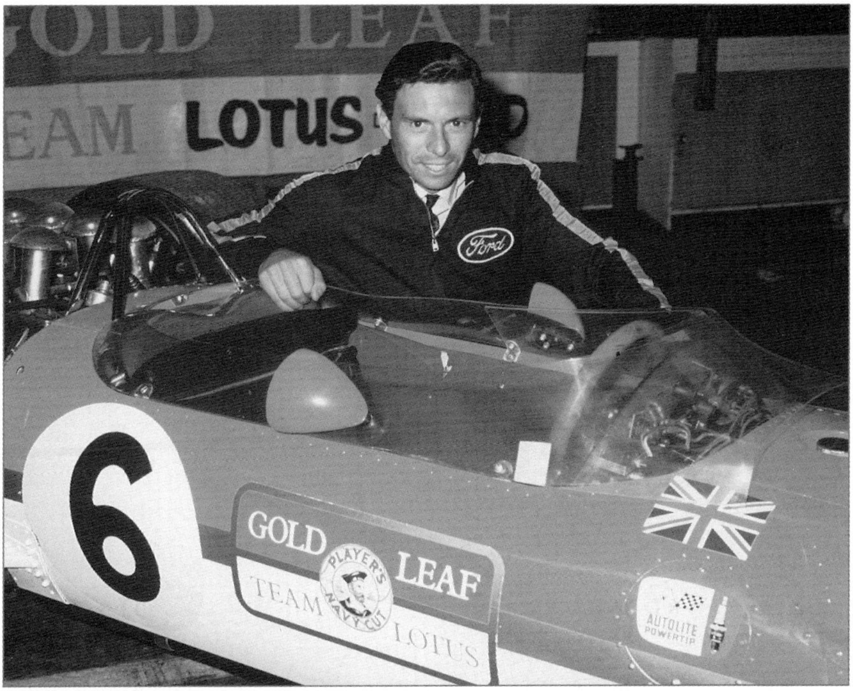

Jim Clark poses with 49/R2 in a garage at Christchurch, New Zealand, during the 1968 Tasman Series. The car had just been painted in the Gold Leaf Colours of Team Lotus's new sponsor.

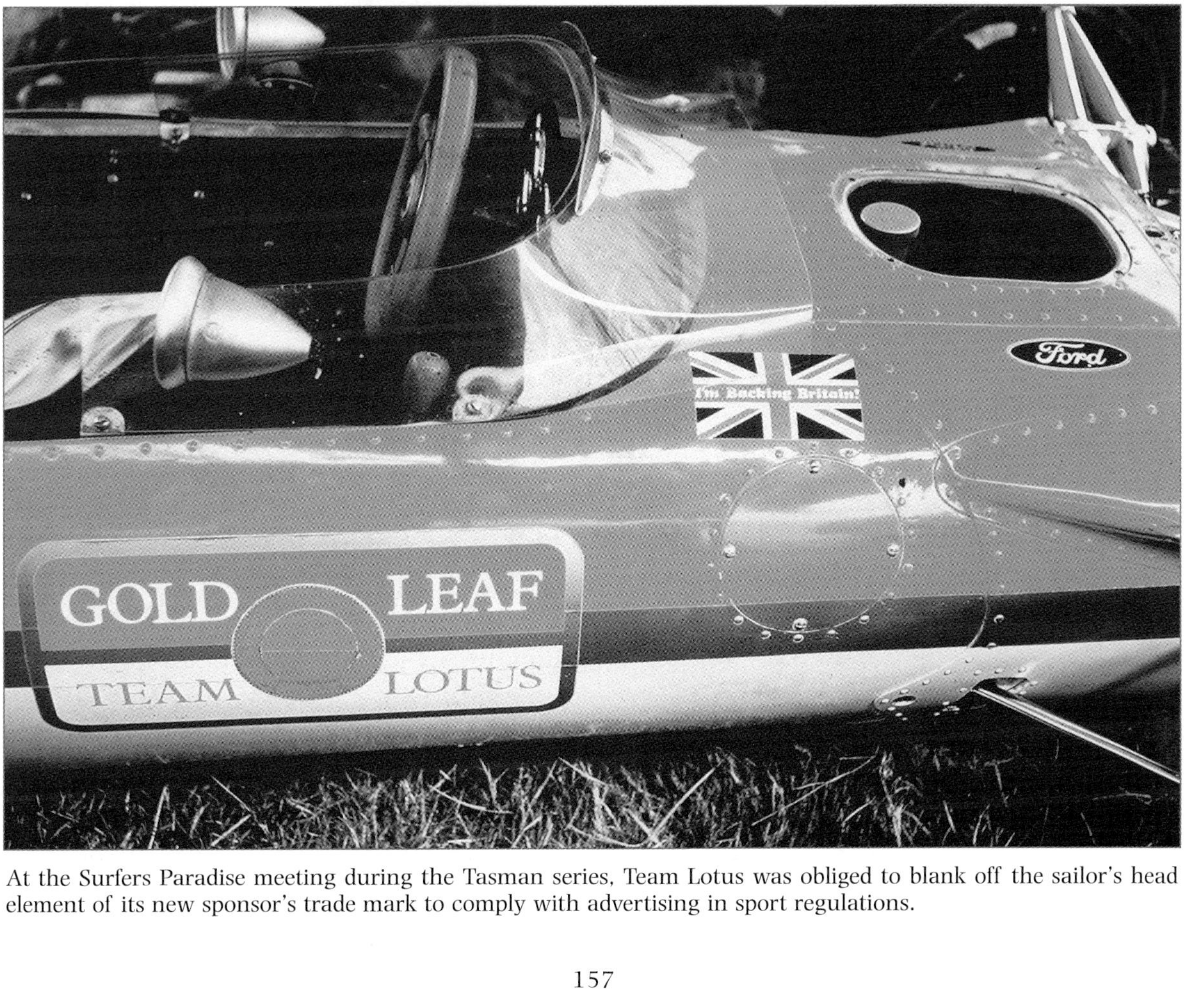

At the Surfers Paradise meeting during the Tasman series, Team Lotus was obliged to blank off the sailor's head element of its new sponsor's trade mark to comply with advertising in sport regulations.

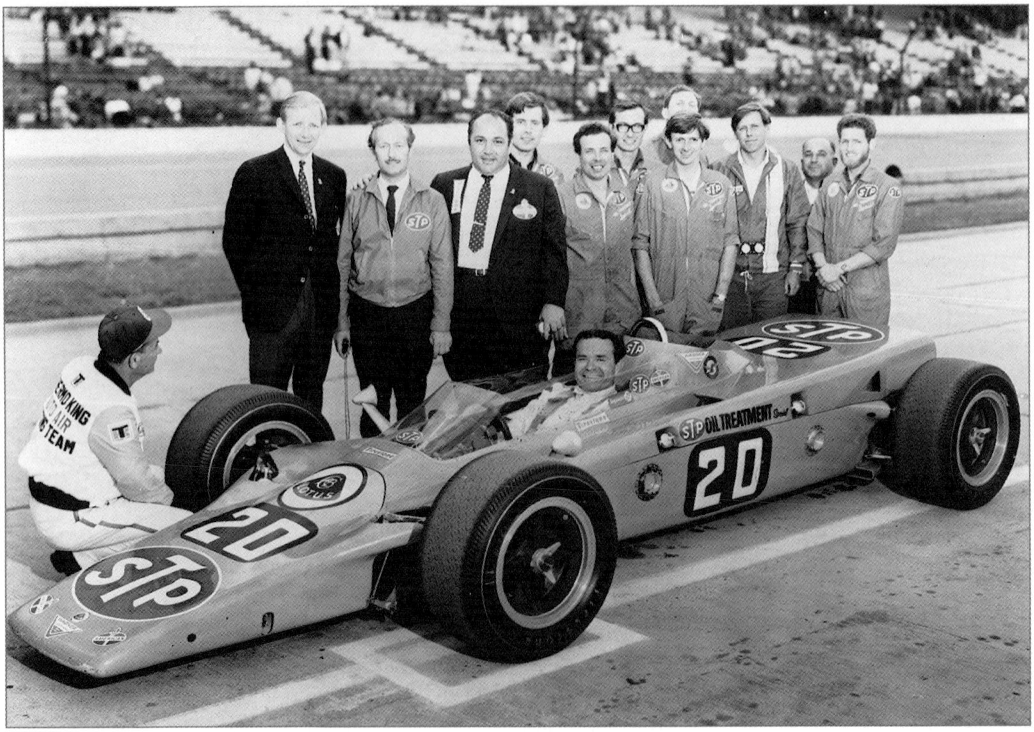

Colin Chapman with STP supremo Andy Granatelli and (mostly Lotus) race crew with the Pratt & Whitney-powered Lotus 56 gas turbine car. In the driving seat is Art Pollard. The 56 was built up on a twin tube ladder-frame stressed skin chassis with wishbone-type suspension. By this time an acrimonious situation was unfolding between Team Lotus and the STP operation, centred on division of prize monies and accreditation of the car's Hethel origins. The car had been tested by Jim Clark before his death, and he was replaced by Mike Spence who perished in qualifying. The 56s were raced by Graham Hill, Joe Leonard – who both started from the front row – and Art Pollard. Graham retired when his suspension broke and a wheel came off, while the Americans were classified twelfth and thirteenth.

Following an F3 test session, Jackie Oliver was brought in, largely on Jim Endruweit's recommendation, to take Clark's place. At the *Daily Express* International Trophy at Silverstone at the end of April, Jo Siffert fielded the new Rob Walker-Jack Durlacher 49. The South African John Love acquired 49/R3 in March 1968. Then, in a race of attrition, Graham Hill won the Spanish Grand Prix at Jarama in May, and lifted Lotus morale immeasurably after the loss of Clark. The following round at Monaco saw the debut of the 49B, which took Team Lotus into further uncharted territory – aerodynamics and high-rise wings.

That's where the story picks up again in Volume Two of *Lotus Racing Cars*. Meanwhile, we leave Lotus on the threshold of the new era of big-business motor sport, which would see a complete transformation come about. At this point, Colin Chapman had achieved a number of his original goals. These included an impressive tally of successes in sports car racing, including class wins at Le Mans. His cars were sought as potential winners in the lesser formulae, and he'd won the two highest accolades in motor racing – the Formula 1 World title and victory in the Indy 500. He had an enviable reputation as the constructor of exciting and trendy road-going sports cars, although he hadn't yet built anything that was in the same class as Ferrari. He was on the verge of moving a stage closer to that objective, though.

ACKNOWLEDGEMENTS

This book has been one of the best projects that I've ever had to do, and I want to thank the following people for their help with information and supplying photographs.

So, I'm most grateful to Clive Chapman and Manning Buckle at Classic Team Lotus for access to their picture archives, to Len Terry, Bob Dance, Eddie Dennis and Jim Endruweit for reminiscences, advice, and loan of pictures from Jim of Indianapolis and the F2 Texaco Stars. One of my colleagues from the JPS days, Ian Catt, delved into his personal Aladdin's Cave for photographs that he took during that era and prior to it. Phillip Stacey allowed full access to the photograph albums of his brother Alan, and Ian Bentall who's restoring Alan's Mk 6 loaned a couple of fine shots of Alan Stacey in action. Thanks also to Bobby Bell for photos of his racing Lotuses, and to Jim Fowler and his colleagues, including Fran Chamberlain, at Ford's Photographic Department for hospitality and photographs of Ford-engined Lotuses. Andrew Davis of Group Lotus PR department supplied some recent Lotus racing shots, while Ted Walker of Ferret Fotographic supplied lots of archive pictures. Another old pal and former BRDC Press Officer Roger Palmer loaned a picture of the Lotus-Cortinas in action, and a couple of the Lotus Seven shots came from Richard Spelberg. The Indy refuelling shot came from Indianapolis Motor Speedway. John Miles and Bill Cowe helped identify cars and people, and Vanna Skelley at Burmah Castrol Photographic Archive supplied some period pix. Thanks also to Sir John Whitmore for supplying and helping to caption a couple of the Lotus-Cortina shots, and Robert Brockington, Mosport International Raceway Track Historian.

I should also like to acknowledge the sources for some of my reference material, including the following books: *Colin Chapman: The Man and his Cars* by Jabby Crombac; *Colin Chapman: Lotus Engineering* by Hugh Haskell; *The Story of Lotus 1947–1960* by Ian H. Smith; *Colin Chapman's Lotus* by Robin Read; *The Story of Lotus 1961–1971* by Doug Nye; *The John Player Motorsport Yearbook* edited by Barrie Gill; *Lotus: The Indianapolis Years* by Andrew Ferguson; *The Lotus 49* by Michael Oliver; *Lotus & Caterham Sevens: Racers for the Road* by John Tipler; and *Lotus Elise: the Complete Story* by John Tipler. I also have a set of *Motor Sport* magazines going back to 1951, which are invaluable for sorting out race results, mechanical mishaps and driver pairings from long ago.

Volume 2 acknowledges some of the above names, with additional sources also credited.

PHOTOGRAPH SOURCES

Bobby Bell: 108, 136
Ian Bentall: 67, 77
Burmah-Castrol (Vanna Skelley): 82, 85, 89, 114, 123
Ian Catt: 7, 122, 137, 138, 145
Classic Team Lotus (Clive Chapman) 11, 17, 23, 24, 34, 85, 96, 97, 140, 144, 146, 147, 151, 158
Jim Endruweit: 102
Ferret Fotographic (Ted Walker): front cover, title page, 1, 2–4, 6, 8, 9, 16, 18, 19, 22, 26, 27, 32–4, 39, 40, 45, 50, 52–5, 57, 58, 61, 63, 65, 66–7, 72, 73, 76, 78, 80, 81, 84, 88, 89, 90, 91, 92, 94–5, 97, 99, 104, 105, 106, 107, 110, 114, 115–18, 120, 121, 124, 125, 129, 138, 139, 141, 142, 149, 150, 152–5, 156
Ford Motor Company (Jim Fowler): 3, 100, 101, 107, 109, 113, 115, 123, 125, 131–5, 150, 155, 157
Indianapolis Motor Speedway: 131
Roger Palmer: 12
Richard Spelberg: 49, 50
Phillip Stacey: 5, 20, 29, 30, 31, 34, 36–9, 41–7, 56, 58, 59, 65, 69, 74, 75
John Tipler: 14, 48, 49, 50, 60, 64, 70, 78, 87, 88, 118, 121, 126–8, 145
Sir John Whitmore: 111, 112

INDEX